Glass Integrated Optics and Optical Fiber Devices

Critical Reviews Series

64- to 256-Megabit Reticle Generation (CR51, G. K. Hearn)
Adaptive Computing: Mathematics, Electronics, and Optics (CR55, S.-S. Chen, H. J. Caulfield)
Applications of Electronic Imaging (v.1082, J. C. Urbach)
Applications of High-Power Lasers (v.527, R. R. Jacobs)
The Aviation Security Problem and Related Technologies (CR42, W. H.Makky)
Characterization of Very High Speed Semiconductor Devices and Integrated Circuits (v.795, R. Jain)
Design of Optical Systems Incorporating Low-Power Lasers (v.741, D. C. O'Shea)
Diagnostic Imaging Applications (v.516, E. S. Beckenbach)
Diffractive and Miniaturized Optics (CR49, S. H. Lee)
Digital Image Processing (v.528, A. G. Tescher)
Digital Optical Computing (CR35, R. Athale)
Electro-Optical Imaging Systems Integration (v.762, R. Wight)
Fiber Optic Communication Technology (v.512, C. W. Kleekamp)
Fiber Optic Sensors (CR44, E. Udd)
Fiber Optics Reliability and Sensing (CR50, D. K. Paul)
Flashlamp-Pumped Laser Technology (v.609, F. Schuda)
Free Electron Lasers (v.728 B. E. Newnam)
Geometrical Optics (v.531, R. E. Fischer, W. H. Price, W. J. Smith)
Glass Integrated Optics and Optical Fiber Devices (CR53, S. I. Najafi)
Gradient-Index Optics and Miniature Optics (v.935, J. D. Rees, D. C. Leiner)
Highly Parallel Signal Processing Architectures (v.614, K. Bromley)
Holographic Nondestructive Testing: Status and Comparison with Conventional Methods (v.604, C. M. Vest)
Holography (v.532, L. Huff)
Image Pattern Recognition: Algorithm Implementation, Techniques, and Technology (v.755, F. J. Corbett)
Industrial Optical Sensing (v. 961, J. Carney, E. Stelzer)
Infrared Detectors (v. 443, W. L. Wolfe)
Infrared Detectors and Arrays (v.930, E. L. Dereniak)
Infrared Optical Design and Fabrication (CR38, R. Hartmann, W. J. Smith)
Infrared Thin Films (CR39, R. P. Shimshock)
Integrated Optics and Optoelectronics (CR45, K.-K. Wong, M. Razeghi)
Interferometric Metrology (v.816, N. A. Massie)
Laser Processing of Semiconductors and Hybrids (v.611, E. J. Swenson)
Laser Wavefront Control (v.1000, J. F. Reintjes)
Lens Design (CR141, W. J. Smith)
Machine Vision Systems Integration (CR36, B. G. Batchelor, F. M. Waltz)
Modeling of the Atmosphere (v.928, L. S. Rothman)
Optical Component Specifications for Laser-Based Systems and Other Modern Optical Systems (v.607, R. E. Fischer, W. J. Smith)
Optical Computing, (v.456, J. A. Neff)
Optical Pattern Recognition (CR40, J. L. Horner, B. Javidi)
Optical Inspection and Testing (CR46, J. D. Trolinger)
Optical Specifications: Components and Systems (v.406, W. J. Smith, R. E. Fischer)
Optical Technologies for Aerospace Sensing (CR47, J. E. Pearson)
Optomechanical and Electro-Optical Design of Industrial Systems (v.959, R. J. Bieringer)
Optomechanical Design (CR43, P. R. Yoder, Jr.)
Photorefractive Materials, Effects, and Applications (CR48, P. Yeh, C. Gu)
Photovoltaics (v.543, S. K. Deb)
Radiation Effects on Optical Materials (v.541, P. W. Levy)
Real-Time Signal Processing for Industrial Applications (v.960, B. Javidi)
Remote Sensing (v.476, P. N. Slater)
Robotics and Robot Sensing Systems (v.442, D. Casasent, E. L. Hall)
Spatial Light Modulators and Applications III (v.1150, U. Efron)
Standards for Electronic Imaging Systems (CR37, M. Nier, M. E. Courtot)
Technology of Stratified Media (v.387, R. F. Potter)

Critical Reviews
of Optical Science
and Technology

Volume CR53

Glass Integrated Optics and Optical Fiber Devices

S. Iraj Najafi
École Polytechnique de Montréal
Editor

*Proceedings
of a conference held
24–25 July 1994
San Diego, California*

Sponsored and Published by
SPIE—The International Society for Optical Engineering

SPIE OPTICAL ENGINEERING PRESS

A Publication of SPIE—The International Society for Optical Engineering
Bellingham, Washington USA

Library of Congress Cataloging-in-Publication Data

Glass integrated optics and optical fiber devices / S. Iraj Najafi, editor.
    p. cm. — (Critical reviews of optical science and technology ; v. CR53)
    "Proceedings of a conference held 24–25 July 1994, San Diego, California."
    ISBN 0-8194-1386-0 (hardbound) — ISBN 0-8194-1385-2 (softbound)
    1. Integrated optics. 2. Optical waveguides. 3. Glass fibers.
I. Najafi, S. Iraj. II. Series.
TA1660.G55 1994
621.36'93—dc20
                        94-11385
                            CIP

Published by
SPIE—The International Society for Optical Engineering
P.O. Box 10, Bellingham, Washington 98227-0010 USA
Telephone 206/676-3290 (Pacific Time) • Fax 206/647-1445

Copyright ©1994, The Society of Photo-Optical Instrumentation Engineers.

Copying of material in this book for internal or personal use, or for the internal or personal use of specific clients, beyond the fair use provisions granted by the U.S. Copyright Law is authorized by SPIE subject to payment of copying fees. The Transactional Reporting Service base fee for this volume is $6.00 per article (or portion thereof), which should be paid directly to the Copyright Clearance Center (CCC), 222 Rosewood Drive, Danvers, MA 01923. Other copying for republication, resale, advertising or promotion, or any form of systematic or multiple reproduction of any material in this book is prohibited except with permission in writing from the publisher. The CCC fee code is 0-8194-1385-2/94/$6.00.

Printed in the United States of America.

Contents

vii   Conference Committee
ix   Preface

WAVEGUIDE DESIGN AND FABRICATION

3   **Ion-exchange process for glass waveguide fabrication**
     G. C. Righini, IROE-CNR (Italy)

25   **Critical issues in designing glass integrated optical circuits**
     A. Tervonen, Optonex Ltd. and Nokia Research Center (Finland)

DEVICE FABRICATION

55   **Glass integrated optical devices on silicon for optical communications** [CR53-03]
     M. F. Grant, BNR Europe Ltd. (UK)

SOL-GEL AND RARE-EARTH-DOPED FIBERS AND WAVEGUIDES

83   **Sol-gel process for glass integrated optics**
     J. D. Mackenzie, Y.-H. Kao, Univ. of California/Los Angeles

114 **Fibers from gels and their applications**
     S. Sakka, Kyoto Univ. (Japan)

132 **Integrated optical devices in rare-earth-doped glass**
     K. J. Malone, National Institute of Standards and Technology

COMMERCIAL DEVICES

159 **Ion-exchanged glass waveguide devices for optical communications**
     S. Honkanen, Nokia Research Center and Helsinki Univ. of Technology (Finland)

180 **Ion-exchanged glass waveguide sensors**
     L. Ross, IOT GmbH (FRG)

200 **Commercial glass waveguide devices**
     M. D. McCourt, Corning Opto-Electronic Components (France)

NONLINEAR FIBERS AND WAVEGUIDES I

211   Quantum dot glass integrated optical devices
      N. Peygambarian, H. Tajalli, E. M. Wright, S. W. Koch, Optical Sciences Ctr./Univ. c
      Arizona; S. I. Najafi, École Polytechnique de Montréal (Canada); D. Hulin, ENSTA (F
      J. MacKenzie, Univ. of California

235   Photosensitive glass integrated optical devices
      B. J. Ainslie, G. D. Maxwell, D. L. Williams, BT Labs. (UK)

NONLINEAR FIBERS AND WAVEGUIDES II

253   Fused bitapered fiber devices for telecommunication and sensing systems
      S. Lacroix, École Polytechnique de Montréal (Canada)

DEVICES FOR COMMUNICATION AND SENSORS I

281   Glass waveguides with grating
      M. Fallahi, National Research Council of Canada and Solid State Optoelectronic
      Consortium (Canada)

295   Multimode glass integrated optics
      O. M. Parriaux, P. Roth, G. Voirin, CSEM (Switzerland)

321   Glass waveguides in avionics
      F. A. Blaha, Canadian Marconi Co. (Canada)

DEVICES FOR COMMUNICATION AND SENSORS II

335   Fiber-to-waveguide connection
      J.-F. Bourhis, Alcatel Cable Interface (France)

367   Optical fiber chemical sensors
      M. Shadaram, Univ. of Texas/El Paso

Conference Committee

Conference Chair
- **S. Iraj Najafi,** École Polytechnique de Montréal (Canada)

Session Chairs

1. Waveguide Design and Fabrication
   S. Iraj Najafi, École Polytechnique de Montréal (Canada)

2. Device Fabrication
   Giancarlo C. Righini, IROE-CNR (Italy)

3. Sol-Gel and Rare-Earth-Doped Fibers and Waveguides
   Nasser Peygambarian, Optical Sciences Center/University of Arizona

4. Commercial Devices
   Ari Tervonen, Optonex Ltd. (Finland)

5. Nonlinear Fibers and Waveguides I
   David N. Payne, University of Southampton (UK)

6. Nonlinear Fibers and Waveguides II
   B. J. Ainslie, BT Laboratories (UK)

7. Devices for Communication and Sensors I
   Seppo Honkanen, Optonex Ltd. (Finland)

8. Devices for Communication and Sensors II
   Mahmoud Fallahi, National Research Council of Canada

Preface

Two distinctly different types of guiding structures are used to make glass waveguide devices: optical fibers and integrated optical waveguides.

There has been remarkable progress in optical fiber devices. Optical fibers have been used to study and demonstrate a number of important phenomena such as optical amplification, soliton propagation, and pulse break-up. All-fiber devices have been produced using readily available optical fibers.

Glass integrated optics was rather slow starting, but outstanding progress has been achieved during the past decade or so. High performance integrated optical devices and circuits have been fabricated.[1-3]

Different techniques are used to make glass integrated optical devices. Ion-exchange is the most popular. This technique is simple and can be used to make reproducible and low-cost devices. Recently, a flame hydrolysis technique has attracted attention, probably because the resultant waveguides can be fused to optical fiber, which improves environmental stability of the chip and eliminates back reflection in the fiber-chip interface. Plasma deposition offers the possibility of doping waveguides to achieve nonlinear devices. The sol-gel method is flexible and can be used to make waveguides with different dopants (e.g., rare earths, semiconductors, photosensitive elements). It is also attractive for fabrication of hybrid circuits. Figure 1 depicts a hybrid 1.3 μm/1.55 μm amplifier/splitter.[2] Ion implantation also has been employed to make glass waveguides. In addition, we have used Ge implantation to make waveguides. A simple photoresist mask has been used to produce channel waveguides. However, the waveguides had rather high propagation losses. In addition, the fabrication process is very costly and is not suitable for device fabrication.

Accurate theoretical tools have been developed to design glass integrated optical devices.[4-6] New and complex devices have been proposed, analyzed, and demonstrated. In particular, waveguides with grating have attracted a lot of attention.[7,8] Figures 2 and 3 depict two examples of such devices. In Figure 2 we propose a new rare-earth-doped glass waveguide laser. The grating with variable width is used to diffract a symmetric laser beam perpendicular to the waveguide surface. In Figure 3 we suggest a narrow band wavelength division multi/demultiplexer.[9]

This critical review includes papers, authored by recognized experts, discussing optical fibers and the progress and future potential of glass integrated optical devices.

(continued)

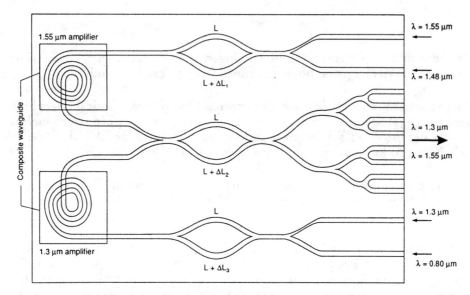

Fig. 1. 1.3 μm/1.55 μm glass integrated optical amplifier/splitter circuit.[2] The composite waveguides can be achieved using rare-earth-doped sol-gel glasses.

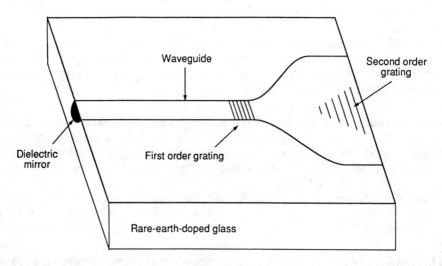

Fig. 2. Rare-earth-doped glass integrated optical symmetric beam surface emitting laser.

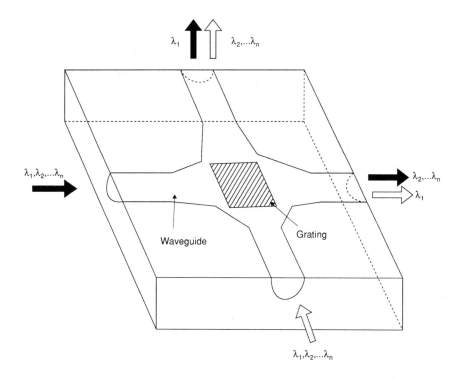

Fig. 3. Integrated optical narrow-band wavelength division multi/demultiplexer.[9]

References

1. S. I. Najafi, *Introduction to Glass Integrated Optics,* Artech House, Boston, 1992.
2. S. I. Najafi and Seppo Honkanen, "Gradient Index Optics," in *CRC Handbook of Photonics,* Editor: M. Gupta, in press.
3. S. Honkanen, A. Tervonen, and S. I. Najafi, "Passive integrated optical components for PONs," *invited paper,* Conf. on Europ. Fiber Optics Communications and Networks, June 1994.
4. Ionex software by Optonex Ltd.
5. P. Auger and S. I. Najafi, "New method to design directional coupler dual wavelength multi/demultiplexer with bends at both extremities," submitted for publication.
6. S. I. Najafi, S. Honkanen, and A. Tervonen, "Recent progress in glass integrated optical circuits," *invited paper,* Conf. on Integrated Optics and Microstructures (SPIE's Annual Meeting), San Diego, July 1994.

7. S. I. Najafi and M. Fallahi, "Circular grating lasers," *invited paper*, International Symposium on Integrated Optics, Lindau, April 1994.
8. S. I. Najafi, "Circular gratings and applications in integrated optics/optoelectronics," in *Nonlinear Optics for High Speed Electronics and Optical Frequency Conversion*, N. Peygambarian et al., Editors, Proc. SPIE 2145, pp. 46–57 (1994).
9. P. Lefebvre, S. I. Najafi, and M. Kavehrad, "Compact narrow-band integrated optical wavelength division multi/demultiplexer," submitted for publication.

S. Iraj Najafi
June 1994

Glass Integrated Optics
and Optical Fiber Devices

Ion exchange process for glass waveguide fabrication

Giancarlo C. Righini

Istituto di Ricerca sulle Onde Elettromagnetiche (IROE-CNR)
Gruppo Tecnologie Optoelettroniche
Via Panciatichi 64, 50127 Firenze, Italy

ABSTRACT

Ion-exchange is a relatively mature technology for the production of optical waveguides in glass substrates, and it appears as the most convenient approach to the development of a number of passive integrated optical devices, due to the low loss and process robustness. The purpose of this paper is to give a quick overview of the status of this technology, with particular attention devoted to the problems of manufacturing tolerances and of fabrication and characterization reproducibility.

1. INTRODUCTION

1.1 Glass and ion-exchange: historical notes

Among optical materials, glass has always played a leading role. In the antiquity glasses were first used in the Mesopotamian region for decorative objects, mainly as a colored glaze on stone or pottery beads, definitely before 3000 BC.[1] The art of making glass was perfected about 1500 BC in Egypt and the Near East. It is not clear how the idea of making melts which would cool to a glass emerged; according to a tale told by the Roman historian, Pliny the Elder, in his *Natural History*, the discovery of glass occurred when a group of Phoenician seamen made a fireplace on a beach and used blocks of soda (that they had aboard their ship for trading) to support the pot. Later, due to the soda melting together with the sand, they could observe a novel, transparent liquid which eventually cooled down to a transparent frozen slag: this was the origin of glass. Such a tale is undoubtly historically wrong, since Phoenicians learned glassmaking techniques from Egyptians or Mesopotomians, but it is interesting since it confirms the nature of the early constituents of glass: sand, which contains silica (as quartz grains) and lime (as shells and skeletons of sea animals), and soda, taken from the dry lakes of Middle East, which contains sodium carbonate, lime, magnesium oxide and traces of sodium sulfate. Accordingly, early Middle Eastern glass was soda-lime-silica, a composition not too different from those of today's bottle and window glasses.

About 50 B.C. the notion of glass blowing was developed, and this resulted in much easier production of a larger variety of glass containers.[2] The art of glassblowing spread rapidly throughout the Roman Empire, and special centers of glassmaking were established in Phoenicia, Egypt, Rome, the Rhineland, and the Rhone Valley. For many centuries after the fall of of Rome in 475 A.D. and of the Roman Empire, glassmaking decreased in importance in Western Europe, as did many other technologies and arts. In Byzantium, however, Greek and Syrian glass centers continued to prosper, and activity in science was kept alive mainly in the Islamic world.

Beginning in the 11th century, several new centers of glassmaking arose in Western Europe. In Bohemia, ash from plants (potash, which is high in potassium) was used as a raw material to make a glass with a lower melting point. A novel impetus to glass development was given by Italian centers: here, glass making was known as early as 900 A.D. and eventually, by 1400 A.D., Venice, where new compositions, colors, forming techniques, and artistic skills were developed, became the most important European center of glass manufacturing.

The importance of glass as an *optical material*, in the modern meaning of this term, was certainly acknowledged by Galileo Galilei at the beginning of XVII century, when, to fabricate his telescopes, he cared very much about selecting a glass with high transparency and homogeneity (for his very first lenses he chose from many pieces of the clearest glass he could get in Murano and tried to isolate the best parts) and also about improving the processing techniques (he also perfected lathes and tools to grind and polish lenses to a perfect shape).[3] Since then, the progress in glass manufacturing and processing has been continuous, but pretty slow. Only the real technological advances in the last 30 years have significantly changed the complexion of glass manufacturing.

Despite its long history and its widespread usage, glass still it is not a material we can say to know completely. This could also be in some way expected since under the name of glass we have to cope with a very wide class of materials, often made up from a large number of constituents, which can result semiconductors or isolators, very transparent or almost opaque, porous or compact, and so on .. What they have in common is that a glass can be defined as a "frozen undercooled liquid" (it is said to be a solid since viscosity at room temperature is higher than 10^{15} poise) and as such it lacks long range order, namely the correlation of position and orientation is vanishing at distances greater than about 20 Å. Due to the random orientations of atomic groups, in absence of residual stress glasses are isotropic; composition fluctuations, however, are unavoidable and generally limit the available homogeneity to index of refraction variations smaller than 10^{-6}. Moreover, since its enthalpy is higher than that of the corresponding crystal, glass is not a perfectly stable solid; its structure depends largely on the cooling rate from the melt, and its state is not only function of the current pressure and temperature but also depends on pressure and temperature history.

The analysis of the structure of various glasses is therefore a complex task and to develop proper models it was necessary to get data from diagnostic techniques such as the X-ray diffraction: one of the models which is still fundamental to the understanding of glass structure is the one proposed by Zachariasen and published in 1932.[4] More recently, the availability of further techniques like Extended X-ray Absorption Fine Structure (EXAFS), Nuclear Magnetic Resonance (NMR) and Small Angle Neutron Scattering (SANS) has allowed the researchers to get more data at selected sites and on the order at different ranges, so that corrections and variations to Zachariasen model have been proposed.[5]

On the other hand, the ion-exchange in glass itself is a very old technique. At its beginning, it was used for coloring glasses. It seems that Egyptians in the VI century already used it to decorate dishes and pots in brownish yellow. Moors used this technique to stain the window glass of their palaces in Spain. In the West the earliest still existing stained glass dates from the XI century, but the finest stained-glass windows were produced in conjunction with gothic art and architecture from about 1130 to 1330. The staining of glass by silver compounds to produce yellow or amber colors was known in France since about the XIV century as *jaune d'argent* (silver yellow), and certainly it was a widely used method to fabricate colored church windows since the early Middle Ages. Silver was spread over the glass as a paste consisting of an inert carrier like clay and of silver chloride or sulfide; when the glass was heated, silver ions diffused into it taking the place of the alkaline ions which, as we said, were largely present in early glasses. Due to the presence of impurities like iron, the silver was subsequently reduced and it was the metal silver to produce the yellow-amber colour.[6,7] Red copper stains were produced in a similar way. Colour intensity was depending on the thickness of the spread paste, and thus it was possible to the artists to produce shaded effects as well.

The first industrial application of the ion-exchange technique came out only in 1913, as a mean to produce chemical surface tempering of glass[8]: a compressive stress was resulting from the introduction into the glass of K^+ ions which have a quite larger size than the substituted Na^+ ions. A few years later, in 1918, Schott researchers observed that ion-exchange was also producing an increase of the refractive index in the diffused layer. Thus, the basic elements of the ion-exchange process were already known almost 80 years ago. Strengthening of glass surfaces by ion-exchange (sometimes also called *ion stuffing*), however, became a standard industrial process only in the 1960s[9-12].

1.2 Ion exchange in glass for integrated optics

Soon after the beginning of integrated optics at the end of 1960s, a research field concerned with the development of optical components like couplers, lenses, modulators and so on in thin-film form and with the interconnection of these components onto a single chip,[13] ion-exchange emerged as one of the most viable and promising techniques to produce low-loss and rugged optical waveguides into glass substrates.[14] The pioneering works by Izawa and Nagakome[15] and by Giallorenzi and coworkers[16] in early 1970s opened the way to a large activity of research and development which have already led to the mass production of commercial integrated optical components and devices.[17]

A number of review papers [18-20] and a book [21] are available to provide the interested reader an overview of the state of the art and a rich bibliography, updated till 1991. As a consequence, here we will not deal with general items but, after a very brief summary of the fundamentals of ion-exchanged optical waveguides, the attention will be focused onto a specific subject: the reproducibility of the fabrication technique and of the characterization measurements, which is also being the main aim of an international cooperative project..

2. FUNDAMENTALS OF ION EXCHANGE FOR INTEGRATED OPTICS

2.1 The diffusing ions

When the glass containing a monovalent cation A is put in contact with the source of the diffusing monovalent cation B, the ion B is driven into the glass by an interphase chemical potential gradient and, in order to maintain charge neutrality, the ion A is released into the melt:

$$A_{glass} + B_{source} \leftrightarrow A_{source} + B_{glass} \qquad (1)$$

Equilibrium conditions and kinetics of the reaction have been studied in detail for different ion pairs.[22-27] The local modification of glass composition made in this way produces in turn a variation of density and, since the volume electronic polarizability is also changed, according to Clausius-Mosotti law it produces a variation of refractive index as well.[28] A contribution to the index variation cames also from the difference in size of the two exchanging ions, which produces elastic stress in the diffused glass layer. As already mentioned, replacement of Na$^+$ by K$^+$ causes the glass to swell (glass network expands to a less packed structure, characterized by a lower refractive index) and induces birifringence: the effect of the higher polarizability of K$^+$, however, is prevailing and the net result is an increase of the refractive index. On the contrary, replacement of Na$^+$ by Li$^+$ makes the glass network to collapse around the smaller Li$^+$ ion and to produce a more densely packed structure having an increased refractive index, despite the lower polarizability of Li$^+$. For convenience of the readers, the electronic polarizability and ionic radius of the most common diffusing ions are given in Table 1 (polarizabilities are at the frequency of the D lines of sodium, λ = 589.29 nm). To give an idea of the toxicity of the different products, which can represent a problem especially for use in a physics or engineering research laboratory, the lethal dose LD50 is also indicated. LD50, in mg/Kg, is the dose in mg per Kg of weight of the animals tested (rats) which is lethal to 50% of them.[29] It has to be noted that there is no a complete agreement on the values of relevant parameters in different reference sources; here we give the values of parameters according to ref.18 except data on LD50 which are taken from ref.29.

TABLE 1. Relevant parameters of monovalent ions for ion-exchange in glass

ION	Electronic polarizability Å^3	Ionic radius Å	Salt	Melting point °C	Decomposition point °C	Lethal dose LD50 mg/Kg -salt
Na$^+$	0.41	0.95	NaNO$_3$	307	380	1955 NaNO$_3$
Li$^+$	0.03	0.65	LiNO$_3$	264	600	710 LiCO$_3$
Ag$^+$	2.4	1.26	AgNO$_3$	212	444	2820 Ag$_2$O
K$^+$	1.33	1.33	KNO$_3$	334	400	1894 KNO$_3$
Rb$^+$	1.98	1.49	RbNO$_3$	310	-	1200 RbCl
Tl$^+$	5.2	1.49	TlNO$_3$	206	430	25 Tl$_2$SO$_4$
Cs$^+$	3.34	1.65	CsNO$_3$	414	-	1200 CsNO$_3$

The most common practice to fabricate optical waveguides by ion exchange is to immerse the glass substrate in a molten salt bath at a high temperature: the salts generally used, with their melting and decomposition temperatures, are also listed in Table 1, while Figure 1 shows the schematic model of this process, with reference to the exchange between the Na$^+$ ion contained in the glass (ion A) and a generic B ion (K$^+$, Ag$^+$, ...) provided by the nitrate melt.

An alternative approach, however, can also be followed to produce ion-exchanged waveguides: a metal film is deposited onto the glass surface, and diffusion is performed by applying an external electric field. Since the early work by Chartier et al.[30], most of the activity in this area has been concerned with silver-film ion exchange, either using a totally dry process or using molten salts as electrodes.[31] Other metal films as well could be used as ion sources, but so far, at least at the author's knowledge, only copper ion-exchange has been demonstrated.[32]

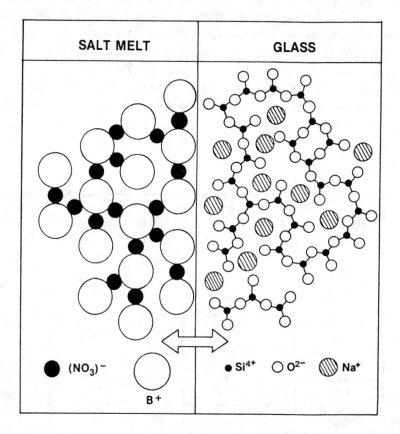

Figure 1. Diagram of the ion-exchange process in a molten salt. (from ref.21)

2.2 The choice of the glass substrate

The ion-exchange process significantly depends on the glass composition, and various glasses have been specially designed, in particular to optimize the fabrication of commercial components and devices[33-37]. The primary requirement is obviously related to the high content of an alkali ion (usually Na^+ or K^+), in order to allow the ion exchange to proceed easily and quickly. Other requirements concern the homogeneity and purity of glass: a glass free from undesirable impurities like iron and arsenic is of particular importance for Ag^+ - Na^+ exchange, since reduction of silver ion to elementary silver and following clustering processes are the main responsible of glass coloration and associated propagation losses.[18] On the other hand, laser-induced coloring of silver-exchanged waveguides has also been suggested as an effect to be exploited in new applications, e.g. to store information in waveguides.[38]

The composition of some common and special glasses for ion-exchange is listed in Table 2; a reference is given for each glass since composition can change slightly from batch to batch even for commercial optical glasses and not all the published data are in mutual agreement. Variations, however, are not large, as it can be seen even for the first three glasses in Table 2, which correspond to commercial microscope slides: these soda-lime glasses are the most used substrates in a large number of laboratories, due to their high Na^+ content, easy availability, and low cost. Schott BK7 and 270 and Corning 0211 and B1664 are optical quality glasses; BK7 and B1664 are widely used borosilicates for conventional optical components such lenses. Schott 270 is particularly suited to be strengthened in a potassium nitrate melt. Last column in Table 1 refers to a glass specially developed in Jena University for multimode waveguides.[36] As to other special glasses, such as the alumoborosilicate developed in Grenoble for thallium-potassium ion exchange[35] or the ones developed by Corning and Schott (e.g., the latter has developed a glass, BGG21, for Cs^+-K^+ exchange and another, BGG31, for Ag^+-Na^+), their exact composition has not been disclosed. The development of special glasses which would be suitable for ion-exchange as well is of critical importance in the area of photonic applications, where nonlinear optical and active materials are being investigated: glasses doped with semiconductor crystallites or with rare earths are deserving particular attention.

2.3 The overall ion-exchange process and device fabrication

Accurate design and modelling tools represent an essential step towards the optimization of any integrated optical device: a continuous feedback is necessary between the design and the fabrication stages to guarantee a high-yield manufacturing process. A number of numerical modelling algorithms for planar and rib (or channel) integrated optical components are already available.[41-44]

Referring to ion-exchange, in order to establish a firm correlation between the process parameters and the required waveguide performance it is necessary to investigate the different steps involved in the overall process; according to Ramaswamy and Srivastava[19], such steps can be categorized in the following areas: i) study of the ion-exchange equilibrium (related to the choice of the melt

TABLE 2. Chemical composition and refractive index of some glass substrates for ion-exchange

Oxide Content % in weight	Chance Propper Select (ref.39)	Chance Propper Gold (ref.39)	Fisher Premium (ref.21)	Schott BK7 (ref.39)	Corning 0211 (ref.21)	Corning B1664 (ref.40)	Schott 270 (ref.41)	Special alumo-boro-silicate (ref.36)
SiO_2	73.24	73.55	72.2	70.36	65	68.3	73.49	mol-% 37.5
Na_2O	14.07	13.86	14.3	9.55	7	9.9	2.61	25.0
K_2O	1.8	0.28	1.2	7.05	7	8.0	12.65	
CaO	6.45	6.07	6.4	0.31			6.65	
MgO	408	3.92	1.2	0.01			2.30	
B_2O_3				11.35	9	11.1		12.5
Al_2O_3	1.46	1.73	1.2	0.28	2		2.30	25.0
BaO				0.43		2.4		
SO_3	0.34	0.43		0.05				
Fe_2O_3	9	0.08		0.01	ZnO 7			
Other components	0.09	0.08	3.5	0.60	TiO_2 3	3		
Refractive index at $\lambda = 633$ nm	1.5111	1.5102	1.512	1.5150	1.522	1.5152 $\lambda = 628.5$ nm	1.5207 $\lambda = 780$ nm	1.5100

composition and of the glass substrate);
ii) solution of the diffusion equation using the appropriate boundary conditions (e.g. related to the use of stirrers in the melt) to predict the ion concentration profile in the glass;
iii) fabrication of the waveguide;
iv) experimental characterization of the waveguide (determination of the refractive index profile and correlation with the theoretical analysis; measurements of the effective indices of the guided modes and/or of the modal field profile, measurements of the propagation losses).

Ion-exchange in glass has thus evolved into a relatively mature technology, and a few commercial components, mostly for telecommunication and sensing applications, are already available on the market. Research, however, is continuing to optimize both the processes and the components design; the demonstrations of some novel structures and devices in gradient-index waveguides are being published at a regular rate.[45-56]

3. REPRODUCIBILITY OF FABRICATION AND CHARACTERIZATION PROCESSES

3.1 Manufacturing tolerances

A high degree of repeatability of the effective refractive index produced by the ion-exchange process is of paramount importance in the batch manufacture of waveguide structures, which may have to be critically phase-matched to a predetermined figure. The problem of manufacturing tolerances which can guarantee a high reproducibility has been investigated since the early developments of ion exchange: the repeatability depends on the temperature and time control imposed in the manufacturing process and on the purity and consistency of the initial materials. It has been shown, for instance, that in the case of $Ag^+ - Na^+$ exchange the melt dilution increases the reproducibility of the processes and relaxes the tight constraints necessary for undiluted melts.[57,58] Typical figures of the standard deviation δn_e of the effective index n_e of planar waveguides obtained in pure $AgNO_3$ are in the range 4 to 7 10^{-4} for undiluted melts when the melt temperature is controlled at ± 0.1 °C and the exchange time at ± 2 s. For diluted melts (with Na^+ mole fraction exceeding 90%) the temperature control can be relaxed to ± 0.6 °C, while keeping the standard deviation almost unchanged. In general, the maximum expected standard deviation $\delta n_e (tot)$ can be expressed as the sum in quadrature of three terms:

$$\delta n_e (tot)^2 = \delta n_e (T_{max})^2 + \delta n_e (t_{max})^2 + \delta n_e (e)^2 \qquad (2)$$

where $\delta n_e (T_{max})$ and $\delta n_e (t_{max})$ are the variations due to the maximum errors in temperature and time, respectively, while $\delta n_e (e)$ is due to the combination of the error in measurement of the synchronous coupling angle and of the standard deviation about the mean substrate refractive index.[57]

Recent tests on waveguides fabricated in the author's laboratory both by K^+ - Na^+ and Ag^+ - Na^+ ion exchange (further details are given in the following subsection) have confirmed standard deviations of n_e in the range $3\ 10^{-4}$ to $7\ 10^{-4}$, depending on the order of mode: these small variations, however, have been obtained with a temperature control not particularly accurate ($\delta T_{max} = \pm 1\ °C$).

3.2 Reproducibility of characterization measurements

The precise experimental determination of the modal propagation constants as well as of the index profile of a fabricated waveguide is in turn affecting all the design and modelling procedure. In the case of ion-exchanged guides, the feedback between modelling and fabrication processes is made more difficult by the fact that the guiding structure presents a graded index profile (most of the times *apriori* unknown), which makes impossible, apart few special cases, an exact treatment of the field propagation inside the waveguide. The modal field has therefore to be numerically evaluated, and the reconstruction of the index profile becomes quite critical for the accuracy of the entire procedure.[59]

For the purpose of assessing the average accuracy and reproducibility of the measurements of the waveguide propagation constants by using prism coupling techniques, a round-robin *Testing of Optical Waveguides* (TOW) project has been recently promoted.[60] Seven laboratories in five countries have agreed to be involved in the measurement of the propagation constants of two sets of K^+/Na^+ and Ag^+/Na^+ ion-exchanged waveguides and the subsequent reconstruction of their refractive index profiles. Different numerical reconstruction methods, namely those routinely used in the laboratories involved, have been compared; for few sample waveguides, a comparison has also been possible between the numerically reconstructed profiles and the profile directly measured through the refracted-near-field (RNF) method.[61]

3.2.1. Waveguide fabrication

Two sets of planar waveguides have been produced by ion-exchange in soda-lime glass microscope slides ("Gold Star" by Chance Propper Ltd., UK): the first set by K^+/Na^+ ion-exchange (100% KNO_3 bath; diffusing time: 9 hours at 395 °C) and the second by Ag^+/Na^+ (20% molar $AgNO_3$ in $NaNO_3$ bath; t = 10 minutes at 305 °C) ion-exchange. Both these exchange times were chosen so that waveguides having respectively 4 and 5 guided modes (TE and TM) were produced: this structure represents a good compromise between the need of a reliable characterisation of the samples and a short measurement time. The waveguides were fabricated in a Techne SBL-2 fluidised-bath oven; the heating of the stainless steel container of the nitrate melt is accomplished in this type of oven by electrical heaters placed at the bottom of the oven and by the "boiling" of the sand kept in movement by a continuos air flow, which guarantees strong heat convection. Good, but not particularly accurate (± 1 °C) temperature stability is achieved through the use of two temperature controls, with thermocouples inside the melt and in the inert sand. The substrates, in a stainless steel holder, were placed in the molten salt horizontally and were separated about 1 cm from each other to allow sufficient solution mobility.

3.2.2. Experimental considerations

A preliminary simple computation of the apriori errors on the effective index of the guided modes as a function of the indetermination of the most relevant experimental parameters has been performed to ascertain which experimental quantities have a stronger influence on the measurement. Assuming typical working conditions for prism coupling as summarised in Table 3, the total error that can be expected on the value of the effective index is of the order of $4 \cdot 10^{-4}$, with the strongest influence coming from any error made on the assumed value of the refractive index and of the base angle of the prism used in the measurement. It is therefore very important to have precise knowledge of these two parameters, since any indetermination affecting them produces a final experimental error of the same order of magnitude.

TABLE 3. Main factors contributing to the error on the measured effective index

	Working condition	Error	Contribution to the error on effective index
Prism base angle (deg)	62°	0.02°	$3 \cdot 10^{-4}$
Prism index	1.750	10^{-4}	10^{-4}
Coupling angle (deg)	4°	0.02°	10^{-5}

3.2.3. Numerical reconstruction of the index profile

Various numerical methods were used by the different groups of the TOW project in order to reconstruct the refractive index profile of the waveguides under test. Most of them were based on the WKB approximation, namely using the dispersion equation for the guided modes in the form:

$$\frac{2}{\lambda_0} \int_0^{x_t} \sqrt{n^2(x) - n_e^2(m)} dx = m\pi + \phi_a + \frac{\pi}{4}, \quad (3)$$

where λ_o is the wavelength of the input light in vacuum, $n_e(m)$ is the measured effective index of the m-th mode, x_t is the turning point and $n(x)$ is the value of the refractive index profile at depth x. This equation must be numerically fitted: in the Forward WKB method a trial index profile is chosen and its parameters fitted to the experimental data; the other methods (Chiang, Inverse WKB, White-Heidrich)[62-65] use recursive algorithms to find empirical profiles starting from the experimental effective indexes. Another method starts from the diffusion equation and retains a more direct link to the physical quantities involved in the ion-exchange process.[66]

3.2.4. Characterization results

The effective indices measured by all groups show a very good agreement, as it can be seen in Table 4, where the data for a typical waveguide of the K-exchanged

TABLE 4. Effective indices measured by all the TOW cooperating groups (1 to 7) on a same K - exchanged waveguide.

Mode	1	2	3	4	5	6	7	Average	Stand. Dev.
TE 0	1.5166	1.5167	1.5172	1.5167	1.5174	1.5166	1.5175	1.5170	0.0004
TE 1	1.5141	1.5143	1.5147	1.5143	1.5149	1.5142	1.5151	1.5145	0.0004
TE 2	1.5122	1.5124	1.5128	1.5123	1.5130	1.5124	1.5132	1.5126	0.0004
TE 3	1.5108	1.5109	1.5113	1.5109	1.5115	1.5109	1.5117	1.5111	0.0004
TE 4					1.5102				
TM 0	1.5179	1.5182	1.5186	1.5181	1.5188	1.5179	1.5189	1.5183	0.0004
TM 1	1.5151	1.5154	1.5158	1.5154	1.5160	1.5151	1.5161	1.5156	0.0004
TM 2	1.5130	1.5131	1.5128	1.5132	1.5138	1.5129	1.5139	1.5132	0.0005
TM 3	1.5113	1.5114	1.5117	1.5114	1.5120	1.5113	1.5122	1.5116	0.0004
TM 4	1.5102	1.5102			1.5108			1.5104	0.0004

set are shown. The maximum difference between the various measurements for the TE_o mode is $9 \cdot 10^{-4}$, which represents a relative error of about 0.05%. The standard deviation of the measurements is of the order of $4 \cdot 10^{-4}$, and the error on the surface index, as resulting from profile reconstruction, is of about the same magnitude. Thus, not surprisingly, the standard deviation resulting from measurements is quite close to the expected error, computed in the previous subsections. Figure 2 shows graphically the the total dispersion of the data for the measurements of all laboratories on all waveguides; different symbols indicate the laboratory (1 to 7) where the related measurement was performed. One can see that there are no important changes between the propagation constants of different waveguides of the same set (a statistical analysis shows that the maximum overall standard deviation is $3\ 10^{-4}$, not significantly different from the value found on a single waveguide). Thus it comes out that also the overall exchange process had been quite reproducible, at least inside the experimental errors.

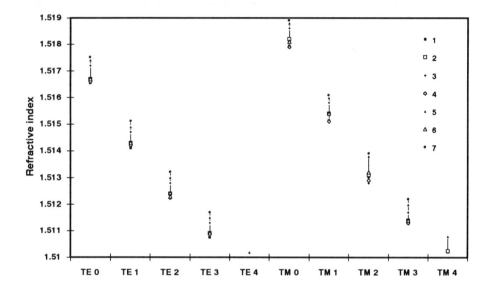

Figure 2 - Dispersion of the effective indices of all 6 K-exchanged waveguides as measured by all groups.

The reconstructed refractive index profiles for the same waveguide considered in Table 4 are drawn in Figure 3 and compared with the result of a direct RNF measurement. In this case slight variations due to the different numerical methods are visible; in particular the profiles produced by Inverse WKB reconstructions tend to be higher than those obtained by starting from trial profile functions. The IWKB method seems therefore not being able to compensate adequately the change of the profile slope near the surface of the waveguide. However, the maximum difference between the values of the surface index is only 0.0018, with a standard deviation of 0.0007 which is of the same order as the deviation of the effective indexes measurements. More problems come out from the observation that the value of the surface index given by the RNF measurement is clearly smaller than all the reconstructed ones, the maximum difference being 0.0031. By considering that the shape and width of the curve for measured and reconstructed profiles are very close, and that the difference between the measured surface index and the average of the reconstructed ones reduces to 0.00212, with a relative error of about 0.1%, we can conclude that the agreement is quite satisfactory anyway. Moreover, it has also to be taken into account that even the direct measurements are generally not able to give exactly the surface index, some kind of extrapolation being usually necessary.

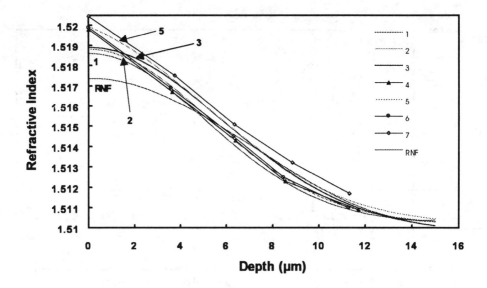

Figure 3 - Refractive index profiles of the same waveguide of Table 4 as numerically reconstructed by each group. RNF designates the profile measured by the Refracted Near-Field Method.

Analogous results have been obtained in the characterization of Ag-exchanged waveguides; the dispersion of effective indices measured on all the 6 samples by all groups is essentially of the same order of that observed for K-echanged waveguides. For what concerns the reconstruction of the index profile, the various results are drawn in Figure 4; it can be said that the standard deviation on the average surface refractive index turns out to be $4 \cdot 10^{-3}$, with a maximum difference of $8 \cdot 10^{-3}$, values which are not far from the results obtainable even in the case of step-index waveguides. Again, it is evident that methods using Inverse WKB approximation tend to produce slightly higher values respect to methods based on the use of trial functions.

From these tests, it appears that there are no significant differences in repeatability, both for the fabrication and the characterization of potassium- or diluted-silver-exchanged waveguides.

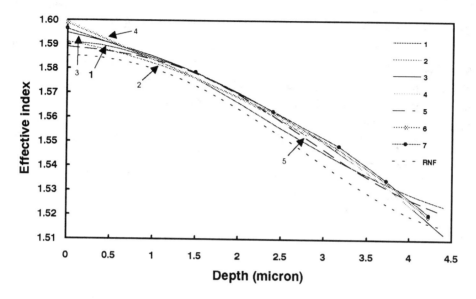

Figure 4. Reconstructed index profiles of a same waveguide produced by diluted-silver ion-exchange.

4. CONCLUSIONS

Ion-exchange in glass substrates is a relatively mature technology, which has already allowed producers to batch manufacture low-loss and rugged components and devices. However, not all the advantages of the ion-exchange technique have been exploited so far, and that depends also on the limited availability of glasses specially designed for this purpose.

Some R&D areas still deserve great attention, and further efforts have to be pursued onto the following targets:
➥ Improvement of the design and modelling techniques for what concerns both the actual ion-exchange process and the characteristics of the graded-index waveguide resulting from it; particular attention can be paid to the problems of buried waveguides[67-70] and to multiple-step processes which can allow a better control of the index profile and/or of the confinement properties of the waveguide.[71, 72]
➥ Improvement of the characterization methods: the preliminary results of the TOW project have shown that there is a quite good agreement among the measurements

routinely performed in different research laboratories, but a further step towards a standardization of the techniques, especially for what concern the measurement of propagation losses, would be desirable. More attention should also be paid to the standardization of measurements concerning channel waveguides.[73, 74]

➥ Investigation of post-fabrication processes, such as annealing, with the aim of developing techniques for a fine tuning of the fabricated devices.[75]

➥ Development of easy, rugged and low-cost packaging techniques: packaging, in fact, appears to be a critical issue in the device manufacturing process. Advanced assembly and interconnection technologies should be used, e.g. to guarantee a high-efficiency fiber-chip coupling with monomode optical fibers; an example of such technologies is represented by the LIGA technique, which allows to create three-dimensional microstructures by using a combination of lithography, electroforming and plastic moulding.[76] An example of fiber-chip coupling structure, where the integrated optical chip can be inserted, is shown in Figure 5 (courtesy of MicroParts GmbH, Dortmund).

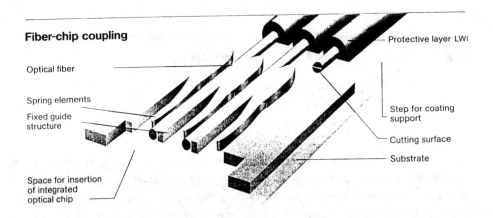

Figure 5. Sketch of a structure fabricated by the LIGA process for connection of monomode fibers to an external integrated optical chip.[76]

➥ Exploitation of the potentialities of ion-exchanged structures and techniques for nonlinear optical devices and optical amplifiers and lasers. Besides the fabrication of optical waveguides in doped-glass substrates, which has been already succesfully demonstrated,[77-85] some peculiar approaches should be further explored.[86] Just to mention few examples, ion-exchanged waveguides have potential as nonlinear optical structures[87-89] and direct exchange of rare earths like Erbium into a glass,[90] once reproducibility problems are solved, can represent an attractive approach to fabricate optical waveguide amplifiers.

5. REFERENCES

1. S. Franck, Glass and Archeology, Academic Press, New York, 1982.
2. R.W. Douglas and S. Frank, A History of Glassmaking, G.T. Foulis & Co. Ltd., Henley-on-Thames (U.K.), 1972.
3. F. Scandone, Galileo and the Telescope, Officine Galileo, Firenze, 1967.
   Related topics are treated, for instance, in:
   A.R. Cooper, "Glass and optics: a historical perspective", in *From Galileo's "occhialino" to optoelectronics*, P. Mazzoldi Ed., pp.50-73, World Sci., Singapore, 1993.
   V. Greco, G. Molesini, and F. Quercioli, "Modern optical testing on the lenses of Galileo", *ibidem*, pp.110-121.
4. W.H. Zachariasen, "The atomic arrangement in glass", J. Amer. Chemical Soc. **54**, 3841-3851, 1932.
5. see for instance:
   C.H.L. Goodman, "The structure and properties of glass and the strained mixed cluster model", Phys. & Chem. of Glasses **26**, 1-10, 1985.
   G.N. Greaves, "EXAFS and the structure of glass", J. Non-Cryst. Solids **71**, 203-217, 1985.
   M.D. Ingram, "A new mechanism of ionic conduction in glass", Material Chem. & Phys. **23**, 51-61, 1989.
6. W.A. Weyl, Coloured Glasses, p. 415, Society of Glass Technology, Sheffield, 1954.
7. H. Rawson, Properties and Applications of Glass, pp. 216-221, Elsevier, Amsterdam, 1980.
8. G. Schulze, "Versuche über die diffusion von silber in glas", Angew. Physik **40**, 335-367, 1913.
9. S.S. Kistler, "Stresses in glass produced by nonuniform exchange of monovalent ions", J. Am. Ceram. Soc. **45**, 59-68, 1962.
10. R.H. Doremus, "Exchange and diffusion of ions in glass", J. Phys. Chem. **68**, 2212-2218, 1964.
11. A.J. Burggraaf and J. Cornelissen, "The strengthening of glass by ion exchange; Part 1, Stress formation by ion diffusion in alkali aluminosilicate glass", Phys. Chem. Glasses **5**, 123-129, 1964.
12. J.A. Marinsky, Ion exchange - A series of advances, Dekker, New York, 1969.
13. G.C. Righini, "25 years of integrated optics: where we are and where we will go", in *Linear and Nonlinear Integrated Optics*, Proc. SPIE **vol. 2212**, 1-6, 1994.
14. G.C. Righini, "Passive and active glasses for integrated optics", in *From Galileo's "occhialino" to optoelectronics*, P. Mazzoldi Ed., pp.272-294, World Sci., Singapore, 1993.
15. T. Izawa and H. Nagakome, "Optical waveguide formed by electrically induced migration of ions in glass plates", Appl. Phys. Lett. **21**, 584-586, 1972.
16. T.G. Giallorenzi, E.J. West, R. Kirk, R. Ginther, and R.A. Andrews, "Optical waveguides formed by thermal migration of ions in glass", Appl. Opt. **12**, 1240-1245, 1973.
17. M. Mc Court, "Status of glass and silicon-based technologies for passive components", Europ. Transact. on Telecom. ETT **4**, 685-689, 1993; - , "Commercial glass waveguide devices", in this Volume.
18. T. Findakly, "Glass waveguides by ion-exchange: a review", Opt. Engin. **24**, 244-250, 1985.

19. R.V. Ramaswamy and R. Srivastava, "Ion-exchange glass waveguides: a review", IEEE J. Lightwave Techn. **LT-6**, 984-1002, 1988.
20. L. Ross, "Integrated optical components in substrate glasses", Glastech. Ber. **62**, 285-297, 1989.
21. S. Iraj Najafi, Ed., <u>Introduction to Glass Integrated Optics</u>, Artech House, Boston, 1992.
22. for what concerns relevant papers published before 1991, see references 17 to 21 and references quoted therein.
23. H. Kahnt and J.M. Reau, "Effective medium approach to the mixed mobile ion effect", J. Non-Cryst. Solids **125**, 143-150, 1990.
24. H. Dunken, "Surface chemistry of optical glasses", J. Non-Cryst. Solids **129**, 64-75, 1991.
25. "Glassy Ionics", in J. Non-Cryst. Solids **131-133**, 1113-1122, 1991.
26. M.D. Ingram, "Relaxation processes in ionically conducting glasses", J.Non-Cryst. Solids **131-133**, 955-960, 1991.
27. S.N. Houde-Walter, J.M. Inman, A.J. Dent, ".Sodium and silver environments and ion-exchange processes in silicate and aluminosilicate glasses", J. Physical Chem. **97**, 9330-9336, 1993.
28. H.L. Anderson, Ed., <u>Physics Vade Mecum</u>, pp. 294-295, Am. Inst. Physics, New York, 1981.
29. L. Ross, "Buried waveguides for passive integrated optics by Cs^+ ion-exchange", in *Integrated Optical Circuit Engineering III*, R.T. Kersten Ed., Proc. SPIE **vol.651**, 32-34, 1986.
30. G.H. Chartier, P. Jaussaud, A.D. de Oliveira and O. Parriaux, "Optical waveguides fabricated by electric-field controlled ion exchange in glass", Electron. Lett. **14**, 132-134, 1978.
31. S. Honkanen, "Silver-film ion-exchange technique", Chap. 3 , pp.39-71, in Ref.21.
32. S.S. Gevorgyan, "Single-step buried waveguides in glass by field-assisted copper ion-exchange", Electron. Lett. **26**, 38-39, 1990.
33. A. Beguin, T. Dumas, M.J. Hackert, R. Jansen and C. Nissim, "Fabrication and performance of low loss optical components made by ion exchange in glass", IEEE J. Lightwave Techn. **LT-6**, 1483-1487 , 1988.
34. L. Ross, H.-J. Lilienhof, H. Holscher, H.F. Schlaak, and A. Brandenburg, "Improved substrate glass for planar waveguide by Cs-ion exchange", in *Techn. Digest Topical Mtg. on Integrated and Guided-Wave Optics* (Atlanta, Georgia, 1986), paper ThBB2.
35. J.F. Bourhis, <u>Contribution à l'étude des mecanismes de l'échange d'ions thallium/potassium dans un verre aluminoborosilicate. Application à la réalisation de composants en optique integrée unimode</u>, PhD Dissertation at Institut National Polytechnique de Grenoble, 1988.
36. C. Kaps, W. Karthe, R. Muller, T. Possner, and G. Schreiter, "Ion exchange in an alumo-boro-silicate glass especially developed for multimode integrated optical elements", in *Glasses for Optoelectronics*, G.C. Righini Ed., Proc. SPIE **vol.1128**, 132-137, 1989.
37. F. Rehouma, D. Persegol, A. Kevorkian, and F. Saint André, "Fabrication and characterization of a buried coupler made by ion-exchange in a special glass", in *Proc. 6th Europ. Conf. on Integrated Optics* (CSEM, Neuchatel), pp. 4-19 to 14.21, 1993.
38. L.B. Glebov, N.V. Nikonorov, and G.T. Petrovskii, "Optical breakdown and laser

coloring of diffused waveguides containing silver", Sov. Phys. Tech. Phys. **28**, 1477-1478, 1983.
39. G.C. Righini, R. Shen, A.M. Losacco, C. Esposito, G. Belli, and R. Saracini, "Ion-exchanged glass waveguides for integrated optics", in *Quantum Electronics and Plasma Physics - 5th Italian Conference*, G.C. Righini Ed. (Italian Physical Society, Bologna) Conf. Proc. **vol.21**, 425-429, 1989.
40. C. Visconti, Etude de la birifringence dans des guides optiques fabriqués par échange d'ions sur substrat de verre: separateur de polarisation, PhD Dissertation at Institut National Polytechnique de Grenoble, 1988.
41. I. Schanen Duport, Etude de structures collimatrices en optique integrée sur verre: application à l'interferometrie, PhD Dissertation at Institut National Polytechnique de Grenoble, 1992.
42. G.L. Yip, "Characterization, modeling, and design optimization of integrated optical waveguide devices in glass", in *Glasses for Optoelectronics* II, G.C. Righini Ed., Proc. SPIE **vol.1513**, 26-36, 1991.
43. G. Perrone, A. Gulisano, D. Petazzi, L. Macchia, I. Montrosset, S. Morasca, F. Pozzi, C. De Bernardi, and G. Zaffiro, "Glass integrated optical WDM devices: a comparison between experimental results and modelling by CAOS software", in *Linear and Nonlinear Integrated Optics*, Proc. SPIE **vol. 2212**, paper 2212-45, 1994.
44. M.A. Forastiere, G.C. Righini, G. Tartarini, and P. Bassi, "Transversally multi-modal 1x4 branch coupler in glass: experimental characterization and BPM modelling", in *Linear and Nonlinear Integrated Optics*, Proc. SPIE **vol. 2212**, paper 2212-72, 1994.
45. D. Jestel, A. Baus, and E. Voges, "Integrated-optic interferometric microdisplacement sensor in glass with thermo-optic phase modulation", Electron. Lett. **26**, 1144-1145, 1990.
46. R.A. Betts, F. Lui, and S. Dagias, "Wavelength and polarization insensitive optical splitters fabricated in K^+ / Na^+ ion-exchanged glass", IEEE Photon. Techn. Lett. **2**, 481-483, 1990.
47. M.-J. Li and S. I. Najafi, "Fully planar ion-exchanged glass channel waveguides with grating taps", Intl. J. Optoelectron. **6**, 575-577, 1991.
48. Zheng Chen and Ji Zhi Dai, "K-ion exchange waveguide directional coupler sensor", in *Optical Fibre Sensors in China*, Proc. SPIE **vol.1572**, 129-131, 1991.
49. A. Tervonen, P. Poyhonen, S. Honkanen, M.T. Tahkokorpi, "Channel waveguide Mach-Zehnder interferometer for wavelength shifting and combining", in *Glasses for Optoelectronics II*, G.C. Righini Ed., Proc. SPIE **vol.1513**, 71-75, 1991.
50. T. Poszner, G. Schreiter, R. Muller, "Stripe waveguides with matched refractive index profiles fabricated by ion exchange in glass", J. Appl. Phys. **70**, 1966-1974, 1991.
51. P.C. Noutsios, G.L. Yip, and J. Albert, "Novel vertical directional coupler made by field-assisted ion-exchanged slab waveguides in glass", Electron. Lett. **28**, 1340-1342, 1992.
52. I. Duport, P. Benech, D. Khalil, "Study of linear tapered waveguides made by ion exchange in glass", J.of Physics. D **25**, 913-918, 1992.
53. S.I. Najafi, P. Lefebvre, J. Albert, "Ion-exchanged Mach-Zehnder interferometers in glass", Appl. Opt. **31**, 3381-3383, 1992.
54. S.I. Bozhevolnyi, K. Pedersen, "Second-harmonic generation in channel glass

waveguides", Appl. Opt. **31**, 5813-5815, 1992.
55. G. Voirin, L. Falco, O. Boillat, O. Zogmal, P. Regnault, and O. Parriaux, "Monolithic double interferometer displacement sensor with wavelength stabilization", in *Proc. 6th Europ. Conf. on Integrated Optics* (CSEM, Neuchatel), pp. 12-28 to 12.29, 1993.
56. S. Honkanen, P. Poyhonen, A. Tervonen, "Waveguide coupler for potassium- and silver-ion-exchanged waveguides in glass", Appl. Opt. **32**, 2109-2111, 1993.
57. C.A. Millar and R.H. Hutchins, "Manufacturing tolerances for silver-sodium ion-exchanged planar optical waveguides", J. Phys. D **11**, 1567-1576, 1978.
58. C.A. Millar, G. Stewart, and R.H. Hutchins, "Improved repeatability of ion-exchanged waveguides by melt dilution", *Conference presentation (1978)*.
59. A. Tervonen, "Theoretical analysis of ion-exchanges glass waveguides", in ref.21, Chap.4, pp. 73-105, 1992.
60. S. Pelli, G.C. Righini, A. Scaglione, G.L. Yip, P. Noutsious, A. Brauer, P. Dannberg, J. Linares, C. Gomez Reino, G. Mazzi, F. Gonella, R. Rimet, and I. Schanen, "Testing of optical waveguides (TOW) cooperative project: preliminary results of the characterisation of K-exchanged waveguides", in *Linear and Nonlinear Integrated Optics*, Proc. SPIE **vol. 2212**, paper 2212-13, 1994.
61. R. Göring, M, Rothardt, Application of the Refracted Near-Field Technique to Multimode Planar and Channel Waveguides in Glass, J. Opt. Commun. **7**, 82-. 1986.
62. K. S. Chiang, Construction of refractive-index profiles of planar dielectric waveguides from the distribution of effective indexes, J. Lightwave Tech. **LT-3**, 385- , 1985.
63. S. Pelli and G.C. Righini, "Introduction to integrated optics: characterisation and modelling of optical waveguides", in: *Advances in Integrated Optics*, S. Martellucci Ed., Plenum Publishing Corporation, New York (in press).
64. G.L. Yip, J. Albert, "Characterization of planar optical waveguides by K^+-ion exchange in glass", Opt. Lett. **10**, 151, 1985.
65. J.M. White and P.F. Heidrich, "Optical waveguide refractive index profiles determined from measurements of mode indices: a simple analysis", Appl. Opt. **15**, 151, 1976
66. J. Liñares, X. Prieto, C. Montero, "A Novel Refractive Index Profile for Optical Characterization of Nonlinear Diffusion Processes and Planar Waveguides in Glass", Optical Materials (in press).
67. J.Albert and J.W.Y. Lit, "Full modeling of filed-assisted ion exchange for graded index buried channel optical waveguides", Appl. Opt. **29**, 18, 1990.
68. P.C. Noutsios and G.L. Yip, "Diffusion and propagation characteristics of buried single-mode waveguides in glass", IEEE J. Quantum Electron. **QE-27**, 549-555, 1991.
69. J. Liñares, C. Montero, X. Prieto and R. de la Fuente, "Simultaneous characterization of surface and buried waveguides produced by ion-exchange in glass", J. Modern Opt. **41**, 5-9, 1994.
70. J. Liñares, X. Prieto, C. Montero, C. Gomez-Reino, G.C. Righini, and S. Pelli, "Buried waveguides fabricated by a purely thermal ion back diffusion in glass and assisted by electric field: a new model", in *Linear and Nonlinear Integrated Optics*, Proc. SPIE **vol. 2212**, paper 2212-15, 1994.
71. M.J. Li, S. Honkanen, W.J. Wang, R. Leonelli, J. Albert, and S.I. Najafi,

"Potassium and silver ion-exchanged dual-core glass waveguides with gratings", Appl. Phys. Lett. **58**, 2607-2609, 1991.
72. A. Tervonen, P. Poyhonen, S. Honkanen, M. Tahkokorpi, and S. Tammela, "Examination of two-step fabrication methods for single-mode fiber compatible ion-exchanged glass waveguides", Appl. Opt. **30**, 338-343, 1991.
73. K. Morishita, "Index profiling of three-dimensional optical waveguides by the propagation-mode near-field method", J. Lightwave Technol. **LT-4**, 1120, 1986.
74. A. Sharma and P. Bindal, "Analysis of near field data of diffused channel waveguides: an accurate method to obtain Δn and the index profile", in *Linear and Nonlinear Integrated Optics*, Proc. SPIE **vol. 2212**, paper 2212-17, 1994.
75. G. Zhang, S. Honkanen, S.I. Najafi, A. Tervonen, and P. Katila, "Effect of thermal post-annealing on spectral transmission of 0.807/1.3 μm, 1.48/1.55 μm and 1.3/1.55 μm ion-exchanged Mach-Zehnder interferometer WDMs", in *Nanofabrication Technologies and Device Integration*, W. Karthe Ed., Proc. SPIE **vol.2213**, paper 2213-27, 1994.
76. J. Mohr, "The LIGA technique - what are the new opportunities", paper presented at MME'92 (3rd Workshop on Micromachining, Micromechanics and Microsystems, Leuven, Belgium, 1992).
77. N. Finlayson, W.C. Banyai, C.T. Seaton, G.I. Stegeman, M.O'Neill, T.J. Cullen, and C.N. Ironside, "Optical nonlinearities in CdS_xSe_{1-x} doped glass waveguides", J. Opt. Soc. Am. B **6**, 675-684, 1989.
78. S.K. Han, Z. Huo, R. Srivastava, and R.V. Ramaswamy, "Thermal nonlinear absorption and power limiting in waveguides", J. Opt. Soc. Am. B **6**, 663-667, 1989.
79. G.C. Righini, G.P. Banfi, V. Degiorgio, F. Nicoletti, and S. Pelli, "Semiconductor doped glasses: structural and waveguide characterization", Mater. Sci. & Engin. **B9**, 397-403, 1991.
80. B.P. Howell and T. Beerling, "Evaluation of ion exchange for fabrication of rare-earth doped waveguides", in *Optoelectronic Materials, Devices, Packaging and Interconnects*, Proc. SPIE **vol.836**, 44-47, 1987.
81. S.I. Najafi, W-J Wang, J.F. Currie, R. Leonelli, and J.L. Brebner, "Fabrication and characterization of neodymium-doped glass waveguides", IEEE Photon. Technol. Lett. **1**, 109-110, 1989.
82. H. Aoki, O. Maruyama, and Y. Asahara, "Glass waveguide laser", IEEE Photon. Technol. Lett. **2** 459-460, 1990.
83. A.N. Miliou, X.F. Cao, R. Srivastava, and R.V. Ramaswamy, "15-dB amplification at 1.06 μm in ion-exchanges silicate glass waveguides", IEEE Photon. Technol. Lett. **4**, 416-418, 1992.
84. N.A. Sanford, K.J.Malone, D.R. Larson, and R.K. Hickernell, "Y-branch waveguide glass laser and amplifier", Opt. Lett. **16**, 1168,1170, 1991.
85. T. Feuchter, E.K.Mwarania, J. Wang, L. Reekie, and J.S. Wilkinson, "Erbium-doped ion-exchanged waveguide lasers in BK-7 glass", IEEE Photon. Technol. Lett. **4**, 542-544, 1992.
86. E. Snoeks, G.N. van der Hoven, A. Polman, B. Hendriksen, and M.B.J. Diemeer, "Doping fibre-compatible ion-exchanged channel waveguides with erbium by ion implantation", in *Proc. 6th Europ. Conf. on Integrated Optics* (CSEM, Neuchatel), pp. 3-38 to 3-39, 1993.
87. J.L. Jackel, E.M. Vogel, and J.S. Aitchison, "Ion-exchanged optical waveguides for all-optical switching", Appl. Opt. **29**, 3126-3129, 1990.

88. P.R. Ashley, M.J. Bloemer, J.H. Davis, "Measurement of nonlinear properties in Ag-ion exchange waveguides using degenerate four-wave mixing", Appl. Phys. Lett. **57**, 1488-1490, 1990.
89. A. Quaranta, F. Gonella, G. Mazzi, and A. Sambo, "Optical and compositional characterization of Ag-containing waveguiding systems", in *Linear and Nonlinear Integrated Optics*, Proc. SPIE **vol. 2212**, paper 2212-14, 1994.
90. X.H. Zheng and R.J. Mears, "Planar optical waveguides formed by erbium ion exchange in glass", Appl. Phys. Lett. **62**, 793-795, 1993.

Critical issues in designing glass integrated optical circuits

Ari Tervonen

Optonex Ltd
P.O. Box 128, FIN-02101, Espoo, Finland
and
Nokia Research Center
Helsinki, Finland

ABSTRACT

Different issues in design of passive integrated optics devices based on glass waveguide technologies are reviewed. The established fabrication technologies, ion exchange into substrate glasses, and deposited doped-silica waveguides on silicon wafers, are described in terms of waveguide engineering for differing needs. The status of various optical guided-wave propagation methods and tools are discussed. Design aspects for two most important categories of glass waveguide devices, M x N couplers and wavelength division demultiplexers/multiplexers, are considered. Particularly, designs with insensitivity to fabrication process tolerances and optical signal bandwidth and polarization state are emphasised. A brief look is also taken on design of complicated guided-wave circuits.

1. INTRODUCTION

The typical characteristics of glass waveguide technologies are low optical losses, good compatibility with singlemode optical telecommunication fibers, high environmental stability and passive operation of devices. Currently, these fabrication technologies have established themselves on a level of development sufficient to produce practical integrated optics components, which can compete with conventional micro-optic and fiber-optic solutions.

This paper is an attempt to review some central aspects of glass waveguide device and component design. To start with, there is an introduction of major technologies and viewpoints on waveguide engineering within these fabrication systems. Next, the numerical analysis methods currently in use for modelling of optical propagation in waveguide structures are described. From this basis, two most important passive integrated optics device categories are discussed. Particularly, choice of designs with low sensitivity to inevitable variation in fabrication processes, and also tolerance for changes in optical signal wavelength and state of polarization is emphasised. Finally, there is a brief look at the design of more complicated guided-wave circuits built up from basic integrated optics elements and devices.

2. WAVEGUIDE ENGINEERING

The starting point for modelling of the optical propagation in guided-wave devices is knowledge of the waveguide structure, defined by the distribution of refractive index.

Inaccurate knowledge of the waveguide structure can often be a limiting factor for modelling the optical properties, and it is not useful to employ highly accurate optical simulation tools without well established waveguide profile parameters. For each waveguide fabrication process, knowledge about waveguide properties has to be achieved experimentally, but in finding out the relationship between process parameters and waveguide profiles, theoretical modelling tools for fabrication processes are important.

Waveguide engineering here means design and fabrication of optical waveguides with transverse refractive index profile distributions optimised for achieving suitable guided-mode properties for different purposes. In addition to having singlemode operation in the operational wavelength band, this usually involves minimising fiber-to-waveguide coupling losses and/or waveguide bending losses.

2.1. Ion-exchanged waveguides

Various different substrate glasses, as well as many different fabrication processes are used for ion-exchanged glass waveguide devices.[1,2] In ion-exchanged waveguides, the nature of refractive index profiles is determined by the fabrication process, and tailoring of waveguide profiles is made by complicated multistep processes. Because of this, modelling of the ion exchange processes is important. For linking the ion exchange processes and the optical properties of the waveguides, knowledge about the relationship between ion concentration and refractive index distributions is used. With the known refractive index profiles of the waveguides, theory of optical propagation in waveguides can be applied for modelling the waveguide properties and for design of integrated optics components.

The modelling of fabrication processes is concerned with development of ion concentration profiles in the glass during the waveguide fabrication, taking into account both diffusion and electric field -induced migration of ions. A general partial differential equation describes the binary ion exchange in glass, giving the change of concentration $c(x,y,t)$ of ions exchanged into the glass as a function of time t:

$$\partial C / \partial t = \frac{D\nabla^2 C}{C(M-1)+1} - \frac{D(M-1)(\nabla C)^2 + M\bar{J}_0 \cdot \nabla C}{(C(M-1)+1)^2} . \quad (1)$$

Here $C(x,y,t)$ is relative concentration $C(x,y,t) = c(x,y,t)/c_0$, with c_0 the original concentration of sodium ions in glass, x and y are the co-ordinates at the transverse cross-section of the waveguide. D is the diffusion coefficient of the incoming ions and M is the ratio of this to the diffusion coefficient of the original ions in the glass. $\bar{J}_0 = \bar{j}_0 / c_0$, with $\bar{j}_0$ the electric current density in the glass. The mask pattern geometry of the process and the ion exchange reactions at the interface of the ion source and the glass together define the boundary conditions to be used in solving the ion exchange equation. In electric field -assisted processes, also distribution of ionic current inside the substrate has to be included. The solution of equation is made numerically, typically using finite difference techniques. Detailed descriptions of the ion exchange modelling method are given in References 3-6.

Refractive indices of glass compositions with oxide constituents can be calculated from a model given in References 7 and 8. This model, based on Gladstone-Dale relations, uses a number of constants to calculate from the weight fractions of the different oxide constituents of the glass, first the density of the glass, and then the refractive index as a function of wavelength. The model actually gives a range for density and refractive index, since the value of these quantities depends not only on the composition but also on the annealing history of the glass.

Applied to ion-exchanged waveguides, the model predicts an approximate linear relation between the relative exchanged concentration and the refractive index change. There are two separate contributions to the refractive index change, from the change of ionic polarizability and from the volume change of the glass. Typically in waveguide fabrication refractive-index increasing ions with higher ionic polarizability also have higher ionic radii, which would make the glass expand. The first contribution to the refractive index change would be positive, the second negative. Because the ion exchange only affects the surface of the glass, the expansion is limited. As glass can only expand in the direction normal to the surface, it will stay compressed from its free volume.

An alternative way of looking at the deviation caused by the compression is as a stress-optical effect. For a waveguide, compressive stress exists in the glass substrate. This nonisotropic stress causes birefringence, and the index changes are different for the two polarizations

It is usually not possible for a given glass to get accurate purely theoretical relationship between the concentration distribution and the refractive index increase. However, this relationship is with good accuracy linear, with the proportionality factor that depends on the polarization and optical wavelength. With the proportionality factor determined experimentally for a given wavelength, it can be extrapolated to other wavelength for the known glass composition.

Examples on optimisation of both buried and on-surface ion-exchanged waveguides with respect to the coupling losses to single-mode fibers are discussed in References 9,10.

2.2. Doped-silica waveguides

Doped-silica waveguides are fabricated by deposition with various techniques of silica layers on substrates, which are typically silicon wafers.[11-14] The structures of these waveguides are determined by the thickness and composition of the deposited films and by the etching processes used for patterning. This leads to clearly defined cross-sections described by a few parameters. The availability of simple optical modelling tools for these rectangular waveguides has often been cited as an advantage of this technology. However, this can be questioned, since these modelling methods are not highly accurate ones. Clearly there is some advantage in the fact, that it is easier to achieve accurate knowledge describing profiles of etched waveguide than those of diffused waveguides fabricated by ion exchange. Still, many doped-silica waveguide processes involve reflow of core material after the etching process, in order to remove edge roughness that causes scattering losses, and in order to enable the filling of narrow gaps between waveguides by the subsequently deposited cladding material. Modelling of reflow processes and cladding deposition may thus be necessary, and the geometry of the actual waveguide profiles after reflow is not so simple.[15]

Reference 14 describes two different waveguide structures designed for different uses in waveguide devices. One of these was a singlemode waveguide optimised for fiber-to-waveguide coupling. The other had smaller cross-section and higher refractive index difference to achieve a low bending loss for small curvature radius waveguides. The latter does not satisfy strictly the single-mode propagation condition, but with this quasi-singlemode waveguide low losses and practical singlemode operation was achieved in small radius bends, that are useful in several waveguide devices.

3. OPTICAL GUIDED-WAVE PROPAGATION MODELLING

Optical waveguide problems are typically divided into two distinct classes. For waveguides, whose cross-sectional dimensions and index profiles do not vary along the direction of propagation, called z-invariant waveguides, optical propagation can be described in terms of normal optical modes. The modelling tools for these waveguides are various mode solvers. For waveguide geometries varying along the propagation direction, called z-variant waveguide structures, general modelling tools are beam propagation methods. Practical guided-wave devices are almost in all cases z-variant, but still local modelling with mode solvers is extremely useful and valuable to calculate waveguide bends, tapers etc.

3.1. Mode solvers

Because of the translational symmetry in z-invariant waveguides, the wave equation solutions are optical normal modes fully described by transverse field distributions and longitudinal propagation constants. In scalar approximation, the Helmholtz equation for a monochromatic field Ψ is, with $k_0=2\pi/\lambda$ and transverse refractive index distribution $n(x,y)$ for the waveguide,

$$\frac{\partial^2 \Psi}{\partial x^2} + \frac{\partial^2 \Psi}{\partial y^2} + \frac{\partial^2 \Psi}{\partial z^2} + k_0^2 n^2(x,y)\Psi = 0. \qquad (2)$$

The mode solution

$$\Psi(x,y,z) = \Psi(x,y)e^{-i\beta z}, \qquad (3)$$

is then obtained by solving scalar wave equation

$$\frac{\partial^2 \Psi}{\partial x^2} + \frac{\partial^2 \Psi}{\partial y^2} = (\beta^2 - k_0^2 n^2(x,y))\Psi. \qquad (4)$$

A large number of mode solver tools exist to calculate these mode properties[16,17]. Most widely used is the well-known effective index method which uses planar waveguide solvers for channel waveguides first to transform the two-dimensional refractive index distribution into one-dimensional effective index profile and then to solve the modes of this slab. The popularity of this tool is due to its ease of implementation and rapidity of calculation, but it is not generally an accurate approximation. More accurate numerical method mode solvers fall into two groups: those that solve the full vectorial wave equation and those using the scalar approximation. The former have not still established a wide use, probably because of the complicated implementation and needs for longer computation times. Also, the scalar description is usually quite accurate enough in glass waveguide profiles with low refractive index variations. Between these two groups, there are vector correction techniques[18] for scalar solutions, and semi-vectorial methods[19] for finding quasi-TE and quasi-TM solutions by setting some field component to zero and taking into account boundary conditions at material interfaces resulting from the vectorial nature of the fields. Extensively used computational techniques are:

a) Finite element methods, in which region of transverse waveguide cross-section is divided into subregions, elements, which are usually triangular. Within each element, field is approximated by a function given in terms of values at element corners. The problem is stated in a variational form. Variational statement integrals are written as sums of integrals over the elements. This leads to a matrix eigenvalue equation, with eigenvectors giving the cornerpoint field values.

b) Finite difference methods, in which, for example, field is given as field values in points of a rectangular mesh. For each point, wave equation is written with five-point difference operator, so a matrix eigenvalue equation is obtained for column vector containing the field values. The eigenvectors (giving transverse mode field distributions) and eigenvalues (giving mode propagation constants) for matrix are solved. Alternatively, solutions may be found by iterative algorithms.

c) Rayleigh-Ritz methods: Field solutions are given as expansions in a set of orthogonal basis functions (for example Hermite-Gaussians). Extrema are found for variational statement in terms of expansion coefficients, which leads to a matrix eigenvalue equation.

3.2. Beam propagation methods

Simulation tools collectively named as Beam Propagation Methods (BPM) are used to simulate the optical propagation through guided-wave circuits. The initial condition for this kind of problem is the known input field distribution in a plane at the device input facet, given as field values in discrete grid points. The propagation through the device, which is described by a three-dimensional refractive index distribution, is then calculated step by step by solving the field distributions at adjacent steps separated by a small distance. Usually more or less paraxial propagation of light along an axis normal to the input plane is involved. In most BPM techniques scalar field approximation is used. As the calculated volume has to be limited, suitable boundary conditions are used at the lateral boundaries in order to minimise the unphysical reflection of light that is radiated out, since light reflected back into the waveguide region causes unwanted interference. The power and convenience of BPM tools is in the fact that no solution in terms of normal modes is needed, the optical propagation is flexibly simulated starting from quite arbitrary input field distribution through almost freely chosen distribution of refractive index.

BPM tools can also be used to solve the normal modes of waveguides, since the propagating field contains all the excited modes. There are three techniques to separate the normal modes from the total field distributions:

i) In the spectral method[20] an initial field distribution that excites the modes is propagated through a long waveguide. A correlation function - propagated field product with initial field - is calculated as a function of propagation distance, and this is Fourier transformed to obtain the spectrum with peaks corresponding to mode propagation constants. The modefield distributions are then calculated by integrating with respect to z the product of propagating field and a phase-factor oscillating with the periodicity given by the mode propagation constant.

ii) In the matrix beam propagation method[21] Kronecker delta initial fields are propagated through a single step to form the rows of transfer matrix as the output field. The eigenvectors and eigenvalues of this matrix give the modefields and propagation constants of the modes, respectively.

iii) The third method is used to solve the lowest order mode: the initial field is propagated in the imaginary axial direction, so that mode amplitudes increase and finally the lowest order mode with the highest propagation constant dominates at the output - its propagation constant is then solved from the variational expression in terms of the lateral modefield and refractive index distributions.[22]

Full three-dimensional BPM tools are still rarely used. This is because the algorithms require considerable processing time. Also, the high refractive index variations in direction normal to the substrate in most waveguide structures are difficult to handle, so that usually the effective index method is used to reduce the refractive index distribution into two-dimensional effective index distribution.

In the conventional beam propagation method[23], also called split-step Fast-Fourier-Transform Beam Propagation Method (FFT-BPM), Slowly varying envelope approximation (SVEA) is used by inserting into Helmholtz equation

$$\Psi(x,y,z) = \psi(x,y,z) e^{-ik_0 n_0 z} \,, \tag{5}$$

and by neglecting the derivative $\dfrac{\partial^2 \psi}{\partial z^2}$. Fresnel or paraxial wave equation is obtained

$$2ik_0 n_0 \frac{\partial \psi}{\partial z} = \frac{\partial^2 \psi}{\partial x^2} + \frac{\partial^2 \psi}{\partial y^2} + k_0^2 (n^2(x,y,z) - n_0^2) \psi \,. \tag{6}$$

Propagation of optical field through a sort distance is split into two different operators. The first of these, the homogeneous space diffraction operator, uses fast Fourier transform to calculate propagation of free-space optical field as a superposition of plane waves. The waveguide structure is represented as a perturbation in refractive index, and its effect is accounted for by multiplication with a lens-correction operator.

The limitations in FFT-BPM are the use of scalar field solution and paraxial propagation, problems with high refractive index contrasts. Also, periodic boundary conditions usually necessitate the use of artificial absorbers at boundaries, and selection of proper value for reference index n_0 is often important.

Recently, several alternative beam propagation techniques have been introduced with advantages over the FFT-BPM. Most of these propagation algorithms are based on finite difference methods to solve the wave equation. These varying techniques are collectively described as Finite Difference Beam Propagation Methods (FD-BPM). These include split-step FD-BPM, in which only the homogeneous-space diffraction operator of FFT-BPM is replaced by a finite difference operator, and the lens operator is used with it[24]. More compact FD-BPM techniques combine the propagation step into a single operator. Usually implicit finite difference methods are used, necessitating the solution of a matrix equation (with tridiagonal matrix in two-dimensional FD-BPM) to obtain the field after the propagation step[25,26]. Alternative explicit finite difference methods give the field after the propagation step in explicit form, field distribution after an initial step is needed to start the calculation[27,28].

The most widely used two-dimensional FD-BPM algorithm is formed from Fresnel equation in one transverse dimension, with effective refractive index distribution n_{eff}:

$$2ik_0 n_0 \frac{\partial \psi}{\partial z} = \frac{\partial^2 \psi}{\partial x^2} + k_0^2 (n^2(x,z) - n_0^2)\psi . \qquad (7)$$

This is a parabolic equation, and standard solution methods exist. Implicit Crank-Nicholson finite differencing gives:

$$\psi_{j+1}^{s+1} + a_j^+ \psi_j^{s+1} + \psi_{j-1}^{s+1} = -\psi_{j+1}^{s} + a_j^- \psi_j^{s} - \psi_{j-1}^{s} , \qquad (8)$$

where

$$a_j^\pm = \mp 2 \pm k_0^2 (n_{eff}^2 - n_0^2)(\Delta x)^2 - 4ik_0 n_0 (\Delta x)^2 / \Delta z , \qquad (9)$$

and

$$\psi_j^s = \psi(j\Delta x, s\Delta z) . \qquad (10)$$

Field at the next step is thus given as a solution of a tridiagonal matrix equation, for solving of which very efficient methods exist. For fully three-dimensional algorithm, the same principle leads to solution of more complex matrix equations, which is time-consuming. Alternative possibilities are using alternate-direction-implicit methods, with which only tridiagonal matrices are encountered, or fully explicit methods.

The main advantages of FD-BPM over FFT-BPM are:

a) Better processing efficiency of the algorithm. For given accuracy of the simulation usually longer steps can be used, and also less calculation is involved for the propagation step so that total advantage in calculation efficiency is typically about an order of magnitude.

b) Nonuniform grids may be used for representing the field, so that grid points may be more freely placed according to the geometry of waveguide structure. Grid can be nonuniform not only in the transverse direction but also in the longitudinal direction[29]. Thus accuracy and efficiency may be optimised at the same time.

c) The use of transparent boundary conditions is possible, so that optical power radiated to the boundaries escapes with negligible reflections[30,31].

More recently, also vector BPM techniques have been introduced[32-35]. This development results from the use of FD-BPM algorithms, which allow adaptation to take into account the vectorial nature of the optical field, and for simulation of propagating field distributions that are discontinuous at refractive index interfaces.

3.3. CAD tools for guided-wave device design

With increasing computing powers in even desktop computers and trend toward industrial production of integrated optics components, the interest in computer-aided-design (CAD) tools for optical guided-wave devices is growing. These software tools help in trying out new device concepts and can save time and cost in realising designs. Particularly, useful guided-wave design CAD tools have to be adaptable to a wide range of device structures, be able to provide detailed information of device performance with accuracy that is high enough for practical needs and carry out modelling with convenient speed.

Practical CAD software covering all the design aspects in fabricating guided-wave devices can be based on the methods and techniques presented above. It also has to be flexible enough so that it can be adjusted to various types of waveguide and device structures. For design of practical integrated optics components to be industrially fabricated, the parameters and details of the fabrication technology have to be attached to the simulation in the design phase, so that design gives the information needed for the device realisation. In particular, the lithography mask pattern geometry is related to the geometry of the actual guided-wave circuit, and CAD tools should be able to generate this mask pattern automatically after the optimum design has been found. First steps to this direction have already been taken[36,37] and some commercial software for guided-wave design is available. It is likely that in near future more advanced CAD tools of this kind will be developed for particular fabrication technologies or for more general applications.

4. M x N couplers

Basic integrated optics devices are M x N splitters, which distribute optical signals from M input ports into N output ports. The important requirements from these devices are transparency (i.e. low losses) and uniformity of operation.

4.1. Couplers built up from three- or four-port elements

As 1 x N splitters such as 1 x 4, 1 x 8, 1 x 16, structures based on cascaded y-junctions have been established as integrated optics solutions.[38-41] The advantage is the inherent uniform 3dB splitting of symmetric singlemode y-junctions with large bandwidth and insensitivity to polarization state, which leads to the excellent performance achieved with these devices. However, the optimum design of 1 x N splitter based on y-junctions is quite a demanding task, since the optical losses and nonuniformities from nonideal junctions have cumulative effect on the output. The simple geometry with parallel splitter joined by s-bends is not the optimum one: the key factor here is the device size achieved with a given waveguide radius of curvature. It is important to keep the curvature of waveguides low to minimise bending losses, but also to avoid the nonsymmetric optical field distributions being launched into the y-junctions due to the shift of modefield in the curved waveguide. Other important factors in y-junction design are the minimising of radiation losses resulting from abrupt termination of the waveguide gap because of lithography limitations, and maintaining singlemode operation in the junction. Reference 42 discusses optimised design of y-branch, with narrower waveguides and smaller angle at the beginning of the branch. With field-assisted ion exchange, this kind of narrowing effect is inherent, since the waveguides close to each other have a limiting effect to the ionic current transporting into the glass. In the 1 x 128 splitter demonstrated in Reference 43, narrow singlemode waveguides were used in the branching region. These were then uptapered into wider waveguides in order to separate outputs in a fanout region, to avoid radiation losses with high waveguide curvature.

For values of M different from 1, M x N couplers cannot be based on y-junctions only. In 2 x N couplers, typical structure has two 1 x (N/2) splitters joined at the different outputs of a 2 x 2 coupler. Since large bandwidth and polarization insensitivity is needed in also these devices, this is an important factor in selecting the 2 x 2 coupler structure. Conventional symmetric directional couplers are described by the complementary output powers P_{bar} and P_{cross} coupled through bar- and cross-connections, respectively, from the input power P_{in} into a single waveguide,:

$$P_{bar}/P_{in} = \cos^2[\pi L/2L_c(\lambda)],$$

$$P_{cross}/P_{in} = \sin^2[\pi L/2L_c(\lambda)]. \qquad (11)$$

Here the coupling length $L_c(\lambda)$ is the length with which total coupling from one waveguide to the other occurs in the two-waveguide region having the total length L. (This is a somewhat simplified account of the directional coupler operation, since some coupling also occurs between the bent waveguides at both input and output. The degree of coupling between the bends at one end can be described by an additional coupling length $L_{sb}(\lambda)$, this can be taken into account by replacing L with $L+2L_{sb}(\lambda)$.)

The spectral behaviour of the device is due to the wavelength-dependence of $L_c(\lambda)$. As 3dB splitters, symmetric directional couplers are not very suitable, since they have narrow bandwidth, and can also be polarization insensitive. There are different alternative 2 x 2 3dB coupler structures with varying degrees of insensitivity to operation wavelength, see Figure 1. The trade-off for increased bandwidth is some increase in optical losses due to more complicated coupler structure.

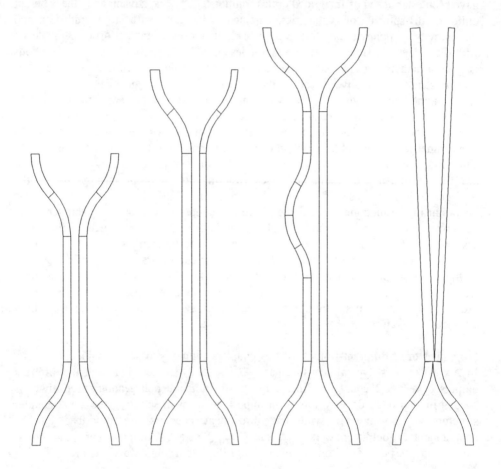

Figure 1. Four-port (2 x 2) 3dB splitter structures. From left: symmetric directional coupler, asymmetric directional coupler, WINC and adiabatic 3dB coupler.

Bandwidth of 3dB directional couplers can be increased by an asymmetric structure, composed of two nonidentical waveguides.[44,45] Difference $\Delta\beta(\lambda)$ arises between the propagation constants of the two waveguides, so that coupling is no longer synchronic, and it is described by a formula:

$$P_{cross} / P_{in} = (L/L_0)^2 \sin^2(\pi/2\sqrt{(L/L_0)^2 + (\Delta\beta \cdot L/\pi)^2})$$

$$/\left((L/L_0)^2 + (\Delta\beta \cdot L/\pi)^2\right) . \qquad (12)$$

Total cross-coupling no longer occurs. When the maximum portion of total power coupled is 0.50, 3dB coupler with increased bandwidth can be made. Alternatively, the coupler region can be divided into two sections of equal length, the second one of which is mirrored along the coupler axis, so that sign of $\Delta\beta$ is reversed. This is alternating $\Delta\beta$ directional coupler, which was used to realise broadband 8 x 8 star coupler by combining twelve 3dB couplers. [45] Asymmetric directional couplers are still relatively sensitive to deviations from the ideal design in fabrication, since their operation depends on getting correct values for two parameters, L_c and $\Delta\beta$.

A wavelength-insensitive coupler (WINC) structure was proposed and demonstrated by Jinguji et al.[46] This has Mach-Zehnder interferometer-type 2 x 2 structure, which consists of two symmetric directional couplers joined by two waveguide arms with a rather small length difference. Different from Mach-Zehnder interferometers, the directional couplers are not 3dB couplers, actually they have different degrees of coupling. At the shorter wavelength limit, the armlength difference is equivalent to one guided-wave wavelength, so that optical phase difference is maintained between the two directional couplers, and their coupling lengths add up to the combined 3dB coupling. The degree of coupling increases with wavelength, but then the armlength difference introduces an out-of-phase factor between the two couplers, which prevents the monotonic increase of coupling. At the longer wavelength limit, the armlength difference is equivalent to half guided-wave wavelength, so that optical phase difference is reversed between the two directional couplers, and the second coupler will couple in reverse direction compared with the first one. Quantitatively, WINC operation is described by formula:

$$P_{cross}/P_{in} = \cos^2(\beta(\lambda)\Delta L/2) \sin^2[\pi(L_1+L_2)/2L_c(\lambda)]$$

$$+\sin^2(\beta(\lambda)\Delta L/2) \sin^2[\pi(L_1-L_2)/2L_c(\lambda)] . \qquad (13)$$

Here $\beta(\lambda)$ and ΔL are the propagation constant and length difference in waveguide arms, L_1 and L_2 are the lengths of two directional couplers.

Still one more wavelength-insensitive 2 x 2 coupler structure is the so-called waveguide hybrid coupler, an adiabatic 3dB coupler.[47-49] This consists of symmetric and asymmetric y-branches of singlemode waveguides joined at a two-mode junction. The two modes of the symmetric branch and the junction are the symmetric (even) and antisymmetric (odd) combinations of the fundamental modes in the two identical waveguides. In the asymmetric branch end, where waveguides are distant from each other, the two modes are localised, corresponding to the individual fundamental modes of the two nonidentical waveguides. If the asymmetric branch angle is small enough, the propagation through the coupler is adiabatic and no power is coupled between the local normal modes. This structure operates inherently as a 3dB splitter, independently of wavelength or polarization, as long as singlemode operation is maintained in all waveguides. An alternative form of this structure, quite identical in function, is the adiabatic directional coupler.[50,51]

4.2. Radiative couplers

Radiative coupler structures have slab waveguide sections between input and output channel waveguide arrays - light radiates from the input waveguides into the slab waveguide, and is captured by the array of output waveguides, which are arranged along a circle, with its center at the center of curvature of farfield phasefront. In the radiative 1 x N couplers, efficiency and uniformity is optimised by varying the output waveguide widths along the array so that the outermost waveguides have the largest width, thus compensating for the decrease of farfield intensity outwards from the center axis.[52,53] The critical factor for achieving a good performance is correct modelling of the farfield distribution from the input waveguide. Since part of the optical power outside the array is inevitably lost, these structures are useful only with larger values of N. However, increasing N does not add much complication to the design.

The design of M x N (typically symmetric N x N) couplers is based on similar principles.[54-57] However, with several input waveguides, variation in the output waveguide widths cannot be used. Instead, the coupling of light between neighbouring input waveguides, when they are brought together before entering the slab area, can be used to modify the farfield pattern so that most of the light is radiated quite uniformly in the angle given by the first Brillouin zone of the input array. With properly designed tapering of the input waveguides before entering the slab, proper sidelobes in the neighbouring waveguides can be excited. Farfield distribution is the Fraunhofer pattern (Fourier transform) of the field at the input array interface to slab, and must be uniform and concentrated over a sector of N output guides. Dummy guides must be included on both sides of arrays, so that all inputs have the similar environment. Using two-dimensional BPM with effective index method, optimised designs have been achieved, also for operation at both 1.30 μm and 1.55 μm wavelengths.[58]

4.3. Multimode interference couplers

Recently, multimode interference couplers based on self-imaging and image-multiplicating properties[59,60] of laterally multimode waveguide sections have attracted a great deal of attention as M x N couplers. The advantages of these structures are low sensitivities and small sizes. So far, however, the possibilities of glass waveguide technologies in fabricating these devices have not been determined. Multimode interference couplers operate best with tight mode confinement in horizontal direction, which is not generally the case in glass waveguides. Also, they are most sensitive to variations in the width of the multimode waveguide section, which may degrade their performance in ion-exchanged waveguides. However, this principle was used in fabrication of 2 x 4 hybrid with phase quadrature outputs by ion exchange.[61] A good overview of this topic is given by Smit.[62]

5. WAVELENGTH DIVISION DEMULTIPLEXERS/MULTIPLEXERS

The second important category of passive glass waveguide components is wavelength division demultiplexers/multiplexers (WDM), which are used to combine signals at different wavelengths into single fiber and to separate these signals at the other end of the fiber. Again, the important characteristic of these devices is transparency, both of the devices (low losses) and between signals (low crosstalk).

5.1. Comparison of dual wavelength WDM devices

This section examines more closely two different designs used for demultiplexing two widely spaced wavelength bands: the symmetric directional coupler and the asymmetric Mach-Zehnder interferometer.

5.1.1. Symmetric directional coupler

For passive integrated optical WDM, operating at the dual 1.30 and 1.55 µm wavelength windows, the symmetric directional coupler has been the typical device.[63-65] It has the advantage of small size and low optical losses due to the simple structure.

The use of directional coupler as a WDM device is based on the wavelength dependence of coupling length, see Eq. (3). Particularly, to separate the two center wavelengths, the odd multiple of coupling lengths at one wavelength must equal the even multiple of coupling lengths at the other wavelength. Since the directional coupler becomes more sensitive with added coupling lengths, it is the best choice to make it one coupling length long at one wavelength and two coupling lengths long at the other wavelength. As the coupling length at the shorter wavelength is longer, the WDM condition is

$$L = L_{1.30} = 2L_{1.55}. \tag{14}$$

Here $L_{1.30}$ and $L_{1.55}$ are the coupling lengths L_c at the two center wavelengths. 1.30 µm wavelength is coupled in a cross-state and the 1.55 µm wavelength is coupled in a bar-state.

Starting from a chosen waveguide profile, there are two parameters to adjust: the waveguide axial separation d and the coupler length L. The condition of Eq. (14) then holds for a single value of d and gives the coupler length L. Reference 66. analyses directional couplers fabricated by thermal ion exchange from $AgNO_3$-melt into Corning 0211 substrate glasses. Solutions for values of d and L were found for waveguide profiles calculated with postbake times t_{pb} at 616 K ranging from 0 s to 2000 s, following the first step ion exchange of duration t_{ie} = 750 s. The 1.30/1.55 µm WDM solution curves are shown in Figure 2. Figure 2a) shows the separation d as a function of postbake time. Figure 2b) shows the coupler length L as a function of postbake time t_{pb}

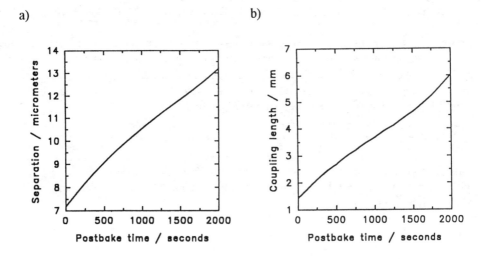

Figure 2. Calculated 1.30/1.55 μm WDM solutions for directional couplers fabricated into Corning 0211 glass substrates by silver ion exchange through 4 μm mask strip openings. The curves show the axial separation d and the coupler length L needed for WDM operation as a function of additional postbake duration at 616 K.

The performance of directional couplers as WDM devices is characterised by the crosstalk, which is defined at a given nominal center wavelength to be the ratio of the optical power in the unwanted output waveguide to the optical power in the wanted output waveguide, calculated from Eq. (3). For 1.30/1.55 μm WDM, -20 dB crosstalk bandwidth is about 30 nm, -30 dB bandwidth is about 10 nm.[64,65] From this the crosstalk due to center wavelength shift from optimum is obtained. The performance can be described in terms of crosstalk at nominal center wavelengths 1300 nm and 1550 nm, since the deterioration of crosstalk is due to the shift of center wavelengths. Thus -30 dB crosstalk corresponds to center wavelength shift of about 5 nm and -20 dB crosstalk to center wavelength shift of about 15 nm. The sensitivities are different at the two center wavelengths. The sensitivity of directional coupler WDM devices is limited by performance at 1.55 μm wavelength.

Figure 3a) shows the calculated tolerances for mask opening widths in a few different devices. It is seen that the WDM coupler with d = 10.18 μm is close to the minimum sensitivity. There are two effects from the variation of strip width: As the gap between waveguides narrows, the coupling gets stronger and the coupling length becomes shorter. However, as the waveguides become wider, the optical fields are also bound more strongly, which decreases the coupling and increases the coupling length. At the optimum solution, these two opposing effects compensate, and the sensitivity is quite low. Far away from this optimum, the reproducible fabrication of WDM couplers is difficult, because of the tolerance requirements on the photolithography.

Crosstalk from inaccuracy of the ion exchange step duration was calculated for different directional couplers. The results are collected in Fig. 3b). The sensitivity increases with increase in waveguide axial separation. These results also give the information about the sensitivity to variation in ion exchange process temperature. The relevant factor is in fact the product Dt. The 60 s change in postbake duration has the same effect as approximately 2.5 K change in temperature.

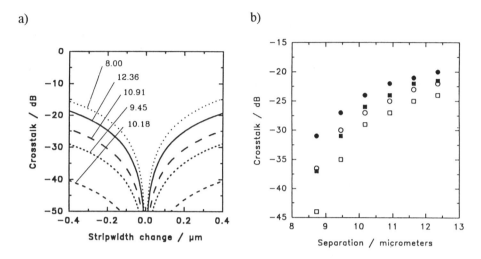

Figure 3. a) Sensitivity of WDM directional couplers to variation of mask opening width. Results showing the crosstalk at 1.55 μm wavelength as a function of change in mask strip width are calculated for five devices with waveguide axial separations d (in micrometers) shown for each curve. b) Directional coupler WDM crosstalks at both wavelengths, when first step ion exchange duration deviates from ideal t_{ie} = 750 s. Open squares: wavelength 1.30 μm, Δt_{ie} = 60 s. Open circles: wavelength 1.30 μm, Δt_{ie} = -60 s. Closed squares: wavelength 1.55 μm, Δt_{ie} = 60 s. Closed circles: wavelength 1.55 μm, Δt_{ie} = -60 s.

Figure 4a) shows the sensitivity of WDM couplers to inaccuracy of postbake duration - it gives crosstalk as a function of postbake time for different couplers, the length of which has been fixed to that given by the optimum WDM condition (14) corresponding to the postbake duration with minimum crosstalk. Again, from knowledge of the temperature-sensitivity of D, the effect of temperature variation can be estimated. For example, with d = 10.18 μm, to achieve crosstalk lower than -20 dB, temperature variation must stay within ±1.3 K.

Figure 4b) presents the crosstalk for different devices due to variation of the separation d from the optimum. The accuracy of d required is very high. However, it is expected that axial separation d can be reproduced very accurately by photolithography, even though the actual gap between mask strip openings varies due to the variation in strip width.

Reference 66. includes also an analysis of coupling at the bends of directional coupler WDM devices. Two-dimensional BPM, with waveguide refractive index distributions transformed into one-dimensional effective index profiles, was used to simulate the propagation of light through directional coupler structures with constant curvature s-bends with no offsets at bend joints, and for each waveguide the bending radius was selected to be large enough to attain low bend losses. Couplers with L optimised for WDM performance were designed. WDM performance with exact 1.30/1.55 µm center wavelengths could not be achieved and for most devices crosstalk was limited to -20 dB level. One reason for this is that somewhat different coupling lengths are obtained using the effective index approximation. More important for practical device design is that the ratio of coupling at the two center wavelengths is not the same in bends as in the straight section. Thus Fig. 3 is no longer the accurate solution for these couplers. For 1.30 µm wavelength 10 - 14 % of the coupling occurred in the bends, and for 1.55 µm wavelength it was 13 - 19 %. Thus the results calculated for straight coupler sensitivity are still quite relevant for estimating realistic device tolerances, though they are not basis for accurate device design.

Since in the couplers with bends, the coupling in bends occurs across increased axial separation, this was compensated for by slightly decreasing the axial separation in the straight section. This way, better WDM operation was achieved.

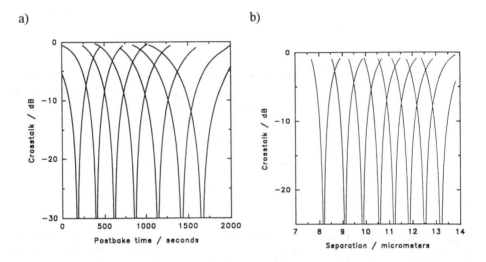

Figure 4. a) The crosstalk for directional coupler WDMs as a function of postbake duration. The different curves according to the position of minimum crosstalk from left to right are for axial separation d = 8.00, 8.73, 9.45, 10.18, 10.91, 11.64 and 12.36 µm.
b) The sensitivity of directional coupler WDM crosstalk to deviation of waveguide separation d from optimum value. Curves are from left to right for postbake durations 250, 500, 750, 1000, 1250, 1500, 1750 and 2000 s.

5.1.2. Asymmetric Mach-Zehnder interferometer

Another dual wavelength device is the guided-wave Mach-Zehnder interferometer[67], which is a more complicated device, but offers the advantage of low sensitivity to variations in fabrication parameters so that very strict control of the fabrication process is not needed. Guided-wave Mach-Zehnder interferometer structures with adiabatic splitters, which eliminate the need for directional coupler splitters, are promising devices with high tolerances for fabrication process deviations. A four-port asymmetric guided-wave Mach-Zehnder interferometer[68] is illustrated in Fig. 5. The device has combinations of asymmetric adiabatic branch and symmetric y-branch at both input and output side. The y-branches are connected to each other by two waveguides having different length. This length-difference is the source of asymmetry in the actual interferometer part of the device and determines its wavelength-dependent operation. With ideal asymmetric adiabatic branch at the input all power coupled through wider waveguide **I** goes into fundamental mode of the two-mode input junction, all power coupled through narrower waveguide **II** goes into the higher-order, antisymmetric mode of the two-mode input junction. The output branch has the reverse function: all power from fundamental mode of the output junction goes into waveguide **III**, all power from higher-order mode of the output junction goes into waveguide **IV**.

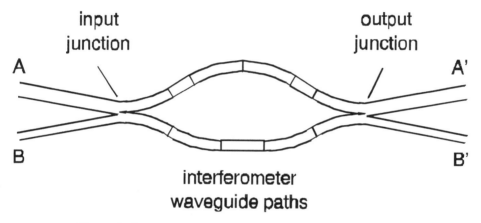

Figure 5. Four-port asymmetric Mach-Zehnder interferometer

The distribution of light into the upper and lower waveguides is given by a vector with the two mode amplitudes of the separate waveguides. At the input and output, this vector is

$$E_{in} = \begin{bmatrix} A \\ B \end{bmatrix}, \quad E_{out} = \begin{bmatrix} A' \\ B' \end{bmatrix}, \tag{15}$$

respectively. The division of light into interferometer waveguide paths is represented by writing the fundamental mode of junction as a combination of amplitudes of two interferometer waveguide modes

$$E_0 = \begin{bmatrix} \sqrt{2}/2 \\ \sqrt{2}/2 \end{bmatrix}. \tag{16}$$

The higher-order-mode has an antisymmetric representation

$$E_1 = \begin{bmatrix} \sqrt{2}/2 \\ -\sqrt{2}/2 \end{bmatrix}. \tag{17}$$

The transfer matrix from the input to the beginning of the interferometer waveguide paths is

$$M_{ij} = \begin{bmatrix} \sqrt{2}/2 & \sqrt{2}/2 \\ \sqrt{2}/2 & -\sqrt{2}/2 \end{bmatrix}. \tag{18}$$

This transfers the input from port **I** into fundamental mode and the input from port **II** into higher-order mode. The transfer matrix from the end of the interferometer waveguide paths to the output is similar

$$M_{oj} = \begin{bmatrix} \sqrt{2}/2 & \sqrt{2}/2 \\ \sqrt{2}/2 & -\sqrt{2}/2 \end{bmatrix}. \tag{19}$$

All the matrix coefficients are real. The phase coefficients can be thus defined by fixing the actual positions of input and output waveguide endpoints within range of one wavelength of the guided wave.

After travelling through the interferometer waveguide paths, light from the fundamental mode is combined into the modes of output junction. The transfer matrix for interferometer waveguide paths is:

$$M_{wp} = \begin{bmatrix} e^{-i(\varphi+\Delta\varphi)} & 0 \\ 0 & e^{-i(\varphi-\Delta\varphi)} \end{bmatrix}. \tag{20}$$

Here $\varphi+\Delta\varphi$ is the optical path-length of the waveguide **a** and $\varphi-\Delta\varphi$ is the optical path-length of the waveguide **b**. The transfer matrix for the whole 4-port Mach-Zehnder interferometer is obtained by multiplication of the three transfer matrices:

$$M_{mz} = M_{oj} M_{wp} M_{ij} = \begin{bmatrix} \cos\Delta\varphi & -i\sin\Delta\varphi \\ -i\sin\Delta\varphi & \cos\Delta\varphi \end{bmatrix} e^{-i\varphi}, \tag{21}$$

$$E_{out} = M_{mz} E_{in}. \tag{22}$$

Again, the phase coefficient $e^{-i\varphi}$ may be eliminated, and the transfer matrix has the same form as that for a symmetric directional coupler. Also for the power coupled through bar- and cross-connections, the formula is

$$P_{bar}/P_{in} = (1+\cos(2\Delta\Phi))/2 = \cos^2(\Delta\Phi),$$
$$P_{cross}/P_{in} = (1-\cos(2\Delta\Phi))/2 = \sin^2(\Delta\Phi). \qquad (23)$$

As practical WDM devices, three-port Mach-Zehnder interferometers[67] are often used, having one of the asymmetric adiabatic branches replaced by a single waveguide.

In the guided-wave Mach-Zehnder interferometer structure, the splitting of the wavelengths is based on the optical pathlength difference in the two waveguide arms. Since for the design shown in Figure 5., the waveguide curves are similar for both paths, the pathlength difference is due to the straight sections, and straight waveguide propagation constants may be used in the analysis. The phase difference of the two paths, with pathlength difference ΔL and the propagation constant β, is $2\Delta\varphi = \beta\Delta L = 2\pi N\Delta L/\lambda$, with mode effective index N. If the paths included different curves, as in special short device designs[69,8], the effect of curves on optical pathlengths would have to be taken into account[70]. The y-junction and adiabatic branch elements are known to be achromatic and insensitive when properly designed[40,48]. Nonideality of these elements would mainly limit the minimum crosstalk achieved without shifting the center wavelengths.

There are less design variables available for fixing the center wavelengths than in the case of the directional coupler, as for the given waveguide profile only the pathlength difference may be changed. The optical pathlength difference for ideal device is $2\Delta\varphi = 2n\pi$ for one center wavelength, $2\Delta\varphi = (2n\pm1)\pi$ for the other, according to Eq. (23). Based on the observation that 1.55/1.30 is quite close to 6/5, device design with the aim $2\Delta\varphi_{1.30} = 6\pi$, $2\Delta\varphi_{1.55} = 5\pi$ can be made. From this the condition is obtained: $N_{1.30}/N_{1.55} = (6/5)(1.30/1.55) \approx 1.00645$. This is close to unity, but in the silver-diffused singlemode waveguides in Corning 0211 glass, the ratio $N_{1.30}/N_{1.55}$ is 1.0046 after first step and approaches 1.002 during postbake[66]. Thus there is actually no solution with exactly the design center wavelengths. However, very good performance can be obtained. Figure 6a) shows the calculated center wavelength crosstalks as a function of pathlength difference with postbake time of 1000 s. It shows that better than -30 dB crosstalks are achieved, if the pathlength difference variation is less than ±5 nm. This seems to be a very strict requirement, but actually the pathlength difference can be maintained with very high accuracy. It is easy to see that for any practical s-bend geometry, see Figure 7., when the length L of bend is much higher than the transverse shift s of the bend, the derivative can be approximated $dL/ds \approx s/L$. It is actually the difference of shifts s in the two interferometer waveguide paths, s_a-s_b, that determines the $\Delta L = L_a$-L_b. As $dL/ds \approx s/L$ can easily be designed to have the value of about 30, it is seen that lithographically the pathlength difference is maintained with high accuracy.

With pathlength difference fixed at $\Delta L = 2570$ nm, Figure 6b) shows the calculated center wavelength crosstalks as a function of postbake time. Crosstalk stays at below the -30 dB for the whole postbake range. This is high insensitivity in comparison with directional couplers - see Figure 4a)., which is actually drawn without including the crosstalk below -30 dB. Variations in mask strip width and ion exchange duration have even smaller effects, since there is less change in the propagation constants. This insensitivity of Mach Zehnder interferometer WDM devices was also confirmed by simulations with two-dimensional FD-BPM, including transparent boundary conditions. Simulated propagation through a device is illustrated in Fig. 8.

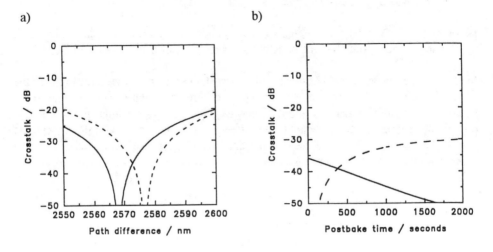

Figure 6. a) Mach-Zehnder interferometer crosstalk at the wavelengths of 1.30 μm (dashed line) and 1.55 μm (solid line) as functions of interferometer path length difference. b) Mach-Zehnder interferometer crosstalk at the wavelengths of 1.30 μm (dashed line) and 1.55 μm (solid line) as functions of waveguide postbake duration.

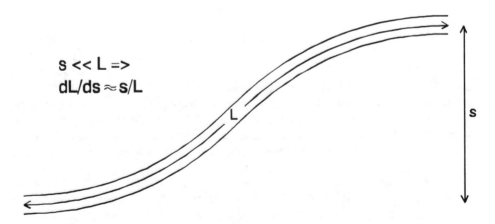

Figure 7. The geometry of waveguide s-bend.

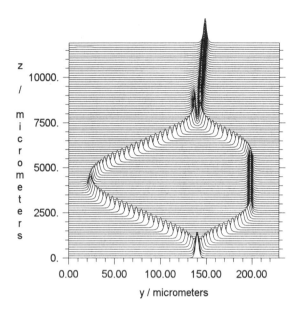

Figure 8. Optical propagation through asymmetric Mach-Zehnder interferometer at wavelength of 1.55 µm simulated by FD-BPM.

5.2. Wide rejection band WDM devices

One disadvantage of the WDM elements described above is the narrow rejection band due to the sinusoidal spectral distribution of optical power at the two complementary outputs[71]. Satisfactorily wide rejection bands can be achieved by additional filtering. Integrated optics technology allows cascading of additional elements after the two outputs. These may be similar to the actual WDM elements, but since the function is to filter out crosstalk left after the first stage, they need to have only single input and output - thus filtering Mach-Zehnder interferometer design can be simplified to have only y-branches between the two waveguide arms. The problem with additional filtering is in maintaining at the same time low losses over the suitably wide passband.

The spectral response of two serial WDM devices is the square of a single element response. With a three-waveguide coupler[72], spectral response is square of that of the symmetric directional coupler. Barbarossa and Laybourn[73] demonstrated the use of three-waveguide couplers, and the simple cascading of these devices to achieve even wider rejection bands. The design of three-waveguide couplers is quite similar to that of two-waveguide couplers, except that the coupling between input and output bends need not to be taken into account.

Another example of a wide rejection band WDM structure, based on a Fresnel mirror etched into doped silica slab waveguide, was demonstrated by Aarnio et al.[74]

5.3. Dense WDM devices

Integrated optical asymmetric Mach-Zehnder interferometers[75,76] make available a wide range of wavelength channel spacings. The channel spacing is inversely proportional to interferometer armlength difference, channel spacings down to 0.1 Å (1 Ghz) have been achieved for frequency division multiplexing.[77] Most of these interferometer structures have used symmetric directional couplers as 3dB couplers at both input and output. For achieving an equal power division in these couplers, precise control in fabrication is necessary. However, equal power division is needed only for wavelength(s) coupled in a cross-state in the Mach-Zehnder interferometer, since for the bar-state wavelength, the function of the output 3dB coupler is to exactly reverse the splitting of the input 3dB coupler, as long as the two couplers are identical. A critical factor is the unequal losses in the two waveguide arms having large differences in lengths, which limits the crosstalk level achievable.

In Mach-Zehnder devices for dense WDM, wavelength spacing can be designed very accurately, but it is very difficult to set the position of center wavelength without additional tuning. For this tuning, thermo-optic effect has been used by adding small heaters on the chip.[78,79] Alternatively, irradiation with high-intensity laser beam has been used to tune the center wavelengths. The number of wavelength channels can be increased by integrating several interferometers in a cascading configuration.[75,80]

Another, different structure for nanometer resolution WDM, the waveguide array multiplexer, was first demonstrated by Vellekoop and Smit.[81] In the radiative MxN star-couplers previously discussed, waveguides in the output array received the light as a divergent beam. Reversing the propagation direction of light, excited array waveguides arranged on a circle will radiate a converging beam that will couple to a single waveguide. Looking back to the star coupler operation, the main difference at the output array excitation from different input waveguides was in the tilt of the optical phasefront. So, again reversing the optical propagation, if controlled tilt can be achieved in the phasefront of light exiting from the excited waveguide array, the single waveguide, into which the converging beam is focused, can be selected.

Connecting two star-couplers with the same waveguide array, so that the length of array waveguides increases linearly with their rank number - that is, there is a constant length difference between neighbouring waveguides - makes an interesting device. The tilt of phasefront at the array output (the input to the second star-coupler) depends on two factors: on the phasefront tilt at the array input (the output of the first star-coupler), and on the wavelength-dependent additional tilt achieved in travelling through the array of waveguides with linearly increasing lengths. This means, that the waveguide that is excited at the output of the total device is selected both by the waveguide excited at the device input waveguide and by the excitation wavelength. The waveguide array with this geometry acts both as a focusing element and as a grating. With this structure, densely spaced NxN multiplexers have been realised.[82-84] It can be used also for broadband two-wavelength WDMs.[85]

The design of array waveguide multiplexers consists of the design of star-couplers and the geometry of waveguide paths in the array to achieve the correct optical pathlength differences. In the design of star-couplers, it should be emphasised that uniform distribution of light into the array is not necessary, in fact it may cause sidelobes in the spectra, thus degrading crosstalk performance. The design of waveguide geometries has similarities to the Mach-Zehnder interferometer arm path design.

6. GUIDED-WAVE CIRCUIT DESIGN

Design of complicated guide-wave circuits (see e.g. Refs. 86-88) can be made by simulation of the total circuit by BPM. This, however, easily leads to very lengthy calculations and large needs of computer resources. Additionally, propagation through all of the circuit may not be paraxial enough or otherwise be suitable for BPM. Even more relevant is, that in this kind of analysis, it is difficult to separate the effects of individual elements and thus optimise the circuit.

Better approach in guided-wave circuit analysis is modelling first the operation of individual elements in terms of response at the output ports to the optical signal at the input ports. Signals at both input and output ports are generally described by excitation of different waveguide modes, given by mode amplitude and phase coefficients. When responses of each elements have been described, including optical waveguide transmission lines between couplers, the totality can be simulated and individual elements optimised in order to achieve the wanted circuit performance.

Some of the guided-wave devices previously discussed can be described this way as simple guided-wave circuits consisting of basic elements. For example, 1xN splitters are cascaded from simple y-branches, Mach-Zehnder interferometer operation was derived above as a small circuit of waveguide arms between couplers, and design of array multiplexers breaks down to modelling of star couplers and waveguide array paths. Design of ring resonators is also typically based on this method. Even more complex circuits can then be built up hierarchically from these basic devices.

One disadvantage of this approach is that it does not take into account the optical crosstalk between various circuit elements brought about by the scattering into radiation modes.

The design methods and computer tools for guided-wave circuit design are currently not in a very advanced state. It is likely, that this must wait for establishment of technologies for fabrication of highly integrated optical waveguide components.

7. REFERENCES

1. L. Ross, "Integrated optical components in substrate glasses", Glastech. Ber. 62, 285-297 (1989).
2. S. I. Najafi [ed.], *Introduction to Glass Integrated Optics*, Artech House, Boston, 1992.
3. A. Tervonen, "A general model for fabrication processes of channel waveguides by ion exchange", J. Appl. Phys. 67, 2746-2752 (1990).
4. J. Albert and J. W. Y. Lit, "Full modelling of field- assisted ion exchange for graded index buried channel optical waveguides", Appl. Opt. 29, 2798-2804 (1990).

5. A. J. Cantor, M. M. Abou el leil and R. H. Hobbs, "Theory of 2-D ion exchange in glass: optimisation of microlens arrays", Appl. Opt. 30, 2704-2713 (1991).
6. A. Tervonen, "Theoretical analysis of ion-exchanged glass waveguides", in Introduction to Glass Integrated Optics, Ed.: S. I. Najafi, Artech House, Boston, 1992, pp. 73-105.
7. S. D. Fantone, "Refractive index and spectral models for gradient-index materials", Appl. Opt. 22, 432-440 (1983).
8. D. P. Ryan-Howard and D. T. Moore, "Model for the chromatic properties of gradient-index glass", Appl. Opt. 24, 4356-4366 (1985).
9. A. Tervonen, P. Pöyhönen, S. Honkanen, M. Tahkokorpi and S. Tammela, "Examination of two-step fabrication methods for single-mode fiber compatible ion-exchanged waveguides", Appl. Opt. 30, 338-343 (1991).
10. J. Albert and G. L. Yip, "Insertion loss reduction between single-mode fibers and diffused channel waveguides", Appl. Opt. 27, 4837-4843 (1988).
11. P. R. Wensley, R. Bellerby, S. Day, M. F. Grant, S. A. Rosser and G.J. Cannell, "Improved waveguide technology for silica-on-silicon integrated optics", *Proceedings of the 6th European Conference on Integrated Optics ECIO'93*, Neuchatel, Switzerland, April 18-22, 1993, pp. (9-8) - (9-9).
12. S. Valette, "State of the art of integrated optics technology at LETI for achieving passive optical components", J. Modern Opt. 35, 993-1005 (1988).
13. C. H. Henry, G. E. Blonder and R. F. Kazarinov, "Glass waveguides on silicon for hybrid optical packaging", J. Lightwave Technol. 7, 1530-1539 (1989).
14. M. Kawachi, "Silica waveguides on silicon and their application to integrated-optic components", Opt. Quantum Electron. 22, 391-416 (1990).
15. M. Nield, F. MacKenzie and J. Rush, "The importance of cavity elimination in planar silica on silicon optical devices", *Proceedings of the 6th European Conference on Integrated Optics ECIO'93*, Neuchatel, Switzerland, April 18-22, 1993, pp. (9-18) - (9-19).
16. T. M. Benson, P. C. Kendall, M. S. Stern and D. A. Quinney, "New results for rib waveguide propagation constants", IEE Proc. J 136, 97-102 (1989).
17. Working Group I, COST 216, "Comparison of different modelling techniques for longitudinally invariant integrated optical waveguides", IEE Proc. J 136, 273-280 (1989).
18. Z. Mao and W-P. Huang, "Analysis of optical rib waveguides and couplers with buried guiding layer", IEEE J. Quantum Electron. 28, 176-182 (1992).
19. M. S. Stern, "Semivectorial polarised finite difference method for optical waveguides with arbitrary index profiles", IEE Proc. J 135, 56-62 (1988).
20. M. D. Feit and J. A. Fleck, Jr., "Computation of mode eigenfunctions in graded-index optical fibers by the propagating beam method", Appl. Opt. 19, 2240-2246 (1980).
21. D. Yevick and B. Hermansson, "New fast Fourier transform and finite-element approaches to the calculation of multiple-stripe-geometry laser modes", J. Appl. Phys. 59, 1769-1771 (1986).
22. B. Hermansson and D. Yevick, "Numerical analyses of the modal eigenfunctions of chirped and unchirped multiple-stripe-geometry laser arrays", J. Opt. Soc. Am. A 4, 379-390 (1987).
23. P.E. Lagasse and R. Baets, "Application of propagating beam methods to electromagnetic and acoustic wave propagation problems: A review", Radio Science 22, 1225-1233 (1987).

24. D. Yevick and B. Hermansson, "Split-step finite difference analysis of rib waveguides", Electron. Lett. 25, 461-462 (1989).
25. Y. Chung and N. Dagli, "An assessment of finite difference beam propagation method", IEEE J. Quantum Electron. 26, 1335-1339 (1990).
26. R. Scarmozzino and R. M. Osgood, Jr., "Comparison of finite-difference and Fourier-transform solutions of the parabolic wave equation with emphasis on integrated-optics applications", J. Opt. Soc. Am. A 8, 724-731 (1991).
27. Y. Chung and N. Dagli, "Explicit finite difference beam propagation method: application to semiconductor rib waveguide y-junction analysis", Electron. Lett. 26, 711-713 (1990).
28. Y. Chung and N. Dagli, "Analysis of Z-invariant and Z-variant semiconductor rib waveguides by explicit finite difference beam propagation method with nonuniform mesh configuration", IEEE J. Quantum Electron. 27, 2296-2305 (1991).
29. M. Artiglia, P. Di Vita, M. Potenza, G. Coppa, G. Lapenta and P. Ravetto, "Variable grid finite difference methods for study of longitudinally varying planar waveguides", Electron. Lett. 27, 474-475 (1991).
30. R. Accornero, M. Artiglia, G. Coppa, P. Di Vita, G. Lapenta, M. Potenza and P. Ravetto, "Finite difference methods for the analysis of integrated optical waveguides", Electron. Lett. 26, 1959-1960 (1990).
31. G. R. Hadley, "Transparent boundary condition for the beam propagation method", IEEE J. Quantum Electron. 28, 363-370 (1992).
32. W. P. Huang, C. L. Xu, S. T. Chu and S. K. Chaudhuri, "A vector beam propagation method for guided-wave optics", IEEE Photonics Technol. Lett. 3, 910-913 (1991).
33. Y. Chung, N. Dagli and L. Thylen, "Explicit finite difference vectorial beam propagation method", Electron. Lett. 27, 2119-2121 (1991).
34. J. M. Liu and L. Gomelsky, "Vectorial beam propagation method" J. Opt. Soc. Am. A 9, 1574-1584 (1992).
35. W. P. Huang, C. L. Xu, S. T. Chu and S. K. Chaudhuri, "The finite-difference vector beam propagation method: analysis and assessment", J. Lightwave Technol. 10, 295-305 (1992).
36. A. Tervonen, S. Honkanen, P. Pöyhönen, and M. Tahkokorpi, "Design software for ion-exchanged glass waveguide devices", *OE/Fibers'92*, 8-11 September 1992, Boston, Proc. SPIE - Int. Soc. Opt. Eng. (USA) 1794, 264-270 (1993).
37. G. Perrone, D. Petazzi and I. Montrosset, "User-friendly computer-aided integrated optics simulator", *OE/Fibers'92*, 8-11 September 1992, Boston, Proc. SPIE - Int. Soc. Opt. Eng. (USA) 1794.
38. P. Pöyhönen, S. Honkanen, A. Tervonen, M. Tahkokorpi and J. Albert, "Planar 1/8 splitter in glass by photoresist masked silver film ion exchange", Electron. Lett. 27, 1319-1320 (1991).
39. S. Kobayashi, T. Kitoh, Y. Hida, S. Suzuki and M. Yamaguchi, "Y branching silica waveguide 1 x 8 splitter module", Electron. Lett. 26, 707-708 (1990).
40. R. A. Betts, F. Lui and S. Dagias, "Wavelength and polarization insensitive optical splitters fabricated in K^+/Na^+ ion-exchanged glass", IEEE Photonics.Technol. Lett. 2, 481-483 (1990).
41. C. Thevenot, M. McCourt, T. Dannoux, F. Jean, J.-L. Malinge and C. Lerminiaux, "Couplers for fiber-in-the-loop systems", Proc. SPIE - Int. Soc. Opt. Eng. (USA) 1580, 408-419 (1991).

42. S. Kobayashi, K. Nakama, S. Sato, H. Wada, H. Hashizume, I. Tanaka and M. Seki, "Novel waveguide y-branch for low-loss 1 x N splitters", *Technical Digest of the Optical Fiber Communication Conference OFC'92*, OSA Technical Digest Series, Vol.5, (Optical Society of America, Washington D.C., 1992) p.145.
43. H. Takahashi, Y. Ohmori and M. Kawachi, "Design and fabrication of silica-based integrated-optic 1 x 128 power splitter", Electron. Lett. 27, 2131-2133 (1991).
44. A. Takagi, K. Jinguji and M. Kawachi, "Broadband silica-based optical waveguide coupler with asymmetric structure", Electron. Lett. 26, 132-133 (1990).
45. H. Yanagawa, S. Nakamura, I. Ohyama and K. Ueki, "Broad-band high-silica optical waveguide star coupler with asymmetric directional couplers", J. Lightwave Technol. 8, 1292-1297 (1990).
46. K. Jinguji, N. Takato, A. Sugita and M. Kawachi, "Mach-Zehnder interferometer type optical waveguide coupler with wavelength-flattened coupling ratio", Electron. Lett. 26, 1326-1327 (1990).
47. M. Izutsu, A. Enokihara and T. Sueta, "Optical-waveguide hybrid coupler", Opt. Lett. 7, 549-551 (1982).
48. C. P. Hussell, R. V. Ramaswamy, R. Srivastava and J. L. Jackel, "Wavelength and polarization insensitive 3 dB cross-coupler power dividers by ion exchange in glass", Appl.Phys.Lett. 56, 2381-2383 (1990).
49. C. P. Hussell, R. V. Ramaswamy, R. Srivastava and J. L. Jackel, "Adiabatic invariance in GRIN channel waveguides and its use in 3 dB cross-couplers", Appl. Opt. 29, 4105-4110 (1990).
50. Y. Shani, C. H. Henry, R. C. Kistler, R. F. Kazarinov and K. J. Orlowsky, "Integrated optic adiabatic devices on silicon", IEEE J. Quantum Electron. 27, 556-566 (1991).
51. R. Adar, C. H. Henry, R. F. Kazarinov, R. C. Kistler and G. R. Weber, "Adiabatic 3dB couplers, filters and multiplexers made with silica waveguides on silicon", J. Lightwave Technol. 10, 46-50 (1992).
52. S. Day, R. Bellerby, G. Cannell and M. Grant, "Silicon based fibre pigtailed 1 x 16 power splitter", Electron. Lett. 28, 920-922 (1992).
53. H. Takahashi, K. Okamoto and Y. Ohmori, "Integrated-optic 1 x 128 power splitter with multifunnel waveguide", IEEE Photonics Technol. Lett. 5, 58-60, (1993).
54. C. Dragone, "Efficiency of a periodic array with nearly ideal element pattern", IEEE Photonics Technol. Lett. 1, 238-240 (1989).
55. C. Dragone, C. H. Henry, I. P. Kaminow and R. C. Kistler, "Efficient multichannel integrated optics star coupler on silicon", IEEE Photonics Technol. Lett. 1, 241-243 (1989).
56. K. Okamoto, H. Takahashi, S. Suzuki, A. Sugita and Y. Ohmori, "Design and fabrication of integrated-optic 8 x 8 star coupler", Electron. Lett. 27, 774-775 (1991).
57. K. Okamoto, H. Okazaki, Y. Ohmori and K. Kato, "Fabrication of large-scale integrated-optic N x N star couplers", IEEE Photonics Technol. Lett. 4, 1032-1034 (1992).
58. K. Okamoto, H. Takahashi, M. Yasu and Y.Hibino, "Fabrication of wavelength-insensitive 8 x 8 star coupler", IEEE Photonics Technol. Lett. 4, 61-63 (1992).
59. R. Ulrich, "Image formation by phase coincidences in optical waveguides", Opt. Comm. 13, 259-264 (1975).
60. R. Ulrich and G. Ankele, "Self-imaging in homogeneous planar optical waveguides", Appl. Phys. Lett. 27, 337-339 (1975).

61. P. Roth, "Passive integrated optic mixer providing quadrature outputs", Proc. SPIE - Int. Soc. Opt. Eng. (USA) 1141, 169-173 (1989).
62. M. K. Smit, "Branching, radiative and self-imaging elements for use in M x N couplers", *Proceedings of the 6th European Conference on Integrated Optics ECIO '93*, Neuchatel, Switzerland, April 18-22, 1993, pp. (14-1) - (14-4), .
63. A. H. Reichelt, P. C. Clemens and H. F. Mahlein, "Single-mode waveguides and components by two-step Cs^+-K^+ ion-exchange in glass", Proc. SPIE - Int. Soc. Opt. Eng. (USA) 1128, 165-168 (1989).
64. C. Nissim, A. Beguin, P. Laborde, C. Lerminiaux and M. McCourt, "Low-loss single-mode wavelength division multiplexers fabricated by ion exchange in glass", *Proceedings: The Eighth Annual European Fibre Optic Communications & Local Area Networks Conference, EFOC/LAN 90*, 27-29 June 1990, Munich, IGI Europe 1990, pp. 114-116.
65. H. C. Cheng and R. V. Ramaswamy, "Symmetrical directional coupler as a wavelength multiplexer-demultiplexer: theory and experiment", IEEE J. Quantum Electron. 27, 567-574 (1991).
66. A. Tervonen, S. Honkanen and S. I. Najafi, "Analysis of symmetric directional couplers and asymmetric Mach-Zehnder interferometers as 1.30/1.55 μm dual wavelength demultiplexers/ multiplexers", Opt. Eng., 32, 2083-2091 (1993).
67. A. Tervonen, P. Pöyhönen, S. Honkanen and M. Tahkokorpi, "A guided-wave Mach-Zehnder interferometer structure for wavelength multiplexing", IEEE Photonics Technol. Lett. 3, 516-518 (1991).
68. W. J. Wang, S. Honkanen, S. I. Najafi and A. Tervonen, "Four-port guided-wave nonsymmetric Mach-Zehnder interferometer", Appl. Phys. Lett. 61, 150-152 (1992).
69. A. Tervonen, P. Pöyhönen, S. Honkanen and M. Tahkokorpi, "Integrated optics 1.48/1.55 μm wavelength division multiplexer for optical amplifier application", J. Modern Opt. 39, 1615-1618 (1992).
70. A. Tervonen, P. Pöyhönen, S. Honkanen and M. Tahkokorpi, "Channel waveguide Mach-Zehnder interferometer for wavelength splitting and combining", *The International Congress on Optical Science and Engineering, ECO4*, 11-15 March 1991, Hague, Proc. SPIE - Int. Soc. Opt. Eng. (USA) 1513, 71-75 (1991).
71. K. Imoto, H. Sano and M. Miyazaki, "Guided-wave multi/demultiplexers with high stopband rejection", Appl. Opt. 26, 4214-4219 (1987).
72. H. A. Haus and C. G. Fonstad, "Three-waveguide couplers for improved sampling and filtering", IEEE J. Quantum Electron. QE-17, 2321-2325 (1981).
73. G. Barbarossa and P. J. R. Laybourn, "Wide rejection band multidemultiplexer at 1.3-1.55 μm by cascading high-silica three-waveguide couplers on Si", Electron. Lett. 27, 2085-2087 (1991).
74. J. Aarnio, P. Heimala and I. Stuns, "Wide passband wavelength multi/demultiplexer at 1.3/1.55 μm based on etched Fresnel mirror", IEE Proc. J, Optoelectron. 139, 228-236 (1992).
75. K. Inoue, N. Takato, H. Toba and M. Kawachi, "A four-channel optical waveguide multi/demultiplexer for 5-Ghz spaced optical FDM transmission", J. Lightwave Technol. 6, 339-345 (1988).
76. B. H.Verbeek, C. H. Henry, N. A. Olsson, K. J. Orlowsky, R. F. Kazarinov and B. H. Johnson, "Integrated four-channel Mach-Zehnder multi/demultiplexer fabricated with phosphorous doped $SiO2$ waveguides on Si", J. Lightwave Technol. 6, 1011-1015 (1988).

77. Takato, N., Kominato,T., Sugita, A., Jinguji,K., Toba, H. and Kawachi, M., 'Silica-based integrated optic Mach-Zehnder multi/demultiplexer family with channel spacing of 0.01 - 250 nm', IEEE J. Selected Areas in Communications, 8, 1120-1127 (1990).

78. H. Uetsuka, M. Kurosawa and K. Imoto, "Novel method for center wavelength tuning of silica waveguide type Mach-Zehnder multi/demultiplexers", Electron. Lett. 26, 251-253 (1990).

79. Y. Hibino, T. Kominato and Y. Ohmori, "Optical frequency tuning by laser-irradiation in silica-based Mach-Zehnder-type multi/demultiplexers", IEEE Photonics Technol. Lett. 3, 640-642 (1991).

80. M. Semjen and C. Lerminiaux, "An 8 channel 2 nm spacing demultiplexer made by ion exchange", *Proceedings 19th European Conference on Optical Communication ECOC'93*, Sept. 13-16, Montreux, Vol. 2 (Regular papers), pp. 25-28.

81. A. R. Vellekoop and M. K. Smit, "Four-channel integrated-optic wavelength demultiplexer with weak polarization dependence", J. Lightwave Technol. 9, 310-314 (1991).

82. H. Takahashi, S. Suzuki, K. Kato and I. Nishi, "Arrayed-waveguide grating for wavelength division multi/demultiplexer with nanometre resolution", Electron. Lett. 26, 87-88 (1990).

83. C. Dragone, "An N x N optical multiplexer using a planar arrangement of two star couplers", IEEE Photonics Technol. Lett. 3, 812-814 (1991).

84. C. Dragone, C. A. Edwards and R. C. Kistler, "Integrated optics N x N multiplexer on silicon", IEEE Photonics Technol. Lett. 3, 896-899 (1991).

85. R. Adar, C. H. Henry, C. Dragone, R. C. Kistler and M. A. Milbrodt, "Broadband array multiplexers made with silica waveguides on silicon", J. Lightwave Technol. 11, 212-219 (1993).

86. M. Kawachi, "Silica-based planar lightwave circuit technologies", *Proceedings 17th European Conference on Optical Communication ECOC'91*, Sept. 9-12, Paris, Vol. 2 (Invited papers), pp. 51-58.

87. Y. Shani, C. H. Henry, R. C. Kistler, R. F. Kazarinov and K. J. Orlowsky, "Integrated optic front end for polarization diversity reception", Appl. Phys. Lett. 56, 2092-2093 (1990).

88. G. Zhang, S. Honkanen, S. I. Najafi and A. Tervonen, "Integrated 1.3 µm/1.55 µm wavelength multiplexer and 1/8 splitter by ion exchange in glass", Electron. Lett., 29, 1064-1066 (1993).

Device Fabrication

Glass integrated optical devices on silicon for optical communications

Michael F. Grant

BNR Europe Limited, London Road,
Harlow, Essex, England, CM17 9NA

ABSTRACT

Silica-on-silicon integrated optics is a key technology for passive optical components, as required for the next generation of fibre optics communications systems. This paper reviews progress on silica-on-silicon technology and key devices developed to date. Waveguide fabrication technology is described, with particular attention to deposition processes. Fibre pigtailing technology is described, and the relative merits of active and passive alignment are discussed. A range of passive components are detailed, concentrating on couplers, splitters and wavelength division multiplexers. Recent developments in rare-earth doping of silica-on-silicon waveguides are also covered. The paper concludes with some discussion on future directions for silica-on-silicon research and development.

1. INTRODUCTION

Optical fibre is now the transmission medium of choice for long distance and high bit rate transmission in telecommunications networks. The next major thrust is into the subscriber network, aimed at delivery of advanced communication services to the home. For this to become a reality, a new generation of optical components is required. These include low cost transmitters and receivers, connectors, and passive optical components such as multiway splitters, wavelength division multiplexers, optical amplifiers and optical switches.

Silica-on-silicon integrated optics technology offers many attractions for the realisation of passive optical components. Silicon is available in cheap, large area substrates (4 inch to 8 inch) with optical quality surface finish. The large area substrates allow the realisation of very complex optical circuits, or alternatively the fabrication of hundreds of simpler components per wafer. Processes such as photolithography, deposition, etching, annealing etc. are compatible with those already developed in the silicon microelectronics industry, and use the same equipment which has been optimised for use with silicon wafers. The waveguides based on deposited silica can be made very low loss and well matched to optical fibres, thus reducing interface loss. The well defined core shapes are amenable to computer modelling and allow precise process control. Silicon allows the development of passively aligned fibre-waveguide coupling using etched grooves in the silicon substrate for location of the fibres. Finally, silicon can also be used as a substrate for the hybridisation of electronic, optoelectronic and waveguide components together with fibre alignment grooves, thus achieving complex opto-electronic modules.

© Northern Telecom Europe Limited 1994

This paper reviews the development of silica-on-silicon integrated optics technology for communication systems. Section 2. commences with a description of the waveguide fabrication process, with particular attention to deposition processes. Fibre pigtailing technology is then described in Section 3. A range of passive components are detailed in Section 4., concentrating on couplers, splitters and wavelength division multiplexers. Section 5. covers recent developments in rare-earth doping of silica-on-silicon waveguides. The paper concludes with some discussion on future directions for silica-on-silicon research and development.

2. WAVEGUIDE FABRICATION TECHNOLOGIES

2.1. Outline of waveguide fabrication process

A schematic of a typical waveguide fabrication process is outlined in figure 1. A buffer layer is first deposited to separate the waveguide core from the high index (n = 3.8) silicon substrate. The core layer is then deposited, with an increased refractive index. Photolithography is then used to pattern the waveguide core, followed by Reactive Ion Etching (RIE) to remove unwanted core material. The cladding layer is then deposited to complete the waveguide structure. Waveguide design is a compromise between bend loss and interface loss to fibre. A large core, low Δn waveguide (e.g. 8 x 8 µm, Δn = 4 x 10^{-3}) will have very low fibre-waveguide interface loss (<< 0.1 dB), but will require use of a high bend radius ($\approx$ 50 mm), leading to excessively long devices. A smaller core, higher Δn waveguide will permit use of a smaller bend radius, but at the expense of increased fibre-waveguide coupling loss. The thickness of the buffer and cladding layers is dependant on the core index, and ranges from a few microns for high index cores to > 10 µm for waveguide cores well matched to optical fibres.

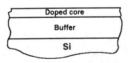

1. Silica based films are deposited on the silicon wafer

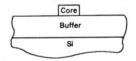

3. Reactive ion etching of unwanted core material

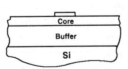

2. Core defined using masking and photolithography

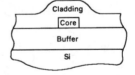

4. Cladding deposited over core

Figure 1 Fabrication process for silica-on-silicon waveguides

Surprisingly, photolithography and reactive ion etching of the thick core layers turn out to be relatively easily achieved using standard equipment, despite the thicknesses involved (up to 10 μm). Masks which can be used are photoresist, amorphous silicon, and metals such as Cr, Ni, Al etc. and etch gases include CHF_3, C_2F_6, SF_6. The development of silica and doped silica deposition processes are more difficult and are therefore described in more detail below.

2.2. Silica waveguide deposition processes

Although silica deposition is well developed in the semiconductor industry, thicknesses used are typically less than a few thousand Angstroms, and the oxide quality has no optical requirements. Silica waveguides require layer thicknesses two orders of magnitude greater than these, with refractive index uniformity around 1×10^{-4}, and low optical attenuation. Waveguide deposition processes are broadly in two classes, namely Flame Hydrolysis Deposition (FHD), similar to the Vapour Axial Deposition (VAD) technique used in optical fibre fabrication, and a range of processes (thermal oxidation, chemical and plasma deposition etc) drawn from processes developed in the semiconductor integrated circuit industry. In both techniques, SiO_2 is deposited and doped to alter the refractive index. Dopants used include TiO_2, GeO_2, P_2O_5, Si_3N_4, and As_2O_3 to increase the refractive index, and fluorine and B_2O_3 to decrease the index. P_2O_5, B_2O_3 and As_2O_3 are network modifiers, and are used to lower the melting point of the deposited layers. This is essential in FHD deposited layers, and also important in other processes to improve cladding conformality and planarisation. The key deposition techniques are described in more detail below.

2.2.1. Flame Hydrolysis Deposition

A schematic of a Flame Hydrolysis Deposition (FHD) system is shown in figure 2. A mixture of gases are burnt in an oxygen/hydrogen torch to produce fine particles which stick onto substrates on a rotating turntable [1,2,3]. The combination of the turntable rotation and traversing of the torch is designed to achieve layer uniformity. Gases used are $SiCl_4$, $TiCl_4$ and $GeCl_4$ to produce SiO_2 doped with TiO_2 or GeO_2 respectively. In addition, small amounts of phosphorous and boron are added using PCl_3 and BCl_3. After deposition, the glass is consolidated by heating to a temperature of around 1200 °C to 1350 °C. During this consolidation, the waveguide layers undergo substantial shrinkage, making final thickness uncertain. A significant advantage of FHD over other methods of depositing waveguides is that very thick layers (many tens of microns) can be built up easily. In addition, the deposition and consolidation process is intrinsically planarising, hence providing excellent cladding conformality over closely spaced cores as used in, for example, directional couplers and Y-junctions. Against these advantages, FHD systems need to be home-built and care must be taken with film uniformity and dust control.

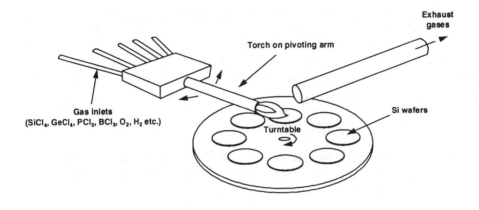

Figure 2 Flame hydrolysis deposition system

2.2.2. Thermal Oxidation

High quality SiO_2 films can be easily grown by placing silicon wafers in a standard tube furnace under flowing O_2 or H_2O vapour at temperatures up to 1300 °C [4]. Typically, this can only produce films a few microns thick, and is therefore only suitable for buffer layers to be used with high Δn waveguides. An extension to this technique is the High Pressure Oxidation (HIPOX), where the silicon is oxidised under high pressure steam (>10 atmospheres) which increases the growth rate [5] and hence the practically attainable thickness. HIPOX SiO_2 up to 15 μm thick has been used as a buffer layer for low Δn waveguides [6, 7].

2.2.3. Plasma Enhanced Chemical Vapour Deposition

Plasma Enhanced Chemical Vapour Deposition (PECVD) is a well known technique for deposition of doped and undoped SiO_2 and Si_3N_4. It is attractive for the deposition of waveguide layers since PECVD can achieve high deposition rates (> 1000 Å/min) [8] with thickness uniformity better than 1% and refractive index uniformity of 0.0001 [9]. A typical parallel plate PECVD system is shown in figure 3. The input gases are fed through holes in the upper electrode to ensure uniform gas distribution in the plasma. Wafers (typically 1 to 10) are placed on the lower electrode, which is heated to around 350 °C. A plasma is generated between the electrodes, at frequencies ranging from around 100 kHz to 13.5 MHz, and at powers of a few Watts up to a few hundred Watts. Deposition occurs on both electrodes and the wafers placed on the lower electrodes. The input gases are SiH_4 and N_2O (diluted by He, Ar or N_2) for the deposition of SiO_2. A range of dopants have been used in PECVD waveguide fabrication, including phosphorous [10], boron and fluorine [8]. PECVD is also suited for the deposition of Si_3N_4 and intermediate oxynitrides, achieved by the addition of NH_3 [11] or by varying the ratio of SiH_4 and N_2O used in the plasma [12,13].

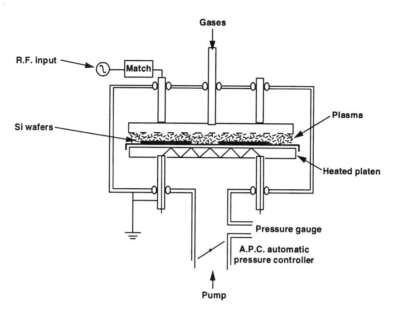

Figure 3   Plasma enhanced chemical vapour deposition system

An alternative system is based on hot wall tube reactor containing many (up to a few hundred) wafers. In this type of system, the gases are introduced from one end and typically, pulsed plasma is used to minimise depletion effects which would cause non-uniformities.

While PECVD has low deposition temperatures (around 350 °C), higher temperatures (800 °C - 1100 °C) are usually required to eliminate hydrogen-related absorption peaks which arise due to the hydrogen based gaseous precursors used. In addition reflow of core and/or cladding at elevated temperatures is also advantageous to improve the conformality and planarisation of the cladding layer over the waveguide core [14].

2.2.4. Low Pressure Chemical Vapour Deposition (LPCVD)

Like PECVD, LPCVD has been developed over many years in the semiconductor IC industry. A mixture of gases is fed into a hot wall tube reactor at low pressure where they react on the silicon wafers and the tube sidewalls. LPCVD has the advantage of depositing on a large number of wafers simultaneously, but suffers the disadvantage of refractive index control along the tube and is not suitable for thick (10 μm) films due to film stress. LPCVD therefore is best used for the nitride type waveguides [15] or in conjunction with another technique such as HIPOX [6,7].

2.2.5. Other deposition techniques

A range of other deposition techniques have been used, including sputtering [16,17], and sol-gel deposition. Sputtering allows a wide choice of glass types, but is not suited for the deposition of very thick layers. Sol-gel deposition also allows a wide choice of glass types, but is conventionally suited only for thin layer depositions. A multiple deposition and rapid thermal annealing scheme has however been demonstrated for deposition of thick silica and doped silica layers [18,19]. This has enabled thick, low Δn waveguides to be demonstrated [20].

3. FIBRE PIGTAILING

To be useful in real fibre optical systems, devices must be interfaced with optical fibres. Fibre pigtailing techniques fall into two categories, where the fibres are bonded to a separate substrate which is then attached to the optical device (figure 4a), or where the fibre-waveguide alignment is achieved using registration features monolithically integrated in the silicon substrate (figure 4b). These techniques are discussed in the following sections.

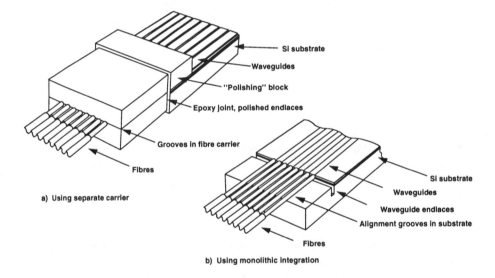

Figure 4 Schematic of fibre-guide coupling techniques

3.1. Fibre pigtailing using separate carrier

Fibre-guide alignment using a separate substrate to carry an array of fibres has been developed over many years for fibre pigtailing of integrated optics devices in LiNbO$_3$ and other technologies [21,22,23]. Typically, an array of fibres is bonded into an array of well defined grooves (often silicon V-grooves). The end face of the integrated optics chip and the fibre array are both polished, and the fibre array is then aligned using micropositioners, and permanently bonded using ultra violet (UV) curing adhesive. The attractions of this scheme are the simplicity of development and its flexibility. In addition low reflectivity (< - 50 dB) is easily achieved by polishing both the waveguide and fibre end faces at an angle. The limitations are that the fibre and chip preparation are time consuming and hence expensive, and that practical considerations usually limit the choice of adhesive used in the interface.

NTT have demonstrated a stable and low loss interfacing technique using a silica glass carrier containing alignment grooves for an 8-way fibre array[24]. An alignment accuracy of better than 1 µm was achieved, resulting in a connection loss of < 0.1 dB/interface, with excellent environmental performance. This was achieved using a humidity resistant UV-curing epoxy resin. Average return loss was < - 46 dB. For pigtailing large numbers of input and output ports, in the case of a 64 x 64 coupler and a 144 x 144 coupler, the coupling was done by sequentially aligning and fixing 8-fibre arrays to the devices [25]. Metal frames were also affixed to the chip and the fibre array, then YAG welded together to achieve a rigid connection.

3.2. Fibre pigtailing using integrated alignment grooves

A number of techniques have been demonstrated to monolithically integrate alignment features into the silicon substrate. This has the major advantage of cost reduction in the long term because it eliminates the expensive fibre and device preparation processes and also the need for active alignment, thus providing a very rapid fibre pigtailing process. This will lead to significant cost reductions in final device costs since fibre preparation and pigtailing dominate device costs. Ruggedisation can also be relatively easily achieved. In the short term however, it requires significantly more development effort.

The use of silicon as a substrate provides the possibility of fabricating V-grooves using anisotropic etching [26]. Because the etching process is almost self-limiting, the dimensions of the V-groove and hence the height of the fibre above the substrate is governed by the width of the etch mask, which is very precisely defined using standard photolithography. BNR Europe Limited have developed a process for integrating preferentially etched V-grooves with waveguides and have demonstrated repeatable fibre-waveguide alignment accuracies better than 1 µm [27]. The resulting mean interface loss is < 0.3 dB (mostly mode field mismatch). Excellent thermal stability has also been achieved using thermally curing epoxies [28].

An alternative process is to use dry etching of silicon to create alignment grooves. This has the advantage that the direction of the grooves is not constrained to be aligned with the crystallographic axis of silicon. Achievement of the required alignment accuracy is however more difficult since the process is not self limiting. LETI have achieved interface losses of less than 0.6 dB/interface (best < 0.25 dB/interface) using dry etching of the silicon substrate [29]. The base of the silicon groove was used for vertical alignment while structures in the waveguide layers on the surface were provided to achieve lateral alignment.

4. DEVICES

4.1. Splitters

The optical splitter is the first large market application for integrated optics devices, driven by requirements of Passive Optical Networks (PONS) systems [30,31,32] as shown in figure 5. A number of structures have been used for power splitting, including

Y-junctions, directional couplers, Wavelength INsensitive Couplers (WINCS) and radiative stars. These are described below.

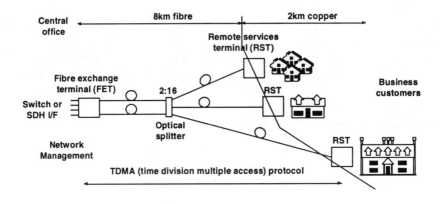

Figure 5 Example passive optical network system

4.1.1. Y-junctions

The Y-junction is the simplest possible power splitting element. A taper is used to gradually double the width of an input waveguide, which is split into two output waveguides which are then fanned out. The basic loss of the Y-junction is low if a gradual angle of divergence is used, however the performance of practical Y-junctions is affected by photolithograpic and etching limitations which can result in a blunt end several microns wide at the apex. The Y-junction has a nearly insensitive wavelength response, due to the intrinsic symmetry used in the design. Wavelength ripple can arise due to imperfect fibre alignment at the input, defects in waveguides or changes in curvatures of the waveguides preceding the Y-junction, all of which can cause the excitation of higher order lateral modes which beat with the fundamental mode prior to the junction.

Higher order 1 x N splitters are built up by cascading Y-junction elements. NTT have reported a pigtailed 1 x 8 splitter module with an excess loss of 1.46 dB, a port-port uniformity of 0.63 dB and a wavelength flatness over the wavelength range 1200 nm - 1600 nm of 2.6 dB [33]. A non-pigtailed 1 x 128 splitter has also been demonstrated, with maximum and minimum insertion losses of 22.2 dB and 25.9 dB respectively, including 21.1 dB splitting loss [34].

4.1.2. 4 port couplers

Many splitter based applications require two input ports to the splitter to allow, for example, system head end redundancy, route diversity or facility for later upgrades in a

PONS network. This cannot be satisfied by Y-junctions alone, so an additional component is required.

The fundamental four port element in integrated optics is the standard directional coupler (figure 6a). In this device, light launched in one input port excites the fundamental and first order lateral mode in the central region of parallel waveguides. These two modes have different propagation velocities resulting in power being exchanged between the waveguides. The power distribution at the output is dependant on the waveguide geometry, refractive index distribution, gap between the waveguides and, importantly, wavelength.

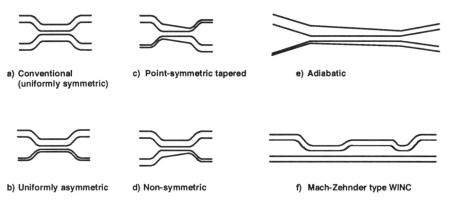

Figure 6   Types of four port coupler

A number of variations on the basic directional coupler design have been developed to improve the spectral uniformity. In fused fibre coupler technology, wavelength flattened couplers are achieved by using two dissimilar fibres in the interaction region [35]. The integrated optics analogue of this is a coupler composed of dissimilar waveguides, usually utilising two different waveguide widths in the interaction region (figure 6b). Asymmetric directional couplers have been used to achieve an 8 x 8 star coupler with wavelength flattened characteristics, variation being reduced from 7.3 dB to 2.1 dB over a 250 nm wavelength range relative to symmetric couplers [36]. Takagi et al examined the wavelength characteristics of these types of directional couplers and two other types which they catagorised as point symmetric tapered and non-symmetric [37] (figures 6c and 6d). The best design was found to be the asymmetric coupler, with which a splitting ratio of 50% ±5% (3 dB ± 0.5 dB) over the wavelength range 1300 nm - 1550 nm was achieved.

An alternative approach which in principle has broader response is the adiabatic 3 dB coupler [38] (figure 6e). The operating principle of this device is as follows. The input to the coupler consists of two waveguides of different width. The separation of these waveguides is reduced very gradually so as to provide an adiabatic transition to the coupling region, where the widths of the waveguides are gradually made equal, before being separated at the output. The fundamental mode of the wider waveguide remains the fundamental (symmetric) mode of the system and gradually evolves to have equal power in the two waveguides at the output. The fundamental of the narrower input waveguide (considered in isolation) is the first order (antisymmetric) mode of the system. This also evolves to have equal power, but opposite phase in the output arms of the coupler. Experimentally, a uniformity of ± 0.6 dB has been achieved over the wavelength range 1200 nm - 1700 nm. This type of device should be less sensitive to fabrication tolerances, but is significantly longer than a conventional directional coupler.

Another way of achieving wavelength flattened coupler is a Mach-Zehnder composed of two different couplers with a differential phase shift in the two waveguides connecting the couplers (figure 6f). - called a WINC (Wavelength Insensitive Coupler)[39]. A WINC is designed such that the phase shift between the couplers stops the coupling ratio monotonically increasing with wavelength. In reference 39, a WINC with a coupling ratio of 20% ± 1.9% over a wavelength range 1250 nm - 1650 nm was achieved. NTT have used a WINC designed as a 3 dB coupler combined with two 1 x 8 Y-junction arrays to achieve a pigtailed 2 x 16 splitter module with a uniformity of ± 0.6 dB over the wavelength range 1250 nm - 1650 nm. The mean loss of the module was 13.0 dB and the worst case loss at 1550 nm (including 12 dB splitting loss) was 13.8 dB.

4.1.3. Radiative stars

An alternative type of M x N coupler is the radiative star coupler, as shown in figure 7. The integrated optics implementation of this type of coupler was first proposed by Dragone [40,41]. This device operates as follows. Light in an input waveguide propagates to the planar waveguide region at the centre, where it is no longer confined laterally but is allowed to radiate freely. The far field diffraction pattern is then coupled into the array of output waveguides which fan out to match the output fibre spacing. Note that this type of device is not constrained to have the same number of input and output waveguides, nor that this number is a power of two. Design of this device is complex, but it offers several important advantages relative to networks of directional couplers or WINCs. The radiative star is more compact than a network of couplers, which is particularly important for large networks, and is rather more insensitive to wavelength, since the far field diffraction pattern is governed by the mode profile, which changes only slowly with wavelength. It is also relatively insensitive to fabrication parameters, for the same reason.

The power distribution in the output waveguide is non-uniform since it is not possible to achieve a perfectly rectangular far field. However coupling in the input waveguide array can produce a nearly rectangular far-field [42,43]. ATT have used this approach to produce a 19 x 19 star coupler with an efficiency of 55%, a uniformity of 2 dB and a worst case loss of 16.3 dB (including 12.8 dB splitting loss) at 1300 nm wavelength [44].

Using the Beam Propagation Method (BPM) [45], NTT designed an 8 x 8 star coupler with an average excess loss of 1.68 dB, with a uniformity of < 2 dB and a worst case loss of 11.6 dB (including 9 dB splitting loss)[46]. BPM modelling, in conjunction with

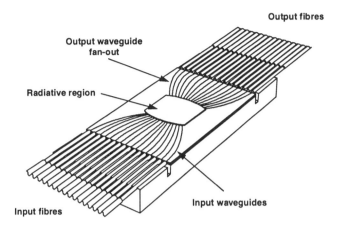

Figure 7   Schematic of radiative star coupler

the well defined square waveguide core, allowed the coupling to neighbouring waveguides to be designed in order to produce an input field with the correct sidelobes to produce a rectangular far field distribution. These results were achieved at 1300 nm wavelength. Further work on tapering the input waveguides prior to the radiative region substantially eliminated the wavelength sensitivity [47], achieving an average excess loss of 1.7 dB at 1300 nm and 2 dB at 1550 nm, uniformity of 1.4 dB at 1300 nm and 1.6 dB at 1550 nm, and corresponding worst case losses of 11.6 dB and 11.9 dB (including the 9 dB splitting loss for the 8 x 8 coupler). NTT have also achieved the largest N x N star couplers yet reported, demonstrating 64 x 64 and 144 x 144 star couplers[48]. For these large scale couplers it was necessary to offset the centres of the planar region surfaces to reduce the size of the offset angles between the peripheral input and output waveguides. These couplers were 60 x 36 mm and 60 x 60 mm for the 64-way and 144-way couplers respectively. For the 64 x 64 coupler average excess loss was 3.16 dB, uniformity was 7 dB and worst case loss was 24.4 dB (including splitting loss of 18. dB). The corresponding losses for the 144 x 144 coupler were 3.4 dB, 11 dB and 30.4 dB (splitting loss was 21.6 dB). The performance of these large scale devices were only degraded by around 1 dB on average when pigtailed as described in reference 25.

BNR Europe Limited has demonstrated 1 x 16 splitters based on the radiative star technique[49]. In this case input coupling does not occur due to the absence of neighbouring input waveguides. The input waveguide field was calculated using the effective index method and Fourier transformed to determine the far field, followed by performing an overlap integral with each of the output waveguides to calculate coupling efficiency. The far field in this case was near-gaussian, resulting in poor uniformity, with excess loss of 5 dB, uniformity of 2.5 dB and worst case loss (including 12 dB splitting loss) of 18.9 dB. Note that these results are for a fibre pigtailed device, in contrast to the results above. The output waveguide widths were then varied to improve this uniformity, with the waveguide widths being altered to ensure uniform power being incident on each waveguide. This improved the excess loss, uniformity and worst case loss values to 3.6 dB, 1.3 dB and 16.8 dB respectively. Very similar results have been obtained for 2 x 16 splitters [50]. Improvements in fabrication technology resulted in pigtailed 2 x 16 splitters with excess loss of 3.0 dB, uniformity of 2.0 dB worst case loss of 16.0 dB[51]. These figures were 3.2 dB and 16.9 dB respectively for a 4 x 16 splitter. A variation in insertion loss with wavelength of +/- 0.5 dB over the range 1100 nm to 1700 nm was also reported. Typical return loss was -35 dB and directivity was < -55 dB. Recently, improvements in waveguide technology allowed the worst case loss for a 2 x 16 splitter to be reduced to 14 dB over the 1300 nm to 1550 nm wavelength range [52].

The above results are summarised in Table 1

Device	Star excess loss *	Uniformity **	Additional losses ***	Splitting loss	Worst case loss	Wavelength (nm)	Pigtailed?	Ref.
19 x 19	2.6	2.0	1.0	12.8	16.3	1300	No	44
8 x 8	1.6	2.0	0.24	9.0	11.6	1550	No	46
8 x 8	1.4	1.4	0.3	9.0	11.6	1300	No	47
	1.7	1.6	0.3	9.0	11.9	1550		
64 x 64	1.7	7.0	1.4	18.1	24.4	1550	No	48
144 x 144	2.0	11.0	1.4	21.6	30.4	1550	No	48
1 x 16	2.6	3.0	2.4	12.0	18.9	1550	Yes	49
1 x 16	2.5	1.3	1.1	12.0	16.8	1550	No	
2 x 16	2.2	1.5	0.8	12.0	16.0	1250	Yes	51
4 x 16	2.4	2.5	0.8	12.0	17.0	→1600	Yes	

All figures in dB unless otherwise stated.
* Excluding interface and waveguide losses   ** Maximum loss - minimum loss
*** Interface and propagation losses

Table 1 Summary of reported radiative star coupler results

4.2. WDM devices

Wavelength Division Multiplexing (WDM) is an attractive way of increasing the information carried by an optical fibre through the use of different data streams on different wavelengths. Multi-wavelength networks are also attracting increasing interest due to the flexibility offered by wavelength drop-and-insert and wavelength routing functions. Another area where Wavelength Division Multiplexers (MUXs) are required is in Er-doped optical amplifiers, for multiplexing pump (980 nm or 1480 nm) wavelength and signal wavelength (around 1530 nm). WDM requirements and devices tend to fall into two classes - multiplexing of two wavelengths (e.g. 1300 nm/1550 nm or 980 nm/1530 nm) and multiplexing many closely spaced channels (e.g. 16 channels at 1 nm spacing in the Er fibre window). Sub-nm WDM is often referred to as Frequency Division Multiplexing (FDM), since the channel spacing is in the GHz or tens of GHz region. A range of device types have been demonstrated for WDM, some of which are described below.

4.2.1. Long Directional Couplers

An obvious way of demultiplexing two wavelengths is a directional coupler with a long interaction region [37], as used in fused fibre coupler MUXs. A significant disadvantage of this type of coupler however is that the filter response is a raised cosine. This means that, although high levels of rejection can be achieved in the centre of the stop band, the level of crosstalk gets worse elsewhere. For example, for a stop-band width of 30% of the channel spacing, the best crosstalk achievable is -12 dB. One way of improving the stop band rejection is by cascading two identical directional coupler MUXs, which gives a transfer function which is the square of that of a single coupler. This has been demonstrated to increase the bandwidth to 100 nm in a 1300 nm/1550 nm MUX, for greater than 20 dB rejection[53], although at the expense of a narrower pass-band. Another difficulty is that a WDM coupler with many interaction lengths requires extremely tight fabrication tolerances. It is for this reason that most workers have preferred the Mach-Zehnder interferometer for demultiplexing two wavelengths.

4.2.2. Mach Zehnder Interferometers

A well known device for demultiplexing two wavelengths is the Mach-Zehnder Interferometer (MZI), which consists of two directional couplers (near-3 dB) connected by waveguides of different length. The phase delay introduced by this differential path length is wavelength sensitive, hence the MZI acts as a wavelength multiplexer/demultiplexer with a periodic raised cosine response. The magnitude of the differential path length ΔL sets the wavelength spacing, e.g. a ΔL of 2.7 µm is used for a 1300 nm/1550 nm MUX while a ΔL of 100 mm is used for a wavelength spacing of 0.008 nm (1 GHz). MZIs with this range of spacings have been demonstrated with losses from 0.5 dB and crosstalks to > 25 dB [54]. The control of the differential path length in the master mask is good enough for devices with the large wavelength spacings, the very close wavelength spacing require thin film heaters, which use the thermo-optic effect to control the absolute phase, and hence the splitting ratio, at the second directional coupler.

WDM devices with greater than two channels can be constructed by cascading MZIs, making use of the periodic response of the MZI. The MZI with the largest ΔL is used for separating two combs of wavelengths, corresponding to the closest wavelength spacing $\Delta \lambda$. The next MZIs with a path length difference $\Delta L/2$ has twice the wavelength spacing $2\Delta \lambda$ and separates the wavelengths into four output channels. Verbeek et al demonstrated a four channel WDM based on cascaded MZIs, with 7.7 nm channel spacing, 2.6 dB insertion loss and -16 dB crosstalk [6]. Takato et al demonstrated a 128 channel frequency selection switch using seven levels of cascaded MZIs [55]. Each MZI had a thin film heater on one arm which were used to shift the response of the MZI. By appropriate choice of switching currents, one channel could be selected from 128 channels at 10 GHz spacing. Insertion loss was 6.7 dB and crosstalk was -13 dB.

Like the directional coupler MUX, the filter response of a MZI is a raised cosine. One way of increasing the stop-band width is integration with a ring resonator [56], which was demonstrated to increase the stop-band width by 1.8 relative to a conventional MZI without a corresponding penalty in reduction in pass band width..

The functions of a Mach-Zehnder 1300 nm/1550 nm multiplexer together with a 3 dB splitting of one wavelength band (a useful component for optical subscriber networks) have been combined in a double Mach-Zehnder configuration [57], with excess losses around 1 dB and length significantly shorter than the same device realised with a single Mach-Zehnder followed by a Y-junction.

4.2.3. Fresnel grating

One attraction of silica-on-silicon integrated optics is that the same RIE processes used to etch the waveguide cores can be used to achieve reflective mirrors by dry etching through the complete three layer structure. LETI have developed a WDM MUX which used a "Fresnel mirror" - a segmented elliptical mirror with combined dispersion and focusing properties - to produce a WDM device as shown in figure 8 [58]. Light from the input channel waveguide is reflected from the Fresnel mirror and focused on the output waveguide array. The position of the image at the output is wavelength dependant, thus forming a demultiplexing operation. A 1 x 4 demultipexer with 20 nm channel spacing, 4 dB insertion loss and < -25 dB crosstalk based on this device structure has been demonstrated [59]. This device did not require metallisation of the Fresnel mirror since it was operated beyond the critical angle. For larger number of channels and/or increased dispersion, a near-normal angle of incidence in the grating is used to operate in higher grating orders and hence increase the grating dispersion. In this case, metallisation is required, and the grating is operated in the equivalent of a Littrow configuration [60].

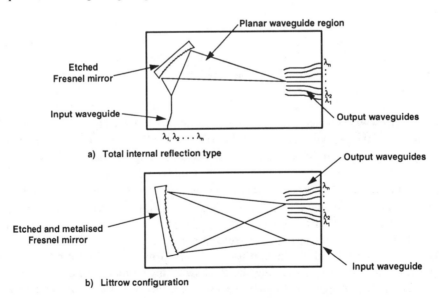

Figure 8 Schematic of wavelength multiplexer based on Fresnel mirror

One difficulty with this approach is that the loss of the device becomes very sensitive to the verticality of the mirror (around 1° verticality is required for < 2 dB insertion loss) and mirror roughness, thus requiring extremely good reactive ion etching technology.

4.2.4. AWG WDM

An alternative and very practical dispersive element is a phased array of waveguides of different lengths as proposed by Smit [61]. This can be achieved most easily by having an array of concentric circular arcs. A broad phase front is coupled into the array of waveguides (typically > 10 waveguides) which are separated to prevent inter-waveguide coupling. The waveguides are then taken through different optical path lengths. The phase distribution across the waveguide array at the output and, hence, the position of the focused spot in the image plane is then a function of wavelength. A channel spacing of 3 nm at 633 nm wavelength with crosstalk of better than 20 dB was obtained.

Sub-nanometer wavelength resolution was demonstrated using 1 mm bend radius waveguides in a non-concentric configuration [62]. This however required external lenses for input and output fibre coupling. Dragone et al integrated the arrayed waveguide grating with radiative star couplers for input and output [63,64] - see figure 9, thus eliminating the need for external lenses, and also demonstrating the potential of the Arrayed Waveguide Wavelength Division Multiplexer (AWG WDM) to operate as an N x N multiplexer/demultipexer. An aberration correction scheme was also described and utilised. For a 7 x 7 multiplexer, the insertion loss was < 2.5 dB and the crosstalk was < -25 dB.

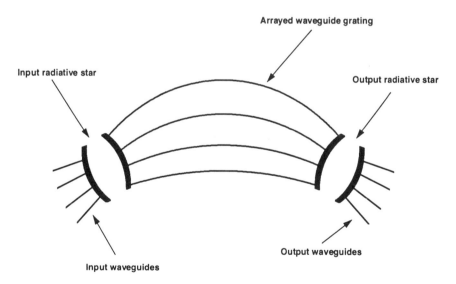

Figure 9   Schematic of arrayed waveguide grating wavelength division multiplexer

Takahashi et al. demonstrated an 11 channel AWG WDM with 10 GHz (0.08 nm) channel spacing [65], thus demonstrating the extremely high resolutions achievable using this device. It was clear from this and earlier work that temperature sensitivity and polarisation sensitivity of devices is a problem at high resolutions. The temperature sensitivity is due to the thermo-optic coefficient of the silica glass ($1.1 \times 10^{-5}/°C$) which results in a coefficient of 0.015 nm/°C and affects all types of WDM structures fabricated in silica. The polarisation sensitivity is due to the stress-induced birefringence of the waveguides, caused by the difference in thermal expansion between the waveguide layers and the silicon substrate following deposition and annealing or reflow. This is usually of the order of 3×10^{-4} to 1×10^{-3}, resulting in the same level of polarisation sensitive shift in centre wavelength (0.3 nm - 1.0 nm). A number of techniques for eliminating this polarisation sensitivity have been demonstrated. The simplest technique was to insert a quartz $\lambda/2$ plate in a 100 µm wide groove in the centre of the arrayed waveguide region [66]. The resulting polarisation rotation from TE to TM and vice versa half way through the grating eliminated the polarisation sensitivity, but at the expense of a 5 dB increase in loss due to the $\lambda/2$ plate. A better technique was to use an amorphous silicon layer to control the birefringence [67]. Here the amorphous silicon reduced the birefringence, and the length difference of amorphous silicon on adjacent waveguides was chosen to eliminate the polarisation sensitivity.

Using a non-symmetric arrangement of waveguides in the arrayed waveguide grating section, Adar et al demonstrated that the AWG WDM was also suitable for greater waveguide separations such as 1310 nm and 1550 nm wavelengths [68]. The advantage of this approach over the directional coupler or Mach-Zehnder WDM is the improved rejection and wider stop bands - about 200 nm at better than -35 dB crosstalk was achieved.

The main results from the above papers are summarised in table 2 below.

Device Type	Channel separation	3 dB passband	Total loss	Crosstalk	TE-TM shift	Ref.
11 x 11	1.5 nm	1.5 nm		< -16 dB	0.6 nm	64
7 x 7	3.3 nm	1.9 nm	2.5 dB	< -25 dB		
11 x 11	0.07 nm (10 GHz)	0.05 nm (6.5 GHz)	7 dB	< -14 dB	0.3 nm	65
1 x 13	1.0 nm	0.3 nm	9.5 dB*	< -30 dB	0.0 nm	66
1 x 16	0.8 nm	0.3 nm	8.5 dB	< -30 dB	0.0 nm	67
1 x 2	1310-1530 nm	90 nm	2.15 dB	< -35 dB	??	68

* Includes 5.0 dB loss from groove containing polarisation compensation plate

Table 2 Summary of reported AWG WDM results

As previously mentioned, one attractive feature of this type of multiplexer is that it operates as a N x N multiplexer. As well as using this feature in wavelength routing systems, this leads to the possibility of new component types. For example, by looping back outputs N to inputs N-1, using a connectorised optical fibre, a wavelength drop and insert component using a single N x N AWG WDM component can be constructed [69]. Using Er doped fibre amplifiers to connect pairs of inputs, together with one pair of outputs being connected, can produce a multi-wavelength ring laser with stable wavelength separation [70].

4.3. Thermo-optic switches

Optical switching is likely to be increasingly important in future transmission systems, both for data switching and for network reconfiguration. This is difficult in silica-on-silicon since the electro-optic effect in silica is extremely small. The best developed method of achieving an optical switching is by using the thermo-optic effect, as pioneered by NTT [71]. This is based on a Mach-Zehnder modulator with a thin film heater on one arm which can create a temperature difference between the two waveguide arms (figure 10a), which causes a phase change via the thermo-optic effect. The power required for switching is typically 0.5 W, the response time is around 2 ms, the crosstalk is -20 dB and the temperature increase of the heated waveguide is 5°C to 10°C with a heater length of 10 mm [3]. This type of device is polarisation insensitive and also insensitive to external temperature, since it is only the temperature difference rather than the absolute temperature which matters. A number of impressive demonstrations of the potential of this device have been made by cascading 2 x 2 thermo-optic switches. An 8 x 8 switch module with driving circuits having TTL interfaces has been fabricated [72]. This required 64 thermo-optic switches for operation, integrated on a 71 mm x 68 mm Si wafer. The insertion loss for the pigtailed 8 x 8 switch ranged from 9 dB - 11 dB, with crosstalk > 20 dB.

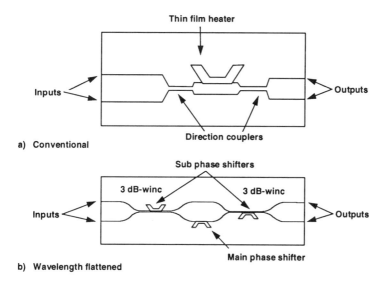

Figure 10 Schematic of thermo-optic switches

The basic Mach-Zehnder switch is wavelength sensitive since the thermo-optic phase change is inversely proportional to wavelength, and the individual couplers from which it is fabricated are themselves wavelength sensitive. Wavelength insensitive thermo-optic switches can be fabricated by replacing the standard couplers by WINCS which incorporate sub-phase shifters (figure 10b). Use of a point-symmetric arrangement of the two WINCS allow these sub-phase shifters to compensate for the wavelength sensitivity of the main phase shifter. An insertion loss of 0.5 dB and an extinction ratio of -17 dB over the wavelength range 1250 nm - 1650 nm has been demonstrated using such a device [73].

An alternative way of realising a switching function is through the hybrid integration of laser diode optical gates (i.e. semiconductor optical amplifiers). A 4 x 4 switch array has been demonstrated using four 1 x 4 splitters followed by sixteen laser diode gates, followed by four 4 x 1 combiners in a full point-multipoint switching configuration[74]. In addition to higher switching speed than the thermo-optic switch, this configuration is suitable for switched-broadcast networks such as video systems, although at the expense of higher insertion losses for the devices fabricated to date.

4.4. Components for signal processing

One of the advantages of using large area silicon substrates is that signal processing circuits incorporating elements such as delay lines or FDM MZIs, which require long lengths of waveguides, become feasible. These of course require very low loss waveguides (<< 0.1 dB/cm), also provided by silica-on-on silicon technology. A few examples of such circuits are given below.

Optical waveguide delay lines, tunable splitters (thermo-optic tuned MZIs) and thermo-optic phase shifters have been combined to produce a coherent optical transversal filter with a tapped delay line structure[75]. Four and eight tap filters were used to achieved specified frequency characteristics, and it was shown that the filter could process high speed broad-band baseband signals over 4 THz. These filters were constructed by looping the splitters, taps and combiners back and forth across the surface of a 3 " silicon wafer.

Delay lines have also been used to convert a 6.3 Gbit/s data stream into a 50 Gbit/s data stream, for use in a time division multiplexing demonstration [76]. This was achieved by using three unbalanced Mach-Zehnder interferometers in series, resulting in a total of eight different paths, and eight corresponding delay times from input to output. Sufficiently short input pulses and appropriate choice of delay times in the MZIs enabled the 6.3 Gbit/s to be multiplexed to 50 Gbit/s. In an extension to this work, a four-stage MZI has also been used as a 16:1 time division multiplexer [77]..

An alternative way of implementing long delay lines is the looped delay line [78], where a few cm of delay line can be fitted into a few square mm of spiralled waveguide. This utilised high Δn waveguides based on arsenosilicate glass to achieve the small bend radii required.

5. RARE EARTH DOPED LASERS AND AMPLIFIERS

Er doped fibre amplifiers (EDFAs) have received much attention over recent years due to the amplification in the 1550 nm wavelength region [79], and are now being used in commercial systems. Er-doped fibres are also applicable to fibre lasers [80]. Both fibre amplifiers and fibre lasers require various passive optical components for operation. These include wavelength multiplexers to mix signal and pump wavelengths (980 nm/1550 nm or 1480 nm/1550 nm), taps to enable pump or signal levels to be monitored, and polarisation multiplexers to provide for pump laser redundancy in high reliability systems. These components are currently realised as fused fibre couplers. Significant long term benefits could potentially be gained by integrating all of these functions together using silica-on-silicon technology. The major technical challenge is achieving the amplification function. Er doped fibre amplifiers typically contain a few hundred ppm of Er doping, and require operating lengths ranging from a few meters to hundreds of meters. Waveguide amplifiers require the same levels of amplification over a few cm or tens of cm, thus requiring much higher Er concentrations, which can lead to clustering and up-conversion problems.

The first demonstration of rare earth doped silica-on-silicon waveguides was a waveguide laser using Nd doping[81]. Lasing action was demonstrated at 1051.5 nm wavelength in a 5 cm long stripe waveguide. Lasing and amplification using Er doping is more difficult since the Er ions act as a three level system, whereas the Nd ions act as a four level system. Lasing has since been demonstrated in Er doped silica-on-silicon waveguides fabricated using flame hydrolysis deposition. A concentration of 8000 ppm of Er was used in a 45 mm long multimode waveguide, which achieved a threshold of 49 mW using 980 nm pump wavelength. Waveguide amplifiers based on Er-doped silica have, to date, required longer waveguide lengths of tens of cm, achieved by looping waveguides back and forth across the waveguide substrate, and using flame hydrolysis [82] or PECVD [83]. High levels of phosphorous co-doping in the core have been used to achieve the required concentrations of Er (about 0.5 wt%), and the demonstrated gain was 0.65 dB/cm. A total gain of 24 dB, saturated output of 11 dBm and a noise figure of 3.8 dB (not simultaneously) were achieved using a pigtailed silica-on-silicon planar waveguide amplifier module[84]. Integration with the 980 nm/1550 nm multiplexer has also recently been demonstrated[85].

One attraction of integrated optics rare earth amplification is the much wider range of host glasses available in comparison with optical fibres. Shmulovich et al demonstrated waveguide amplification based on a sputtered highly Er-doped sodium calcium silicate glass on a HIPOX buffer layer, with a silicone cladding [86]. Using this material system, a gain of 3.3 dB/cm was demonstrated, which enabled 15 dB net gain to be demonstrated in a 4.5 cm long straight waveguide [87].

6. CONCLUSIONS

Following about ten years of research, silica-on-silicon technology is now maturing rapidly. Very low loss waveguides, simple and stable fibre alignment techniques and wafer scale integration have all been developed. A wide range of device types, from simple splitters to wavelength division multiplexers to switch arrays have been demonstrated. Work has commenced on development of rare earth doped waveguide amplifiers.

Current research directions are in optimising the performance and improving the manufacturability of components such as splitters and wavelength multiplexers, in further developing large scale complex circuits such as the AWGWDM, N x N switch arrays and signal processing circuits. Exciting new avenues of exploration are in rare earth doped amplifiers and lasers, and in hybridisation with opto-electronic and electronic components.

Components are now being moved from the laboratory into real production. This is being led by splitters and other components for Passive Optical Networks. Reliability of these components, and cost competitiveness, particularly against fused fibre and ion-exchanged glass technologies, must be proven in the market place. This will then open the way for the deployment of more complex systems, and the silica-on-silicon components on which they depend.

7. ACKNOWLEDGEMENTS

The author would like to thank the Directors of BNR Europe Limited for permission to publish this paper.

© Northern Telecom Europe Limited 1994

8. REFERENCES

1. M. Kawachi, M. Yasu, and T. Edahiro, "Fabrication of SiO_2-TiO_2 glass planar optical waveguides by flame hydrolysis deposition", Electronics Letters, Vol. 19, No.15, pp 583-584, 21st July 1983.
2. N. Takato, M. Yasu, M. Kawachi, "Low loss high silica single mode channel waveguides", Electronics Letters, Vol 22, No. 6, pp 321-322, 13 March 1986
3. M. Kawachi, "Silica waveguides on silicon and their application to integrated-optic components", Optical and Quantum Electronics, Vol. 22, pp 391-416, 1990
4. P.B. Moynagh and P.J. Rosser, "Thermal Oxidation of Si", <u>Properties of Silicon</u>, No. 4, pp 469 - 479, INSPEC, The Institution of Electrical Engineers, London and New York, 1988

References (cont)

5. S.P. Tay, J.P. Ellul and M.I.H. King, "Oxidation rates of Si at high pressure", Properties of Silicon, No. 4, pp 482-490, INSPEC, The Institution of Electrical Engineers, London and New York, 1988

6. B.H. Verbeek, C.H. Henry, N.A. Olsson, K.J. Orlowski, R.F. Kazarinov and B.H. Johnson, "Integrated four-channel Mach-Zehnder multi/demultiplexer fabricated with phosphorous doped SiO_2 waveguides on silicon", Journal of Lightwave Technology, Vol. 6, No. 6, June 1988.

7. C.H. Henry, G.E. Blonder and R.F. Kazarinov, "Glass waveguides on silicon for hybrid optical packaging", Journal of Lightwave Technology, Vol. 7, No. 10, October 1989.

8. R. Bellerby, B. Bhumbra, G.J. Cannell, S.Day, M.Grant and M. Nelson, "Low cost silica-on-silicon single mode 1:16 optical power splitter for 1550 nm", Proc EFOC '90, Munich, June 1990, pp 100-103

9. G. Grand, S. Valette, G.J. Cannell, J. Aarnio, M. del Giudice et al, "Fibre pigtailed silicon based passive optical components", Proc. ECOC'90, Amsterdam, September 1990, pp 525-528

10. G. Grand, J.P. Jadot, H. Denis, S. Valette, A. Fournier and A.M. Grouillet, "Low-loss PECVD silica channel waveguides for optical communications", Electronics Letters, Vol. 26, No. 25, pp 2135-2137, 6 December 1990.

11. F. Bruno, M. del Guidice, R. Recca and F. Testa, "Plasma-enhanced chemical vapour deposition of low-loss SiON optical waveguides at 1.5 µm wavelength", Applied Optics Vol. 30, No. 31, pp 4560-4564, 1 November 1991

12. K. Kapser, C. Wagner and P.P Deimel, "Rapid deposition of high-quality silicon-oxinitride waveguides", IEEE Photonics Technology Letters, Vol 3, No. 12, pp 1096-1098, December 1991.

13. Q. Lai, J.S. Gu, M.K. Smit, J. Schmid and H. Melchior, "Simple technologies for fabrication of low loss silica waveguides", Electronics Letters, Vol 28, No. 11, pp1000-1001, 21st May 1992

14. P.R. Wensley, R. Bellerby, S. Day, M.F. Grant and S.A. Rosser, "Improved waveguide technology for silica-on-silicon integrated optics", ECIO, Neuchatel, April 18-22, 1993, pp 9.8-9.9.

15. C.H. Henry, R.F. Kazarinov, H.J. Lee, K.J. Orlowski and L.E. Katz, "Low loss Si_3N_4-SiO_2 optical waveguides on Si", Applied Optics, Vol. 26, No. 13, 1 July 1987

16. N. Imoto, N. Shimuzu, H. Mori and M. Ikeda, "Sputtered silica waveguides with an embedded three dimensional structure", Journal of Lightwave Technology, Vol. 1, pp 289-293, 1993

17. J.T. Boyd, R.W. Wu, D.E. Zelmon A. Naumann and H.A. Timlin, "Guided wave optical structures using silicon", Optical Engineering, Vol 24, pp 230-234, 1985

18. R.R.A. Syms, "Stress in thick sol-gel phosphosilicate glass films formed on Si substrates", Journal of Non-Crystalline Solids, Vol. 167, pp 16-20, 1994

19. R.R.A. Syms and A.S. Holmes, "Reflow and burial of channel waveguides formed in sol-gel glass on Si substrates", IEEE Photonics Technology letters, Vol. 5, No. 9, pp 1077-1079, September 1993

References (cont)

20. A.S. Holmes, R.R.A. Syms, Ming Li and M. Green, "Fabrication of buried channel waveguides on silicon substrates using spin-on glass", Applied Optics, Vol.32, No. 25, pp 4916-4921, 1 September, 1993
21. E.J. Murphy and T.C. Rice,"Low loss coupling of multiple fibre arrays to single mode waveguides", Journal of Lightwave Technology, Vol. 1, pp 479-482, 1983
22. E.J. Murphy, T.C. Rice, L. McCaughen, G.T. Harvey and P.H. Read,"Permanent attachment of single mode fibre arrays to waveguides", Journal of Lightwave Technology, Vol.3, pp 795-799, August 1985
23. M.F. Grant, A. Donaldson, D.R. Gibson and M.Wale,"Recent progress in lithium niobate integrated optics technology under a collaborative Joint Opto-Electronic Research Scheme (JOERS) programme", Optical Engineering, Vol. 27, No. 1, pp 002-010, January 1988
24. Y. Yamada, F. Hanawa, T. Kitoh and T. Maruno, "Low -loss and stable fiber-to-waveguide connection utilising UV curable adhesive", IEEE Photonics Technology Letters, Vol. 4, No. 8, pp 906-908, August 1992
25. K. Kato, K. Okamoto, H. Okazaki, Y. Ohmori and I. Nishi, "Packaging of large-scale integrated optic N x N star couplers", IEEE Photonics Technology Letters, Vol. 4, No. 3, pp 348-351, March 1993
26. K.E. Bean, "Anisotropic Etching of Silicon", IEEE Transactions on Electron Devices, Vol. ED-25, No. 10, pp 1185-1193, October 1978
27. M.F. Grant, R. Bellerby, S. Day, G.J. Cannell and M. Nelson, "Self aligned multiple fibre coupling for silica-on-silicon integrated optics", Proc. EFOC-LAN, London, June 19-21 1991, pp 269-292
28. M.F. Grant, S. Day and R. Bellerby,"Low loss coupling of ribbon fibres to silica-on-silicon integrated optics using preferentially etched V-grooves", Proc. Integrated Photonics Research, New Orleans, April 13-16 1992, pp 166-167
29. G. Grand, H. Denis and S. Valette, "New method for low cost and efficient optical connections between singlemode fibres and silica guides", Electronic Letters, Vol. 27, No. 1, pp 16-18, 3 January, 1991
30. M.H. Reeve, S. Hornung, L.Bickers, P. Jenkins and S. Mallinson, "Design of passive optical networks", British Telecom Technology Journal, Vol. 7, No. 2, pp 89-99, April 1989
31. P.J. Dyke, "Introduction and evolution of optical fibre systems in the local loop", International Journal of Digital and Analogue Cabled Systems, Vol.2, pp 187-189, 1989
32. W. Weippert, "Chances for optical communication in Germany", <u>ECOC '92</u>, Vol 2, pp 649-659, Berlin und Offenbach, Berlin, 1992
33. S. Kobayashi, T. Kitoh, Y. Hidu, S. Suzuki and M. Yamaguchi, "Y-branching silica waveguide 1 x 8 splitter module", Electronics Letters, Vol. 6, No. 11, pp 707-708, 24 May 1990
34. H. Takahashi, Y. Ohmori and M. Kawachi, "Design and fabrication of silica-based integrated optic 1 x 128 power splitter", Vol 27, No. 23, pp 2131-2133, 7 November 1991
35. D.B. Mortimer, "Wavelength flattened fused couplers", Electronics Letters, Vol. 21, pp 742-743, 1985

References (cont)

36. H. Yanagawa, S. Nakamura, I. Ohyama and K. Ueki, "Broad-band high-silica optical waveguide star coupler with asymmetric directional couplers", Journal of Lightwave Technology, Vol.8, No. 9, pp 1292-1297, September 1990
37. A. Takagi, K. Jinguji and M. Kawachi, "Wavelength characteristics of 2 x 2 optical channel-type directional couplers with symmetric or nonsymmetric coupling structures", Journal of Lightwave Technology, Vol. 10, No. 6, pp 735-746, June 1992
38. Y. Shani, C.H. Henry, R.C. Kistler, R.F. Kazarinov and K.J. Orlowski,"Integrated optic adiabatic devices on silicon", IEEE Journal of Quantum Electronics, Vol. 27, No. 3, pp 556-566, March 1991
39. K. Jingui, N. Takato, A. Sugita and M. Kawachi, "Mach-Zehnder interferometer type optical waveguide coupler with wavelength-flattened coupling ratio", Electronics Letters, Vol. 26, No. 17, pp 1326-1327, 16 August 1990
40. C. Dragone, "Efficient N x N star couplers using Fourier optics", Journal of Lightwave Technology, Vol. 7, No. 3, pp 479-489, March 1989
41. C. Dragone, "Efficient N x N star coupler based on Fourier optics", Electronics Letters, Vol 24, No. 15, 21 July 1988
42. C. Dragone, "Efficiency of a periodic array of a nearly ideal element pattern", IEEE Photonics Technology Letters, Vol. 1, No. 8, pp 238-240, August 1989
43. C. Dragone, "Optimum design of a planar array of tapered waveguides", Journal of the Optical Society of America, Vol. 7, No. 11, pp 2081-2093, November 1990
44. C. Dragone, C.H. Henry, I.P. Kaminow and R.S. Kistler,"Efficient multichannel integrated optics star coupler on silicon", IEEE Photonics Technology Letters, Vol. 1, No. 8, pp 241-242, August 1989
45. J. Van Roey, J. Van Der Donk and P.E. Lagasse, "Beam propagation method: analysis and assessment", Journal of the Optical Society of America, Vol. 71, No. 7, pp 803-810, July 1981
46. K. Okamoto, H. Takahashi, S. Suzuki, A. Sugita and Y. Ohmori, "Design and fabrication of integrated optic 8 x 8 star coupler", Electronics Letters, Vol. 27, No. 9, pp 774-775, 25 April 1991
47. K. Okamoto, H. Takahashi, M. Yasu and Y. Hibino, "Fabrication of wavelength insensitive 8 x 8 star coupler", IEEE Photonics Technology Letters, Vol. 4, No. 1, pp 61-63, January 1992
48. K. Okamoto, H. Okazaka, Y. Ohmori and K. Kato,"Fabrication of large scale integrated optic N x N star couplers", IEEE Photonics Technology Letters, Vol. 4, No. 9, pp 1032-1035, September 1992
49. S. Day, R. Bellerby, G.J. Cannell and M.F. Grant, "Silicon based fibre pigtailed 1 x 16 power splitters", Electronics Letters, Vol 28 No.10 pp 920-921, 7 May, 1992
50. S. Day, R. Bellerby, G. Cannell and M. Grant, "Silicon based fibre pigtailed 1 x 16 and 2 x 16 power splitters", Proc. ECOC '92, Berlin, September 1992, pp. 525-528
51. M. Grant, S. Day, R. Bellerby, P.R. Wensley, S.A. Rosser and G.J. Cannell, "Low cost M x N couplers in silica-on-silicon for passive optical networks", To be published, International Journal of Opto-Electronics, 1994, Vol. 9, No. 2
[52] S. Day, P.R. Wensley, private communication.

References (cont)

53. K. Imoto, H. Sano, M. Miyazaki, Y. Takasaki, "Guided-wave multi-demultiplexers with high stopband rejection", Electronics Letters, Vol. 23, No. 9, pp 472-473, 23 April 1987

54. N. Takato, T. Kominato, A. Sugita, K. Jingui, H. Toba and M. Kawachi, "Silica-based integrated optic Mach-Zehnder multi/demultiplexer family with channel spacing of 0.01 - 250 nm", IEEE Journal of Selected Areas in Communications", Vol. 8, No. 6, pp 1120-1127, August 1990

55. N. Takato, A. Sugita, K. Onose, H. Okazaki, M. Okuno, M. Kawachi and K. Oda, "128-channel polarisation-insensitive frequency-selection-switch using high-silica waveguides on Si", IEEE Photonics Technology Letters, Vol. 2, No. 6, pp 441-443, June 1990

56. K. Oda, N. Takato, H. Toba and N. Nosu, "Wideband guided-wave periodic multi/demultiplexerwith a ring cavity for optical FDM transmission systems", Electronics Letters, Vol. 24, No. 4, pp 210-212, 18 February 1992.

57. A. Takagi, T. Kominato and I. Nishi, "1x3 WDM/Splitter circuits based on guided wave double Mach-Zehnder interferometer configuration", ECOC 92

58. S. Valette, P. Gidon and J.P. Jadot, "New integrated optical multiplexer-demultiplexer realised on silicon substrate", ECIO'87, Glasgow

59. G. Grand, J.P. Jadot, S. Valette, H. Denis, A. Fournier and A.M. Grouillet, "Fiber pigtailed wavelength multiplexer/demultiplexer at 1.55 microns integrated on silicon substrate", Proc EFOC '90, Munich, June 1990, pp ??

60. G. Grand, B. Corselle, J.P. Jadot, E. Parrens, A. Fournier, A.M. Grouillet and S. Valette, "16-channel optical wavelength multipleter/demultiplexer integrated on silicon substrate", Proc. EFOC-LAN, London, June 19-21 1991, pp ????

61. M.K. Smit, "New focusing and dispersive planar component based on an optical phased array", Electronics Letters, Vol. 24, No. 7, pp 385-386, 31 March 1988.

62. H. Takahashi, S. Suzuki, K. Kato and I. Nishi, "Arrayed-waveguide grating for wavelength division multi/demultipexer with nanometer resolution", Electronics Letters, Vol. 26, No. 2, pp 87-88, 18 January 1990

63. C. Dragone, "An N x N optical multiplexer using a planar arrangement of two star couplers", IEEE Photonics Technology Letters, Vol. 3, No. 9, pp 812-815, September 1991

64. C. Dragone, C.A. Wdwards and R.C. Kistler,"Integrated Optics N x N Multiplexer on silicon", IEEE Photonics Technology Letters, Vol. 3, No. 10, pp 896-899, October 1991

65. H. Takahashi, I. Nishi and Y. Hibino, "10 GHz spacing optical frequency division multiplexer based on arrayed waveguide grating", Electronics Letters, Vol. 28, No. 4, pp 380-382, 13 February 1992

66. H. Takahashi, Y. Hibino and I. Nishi, "Polarisation-insensitive arrayed-waveguide grating multiplexer on silicon", Optics Letters, Vol. 17, No. 7, pp 499-501, April 1, 1992

67. H. Takahashi, Y. Hibino, Y. Ohmori and M. Kawachi, "Polarisation-insensitive arrayed waveguide wavelength multiplexer with birefringence compensating film", IEEE Photonics Technology Letters, Vol. 5, No. 6, pp 707-709, June 1993

References (cont)

68. R. Adar, C.H. Henry, C. Dragone, R.C. Kistler and M.A. Milbrodt, "Broad band array multiplerers made with silica waveguides on silicon", Proc. Integrated Photonics Research, New Orleans, pp 12-15, April 13-16 1992
69. Y. Tachikawa, Y. Inoue, M. Kawachi, H. Takahashi and K. Inoue, "Arrayed-waveguide grating add-drop multiplexer with loop-back optical paths", Electronics Letters, Vol. 29, No. 24, pp 2133-2134, 25 November 1993.
70. H. Takahashi, H. Toba, Y. Inoue, "Multiwavelength ring laser composed of EDFAs and an arrayed-waveguide wavelength multiplexer", Electronics Letters, Vol. 30, No. 1, pp 44-45, 6 January 1994
71. H. Himeno, M. Kobayashi, "Single-mode guided-wave optical gate matrix switch using Mach-Zehnder interferometric gates", Electronics Letters, Vol. 23, No. 17, pp 887-888, 13 August 1987
72. R. Nagase, A. Himeno, K. Kato and M. Okuno, "Silica-based 8x8 optical-matrix switch module with hybrid integrated driving circuits", Proc. ECOC '93, Montreaux, September 12-16 1993, pp17-20
73. T. Kitoh, N. Takato, K. Jinguji, M. Yasu and M. Kawachi, "Novel broad-band optical switch using silica-based planar lightwave circuit", IEEE Photonics Technology Letters, Vol. 4, No. 7, pp 735-737, July 1992
74. Y. Yamada, H. Terui, Y. Ohmori, M. Yamada, A. Himeno and M. Kobayashi, "Hybrid-integrated 4 x 4 optical gate matrix switch using silica-based optical waveguides and LD array chips", Journal of Lightwave Technology, Vol. 10, No. 3, pp 383-389, March 1992
75. K Sasayama, M. Okuno and K. Habara, "Coherent optical transversal filter using silica-based waveguides for high speed signal processing", Journal of Lightwave Technology, Vol. 9, No. 10, pp 1225-1230, October 1991
76. S. Kawanishi, H. Takara, K. Uchiyama, T. Kitoh and M. Saruwatari, "100 Gbit/s, 50 km, and nonrepeated optical transmision employing all-optical multi-demultiplexing and PLL timing extraction", Electronics Letters, Vol. 29, No. 12, pp 1075-1077, 10 June 1993
77. S. Kawanishi, H. Takara, K. Uchiyama, M. Saruwatari and T. Kitoh, "Fully time-division-multiplexed 100Gbit/s optical transmission experiment",Electronics Letters, Vol. 29, No. 25, pp 2211-2212, 9 December 1993
78. C.J. Beaumont, S.A. Cassidy, D. Welbourne, M. Nield, A. Thurlow and D.M. Spirit, "Integrated silica optical delay line for signal processing", BT Technology Journal, Vol. 9, No. 4, pp 30-34, 1991
79. R.J. Mears, L. Reekie, I.M. Jauncey and D.N. Payne, "Low noise erbium-doped fiber amplifier operating at 1.54 µm", Electronics Letters, Vol. 23, pp. 1026-1028, 1987
80. L. Reekie, R.J. Mears, S.B. Poole and D.N. Payne, "Tunable single mode fiber lasers", Journal of Lightwave Technology, Vol. 4, pp 956-960, 1986
81. Y. Hibino, T. Kitagawa, M. Shimuzu, F. Hanawa and A. Sugita, "Neodymium-doped silica optical waveguide laser on silicon substrate", IEEE Photonics Technology Letters, Vol. 1, No. 11, pp 349-350, Novmber 1989.

References (cont)

82. T. Kitagawa, K. Hattori, K. Shuto, M. Yasu, M. Kobayashi and M. Horiguchi, "Amplification in erbium-doped silica-based planar lightwave circuits", Electronics Letters, Vol. 28, No. 19, pp 1818-1819, 10 September 1992

83. K. Shuto, K. Hattori, T. Kitagawa, Y. Ohmori and M. Horiguchi, "Erbium-doped phosphosilicate glass waveguide amplifier fabricated by PECVD", Electronics Letters, Vol. 29, No. 2, pp 139-141, 21 January 1993

84. T. Kitagawa, K. Hattori, K. Shuto, M. Oguma, J. Temmyo, S. Suzuki and M. Horiguchi, "Erbium doped silica-based planar amplifier module pumped by laser diodes", Proc. ECOC '93, Montreaux, September 12-16 1993, pp.41-44

85. K. Hattori, T. Kitagawa, M. Oguma, Y. ohmori and M. Horiguchi, "Erbium-doped silica-based planar-waveguide amplifier integrated with a 980 nm/1530 nm WDM coupler", Conference of Optical Fibre Communication, OFC '94 Technical Digest, pp 280-281

86. J. Shmulovich, A. Wong, Y.H. Wong, P.C. Becker, A.J. Bruce, and R. Adar, "Er^{3+} glass waveguide amplifier at 1.5 µm on silicon", Electronics Letters, Vol. 28, No. 13, pp 1181-1182, 18 June 1992

87. J. Shmulovich, Y.H. Wong, G. Nykolak, P.C. Becker, R. Adar, A.J. Bruce, D.J. Muehlner, G. Adams and M. Fishteyn, "15 dB net gain demonstration in Er^{3+} glass waveguide amplifier on silicon", Proc. Optical Fibre Communications 1993, San Jose, paper PD18-1, pp 75-78, February 21-26, 1993

Sol-Gel and Rare-Earth-Doped Fibers
and Waveguides

Sol-gel process for glass integrated optics

John D. Mackenzie and Yu-Hua Kao
Department of Materials Science & Engineering
University of California at Los Angeles
Los Angeles, CA 90024

ABSTRACT

The sol-gel process allows the synthesis of a wide variety of amorphous as well as crystalline materials which can exhibit useful passive or active optical properties. The process offers many advantages, such as low-temperature synthesis, excellent control and flexibility over composition and design. For applications in future integrated optics, the sol-gel process is flexible in making various kinds of optical components which either have been successfully made or have potential in their realization. This paper examines a number of the state-of-the-art optical components fabricable by the sol-gel process for glass integrated optics. Major examples furnished are in glass substrates and waveguides, third-order nonlinear materials, lasers and optical amplifiers, optical fibers, and gradient-index lenses. The benefits as well as limitations by using the sol-gel approach will be critically presented.

1. INTRODUCTION

The motivation for developing integrated optics over electronics is that it offers significant advances in device performance[1]. Photonic devices can operate at much higher speeds than electronic devices. They are designed to switch and process light signals without the need to convert them to electronic form, leading to very fast operating speeds. The fastest electronic components operate at speeds in the picosecond range, which is 1000 times slower than the experimental optical components that can operate in the femtosecond range[2]. Furthermore, in optics, photons essentially do not interact, but electrons interact very strongly. Thus, optical devices are capable of massive parallelism and are free from cross-talk[1].

The development of integrated optical devices based on glass is desirable because of its excellent transparency, high threshold-to-optical damage ratio, thermal, mechanical, and chemical stablility, and small coupling losses between waveguides made on silica-based glasses and optical fibers because of their close match in refractive indices[3]. Moreover, in recent years, glasses and various kinds of amorphous solids are being developed for use as photonic materials. Some of these materials are relatively easy to incorporate with functional entities to make active optical components, such as third-order nonlinear waveguides and solid-state lasers.

In the making of amorphous photonic materials, the gel approach has attracted much attention over the conventional melt-quench method. The sol-gel process offers many inherent advantages such as low-temperature processing; high-purity and homogeneity; ease of chemical control of the processing; and the flexibility of making photonic materials in various configurations such as bulk solids, fibers, film waveguides, and coatings for device applications.

Because the sol-gel process offers flexibility in its processing, various optically-active compounds have been incorporated in gel-derived matrices. They have been proven to be suitable hosts for various optically active compounds. The gel process has allowed the synthesis of third-order nonlinear optical (NLO) materials based on organics and quantum dots, and lasers and optical amplifiers containing organic dyes and rare-earth ions. Furthermore, the gel process has offered specific advantages to the fabrication of passive materials such as glass substrates, optical fibers, gradient refractive index (GRIN) lenses, and protective coatings. For this critical review, in addition to glasses, it will be shown that amorphous solids containing organics exhibit good performance for optical applications.

2. THE SOL-GEL PROCESS

The sol-gel process is a liquid route to make amorphous and crystalline materials via the hydrolysis and polycondensation of metal alkoxides[4,5]. Optically active compounds initially in the form of salts or alkoxides can be incorporated in the sol for making various photonic materials. In addition, a wide variety of optically active organics can be incorporated without thermal degradation via this low-temperature chemical route.

Equations 1 and 2 show how gels and glasses can be prepared by the sol-gel techniques. The metal alkoxide mixed homogeneously in a mutual solvent undergo hydrolysis via water and catalyst. As polycondensation occurs, the viscosity of the sol increases to form a three-dimensional network, i.e. a gel, by polymerization.

Hydrolysis: $M(OR)_n + nH_2O \longrightarrow M(OR)_n + nROH$ (1)

Condensation: $M(OH)_n + M(OH)_n \longrightarrow \begin{array}{c} O \\ | \\ O-M-O-M-O \\ | \\ O \end{array} \begin{array}{c} O \\ | \\ \\ | \\ O \end{array} + nH_2O$ (2)

(M = Si, Ti, Zr, etc..., R = alkyl group)

The gel so-formed is a highly porous solid. It must be dried and heat-treated to become a stable and useful component. For some systems, after high-temperature treatment, a dense oxide glass is formed. For other systems, high-temperature heating will yield a polycrystalline solid. Those containing organics cannot obviously be heat-treated to very high temperatures. In some systems in which the organic forms chemical bonds with the oxide, rubbery solids can be prepared[6]. Since the sol-gel process is based on liquid solutions, the fabrication of films and fibers can be easily achieved.

The sol-gel process offers inherent advantages over the conventional glass-melting method. Because the starting alkoxides are purified then mixed in the liquid state, the sol-gel-derived gels have low impurity levels and excellent homogeneity. These factors are especially important for photonic materials with low loss. Higher concentrations of dopants in the form of salts or alkoxides can be uniformly dispersed in the sol to form doped photonic materials. Low-temperature processing can lead to the fabrication of amorphous materials without the phase separation and crystallization so that special multicomponent compositions can be fabricated. Figure 1 shows that through careful control of viscosity, drying and heating conditions, bulk, fiber and film configurations can be easily fabricated. Furthermore, organically-modified silicate (Ormosil) matrices with hybrid organic-inorganic properties can be synthesized through the addition of an organic at the solution stage. Transparent Ormosils have been demonstrated to be excellent hosts for wide-ranging optical materials[7,8]. Flexibility of the Ormosil matrix has improved crack resistance of thick films and coatings[9].

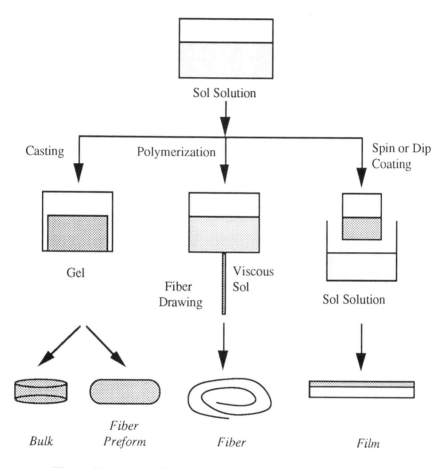

Fig. 1. Material configurations fabricable by the sol-gel process.

3. COMPONENTS FOR GLASS INTEGRATED OPTICS

3.1. Glass substrates and waveguides

Hench et. al. have successfully fabricated fully dense, very high purity and homogeneous silica glasses[10,11,12] which can be used as substrates. The acid-catalyzed alkoxide-derived silica monoliths have excellent physical and optical properties. Compared to other commercial silica glasses, these gel-derived glasses have lower vacuum ultraviolet cut-off, lower dispersion, lower thermal expansion coefficient (0-650°C), higher homogeneity, and less defects[12]. Figure 2 shows the UV transmission curve for a densified gel-derived silica glass and the curves for commercial silica glasses[12]. Due to the elimination of OH⁻ groups from the SiO_2 network, the gel-derived glasses have improved vacuum UV transmission compared to the commercial glasses. Based on optical interferometry, the gel-derived glasses are bubble-free and have a level of homogeneity less than 1 ppm[13,14].

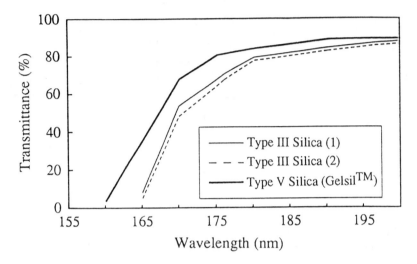

Fig. 2. UV transmission of commercial (Type III) and gel-derived (Type V Gelsil™) silica glasses[12].

Various processes which have been investigated for making waveguides in sol-gel-derived materials are ion exchange[15], sol-gel coating[16,17], sol-gel coating with subsequent pattering[18] and laser densification[19-22]. The ion exchange technique has been widely investigated for glasses containing alkali ions[15]. Active optical glasses containing these ions have been made by the sol-gel process (see Sec. 3.2(B)), providing great potential as functional components in integrated optics.

Planar waveguides of high-index GeO_2-SiO_2 glass films on silica substrate have been fabricated[23]. Double alkoxide solutions of germanium and silicon were used for dip-coating. After multiple dipping-and-firing cycles, the films were fully densified at 850-1110°C, depending on the composition. Films with GeO_2 as high as 67 mol% have been deposited. The thickest film was 2.6 μm which contained 9.5 mol% GeO_2. These waveguides have been demonstrated to have good waveguiding properties. The loss of a 50GeO_2-50SiO_2 glass waveguide was 3.31dB/cm. Subsequent waveguide fabrication has been investigated by pattering or laser densification on gel films. Waveguides can be made by stamping a desired configuration on the as-cast gel film[18]. Laser densification have been explored to make channel and ridge waveguides in porous gels films[19-22].

In the making of films for waveguides, the sol-gel process has two limitations: film thinness and high OH⁻ content. Multiple dip-coating is necessary to make >1μm films needed for waveguide. Thick films crack due to coating/substrate stresses during film shrinkage. However, Menning et. al. have shown that an organically-modified silica sol greatly improves the ability to deposit >1 μm crack-free films[24]. The OH⁻ groups which are generated upon hydrolysis of the alkoxide remain in the film and cause transmission losses in the IR and near IR. With further research, the OH⁻ groups can be removed during the processing of gel-derived materials. Furthermore, the short interaction lengths involved in integrated optics may reduce the importance of losses associated with OH⁻ groups.

Sol-gel-derived materials have low foreign impurity and good homogeneity which are crutial for use as low loss components. High levels of high-index dopants can be incorporated through the sol-gel process to allow the synthesis of high-index films. This provides great flexibility for engineering a wide-range of Δn. Large coated areas on either one or both sides of the substrate can be made easily without complicated equipment. Finally, the sol-gel process has the flexibility of making not only passive but active waveguides with the addition of active entities in the gel (see Sec. 3.2 and 3.3).

3.2. Third-order nonlinear optical materials

Research in nonlinear optical (NLO) materials has become increasingly active because they hold great potential in the development of photonic-based technologies. Third-order NLO processes are fundamentally important to all optical information processing operations. The control of light by light in high-speed photonic switches allows their operation in the femtosecond region. High nonlinearity is needed in order to miniturize the components for integrated optics, and to allow the eventual use of low-intensity lasers (*e.g.* semiconductor diode lasers) in photonic devices. However, desirable NLO materials for integrated optics must not only exhibit large nonlinearity but also have chemical, thermal and mechanical durability, low optical loss, high threshold for optical damage, and be capable of existing in useful device configurations. In the following section, the sol-gel process has been used to make third-order NLO materials in bulk and thin film forms. Organics molecules, semiconductor quantum dots and metal clusters exhibiting high nonlinearities have been incorporated in gels, gel-derived inorganic glasses, and Ormosils.

3.2(A). Organic-doped materials

Numerous third-order nonlinear organic molecules have widely been studied in gel-derived matrices[25-27]. For example, Prasad et. al. have fabricated film waveguides of poly-p-phenylene vinylene (PPV) in silica gel[25]. Up to 50 wt% of PPV has been incorporated in silica by first mixing a polymer precursor with a silica precursor. The organic was then polymerized to form the conjugated structure. The $\chi^{(3)}$ was approximately 3×10^{-10} esu measured by degenerate four-wave mixing (DFWM) at 1.064 μm, and the response time was in the sub-picosecond range[26].

3.2(B). Semiconductor quantum dot-doped materials

Three-dimensionally-confined semiconductor quantum dots exhibit enhanced nonlinearity compared to their corresponding bulk semiconductors[28-30]. Quantum confinement leads to a blue shift in the absorption edge of the bulk semiconductor. Physical confinement of excitons at some radius smaller than the Bohr exciton radius increases the binding energy of excitons at room temperature. Upon irradiation with light, the exciton levels become saturated. This saturation blocks the further absorption of light, resulting in the nonlinear optical behavior of the material[31-33]. The enhancement of nonlinearity in quantum dots for a simple two-level system are shown in Eqs. 3 and 4

$$\Delta\alpha/I = K_1 [R_e/R]^3 \text{Im}[\chi_n^{(3)}] \qquad (3)$$

$$\Delta n/I = K_2 [R_e/R]^3 \text{Re}[\chi_n^{(3)}] \qquad (4)$$

where $\Delta\alpha$ is the enhanced change in absorption due to confinement, Δn is the enhanced change in index due to confinement, R_e/R is the ratio of the exciton Bohr radius to the quantum dot radius, $\chi_n^{(3)}$ is the normalized nonlinearity, I is the intensity of light, and K_1 and K_2 are scaling factors[34]. The material performance is given by the figure of merit (FOM) as shown in Eq. 5

$$\text{FOM} = n_2/(\tau\alpha) \qquad (5)$$

where n_2 is the nonlinear refractive index, τ is the response time, and α is the absorption.

Research on inorganic third-order nonlinear quantum dot materials grew after CdS_xSe_{1-x} filter glasses were discovered to exhibit large resonant nonlinearities and very fast response times of 10^{-9} to 10^{-8} esu and 10^{-12} to 10^{-11} sec, respectively, by DFWM at 532, 580, and 694 nm[28]. Table 1 below shows values of nonlinearity exhibited by different quantum dot systems made by various processing methods.

Table 1. Third-order nonlinearity of semiconductor quantum dots made by various processing methods.

Quantum Dot Materials	$\chi^{(3)}$ [esu]	λ (nm)	Method	Ref.
Sol-gel				
CuCl in glass	10^{-8}	380	DFWM	35
CdS in silica glass	10^{-10}	390	DFWM	36
CdS in Na_2O-B_2O_3-SiO_2 glass	10^{-7} - 10^{-6}	450 - 460	DFWM	37
CdS in Ormosil film	10^{-8}	436	pump & probe	38
Melt-quench				
CdS_xSe_{1-x} filter glasses	10^{-9} - 10^{-8}	532, 580, 694	DFWM	28
CdTe in 50CdO-50P_2O_5 glass	10^{-7}	600	DFWM	39
Sputtering				
CdS in SiO_2 film	10^{-8}	532	DFWM	40
Laser evaporation				
CdTe in SiO_2 film	10^{-7}	580	DFWM	41

This table shows that sol-gel-derived materials exhibit approximately the same order of magnitude in nonlinearity as those made by the other methods considered here. Moreover, the sol-gel process provides the following advantages:

1. Incorporation of high semiconductor content (up to ~ 20wt%)[42]
2. Improved control over the formation and growth of the quantum dots[43-44]
3. Reduced photodarkening effect[45]
4. Control over the chemistry and physical properties of the quantum dots[46-47]

Thus, many investigators have used this process to incorporate semiconductor quantum dots[48-52]. For the sake of brevity, only work at UCLA is described in detail.

Our research has been concentrated in the fabrication and optical properties of CdS quantum dots in sodium borosilicate (NBS) glasses, Ormosils, and SiO_2 gel matrices. The process for the fabrication of fully dense CdS-doped $5Na_2O$-$15B_2O_3$-$80SiO_2$ (mol%, nominal) glasses is shown in Fig. 3[53]. Tetramethylorthosilicate (TMOS) and triethylborate [$B(OEt)_3$] are mixed in the presence of water and the catalyst HCl. Hydrolysis and polycondensation occurs to form the borosilicate gel. Cadmium and sodium salts are doped into the precursor solution. Upon gelation and drying, the Cd salt precipitates on the pore walls as nano-sized particles. The Cd salt is converted to CdO by oxidation in O_2 and subsequently converted to CdS by sulphuration in H_2S. The porous gel is fully densified in vacuum into a transparent optical quality glass. The NBS matrix has a low densification temperature which reduces the sublimation loss of CdS during densification. These CdS quantum dot-doped glasses can contain high concentrations of CdS (10wt%) with dot size well within the quantum confinement range of 7-10 nm[37]. The nonlinearities were large on the order of 10^{-7} to 10^{-6} esu, as shown in Fig. 4[54]. For device applications, ion exchange of sodium for potassium has been performed on these NBS glasses to make planar waveguides.

The Ormosils matrix has also been proven to be an excellent matrix for CdS quantum dots[55]. With Ormosils as the matrix, the tendency for the gel to crack during shrinkage, drying and heat-treatment is reduced. The process for the fabrication of 28 wt% polydimethylsiloxane (PDMS) - 72 wt% tetraethoxysilane (TEOS) Ormosils is shown in Fig. 5[42]. The CdS-doped Ormosils were crack-free with maximum CdS concentration of 20wt%[42]. Quantum dot sizes in Ormosils have been determined by photoluminescence excitation spectroscopy (PLE). Each PLE spectrum in Fig. 6 corresponds to the energies of a specific transition for some distribution of particle sizes in the Ormosils matrix[56]. These spectra were in good agreement with the values calculated from X-ray peak broadening[42,56]. Both the sol-gel-derived CdS-doped Ormosils and NBS materials are much more resistant to photodarkening than the melt-quenched glass. Results from the pump-and-probe experiment in Fig. 7 shows that unlike the melt-quenched glass, the Ormosil sample retained its third-order response after two hours of pumping[45,57]. Table 2 shows a summary of the unique properties of our CdS-doped materials.

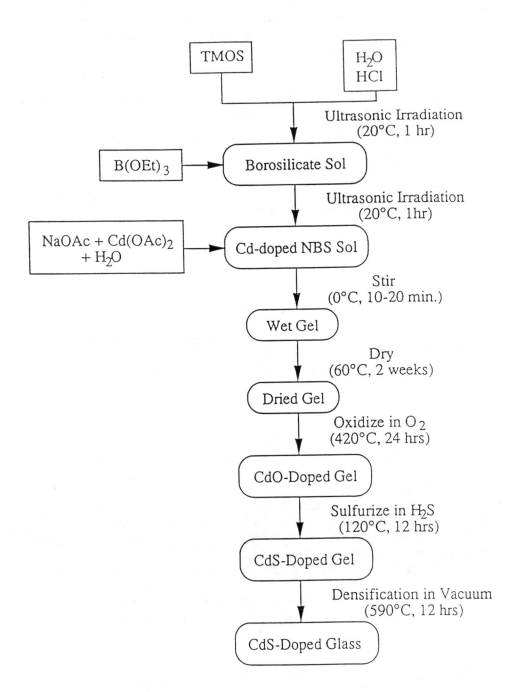

Fig. 3. Sol-gel process for the preparation of CdS-doped Na_2O-B_2O_3-SiO_2 glasses[53].

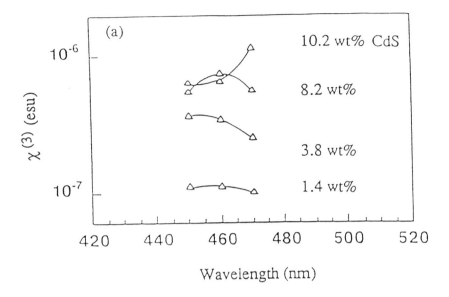

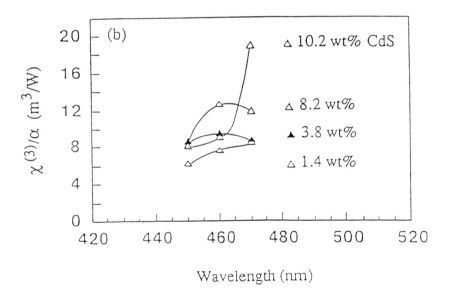

Fig. 4. Third-order nonlinear optical properties of CdS-doped Na_2O-B_2O_3-SiO_2 glasses measured by DFWM with 7 ns laser pulse[54].
(a). $\chi^{(3)}$
(b). $\chi^{(3)}/\alpha$, which eliminates the effect of CdS concentration on $\chi^{(3)}$

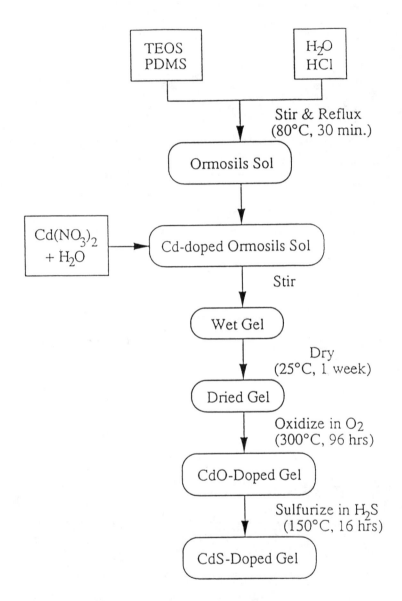

Fig. 5. Sol-gel process for the preparation of CdS-doped Ormosils[42].

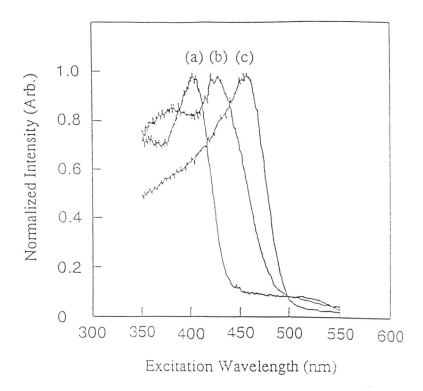

Fig. 6. Photoluminescence excitation spectra (monitored at 680 nm and 4.2K) of CdS-doped Ormosils with different microcrystallite sizes[56].
(a). 2.8 nm
(b). 3.5 nm
(c). 6.0 nm

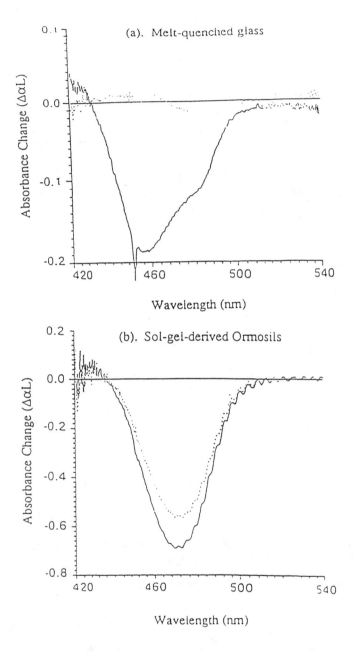

Fig. 7. Room temperature absorbance change by the pump-and-probe technique using laser pulses at 10 Hz and 5 µJ. Solid and dotted curves are before and after two hours of pumping, respectively[42,57].
(a). 0.5 wt% CdS-doped glass made by the melt-quench method
(b). 4 wt% CdS-doped Ormosils made by the sol-gel method

Table 2. Summary of the properties of the CdS quantum dot materials made by the sol-gel method.

Property	Sodium Borosilicate[53,54]	Ormosil[42,55,57]
Maximum CdS content (wt%)	11	20
CdS quantum dot size (nm)	3 to 7	2 to 7
Nonlinearity, $\chi^{(3)}$ On-resonnance (esu) Off-resonnance (esu)	10^{-6} to 10^{-7} 10^{-11} to 10^{-12}	10^{-7} to 10^{-8} 10^{-11} to 10^{-13}
Photodarkening	Resistant	Resistant
Matrix microstructure	Pore-free glass	Porous gel
Mechanical properties	Similar to glass (polishable to 15 μm)	Vickers hardness number of 160
Waveguide fabrication	Ion exchange (bulk glass)	Sol-gel coating (crack-free film)

Although both of the sol-gel-derived CdS-doped NBS and Ormosils materials are superior to the melt-quenched filter glasses in many ways, the quantum dot size and size distribution can be further improved. A novel method to control and further narrow the size distribution of quantum dots has been developed. Cadmium ions in the gel are uniformly distributed by anchoring them to the gel matrix via a bifunctional ligand of an organically-substituted alkoxysilane[43]. A bifunctional ligand, X---Si(OR)$_3$, consists of a polar group (X) capable of coordinating the metal ion, an inert group (---, e.g. $(CH_2)_3$), and hydrolyzable silyl groups (Si(OR)$_3$). The resulting complex, L_mM-X---Si(OR)$_3$, is formed by mixing together the metal moiety (L_mM) and the bifunctional alkoxysilane. This complex is then anchored to the silica gel matrix upon hydrolysis and polycondensation with the metal alkoxide. The reaction in Eq. 6 shows how the bifunctional ligand, 3-aminopropyl triethoxysilane (APTES) can be used to anchor Cd^{2+} ions to the gel matrix.

$$L_mCd\text{---}NH_2(CH_2)_3Si(OC_2H_5)_3 + Si(OEt)_4 \xrightarrow[-EtOH]{+H_2O}$$

$$L_mCd\text{---}NH_2(CH_2)_3\text{---}\underset{\underset{O}{|}}{\overset{\overset{O}{|}}{Si}}\text{---}O\text{---}\underset{\underset{O}{|}}{\overset{\overset{O}{|}}{Si}}\text{---}O\text{---} \qquad (6)$$

When the metal ions have been anchored to the gel, precipitation of the metal salt during drying is prevented. Upon subsequent heat-treatment, a much narrower quantum dot size distribution is expected. Figure 8 shows schematically how this (b) chelating process improves the particle size and size distribution over (a) processes without the use of the chelating agent[44]. Indeed, the transmission electron micrographs of Fig. 9 shows that the CdS-doped SiO_2 gels fabricated with the use of APTES ((a) and (b)) have much narrower sizes and size distributions over the materials prepared without APTES ((c) and (d))[44].

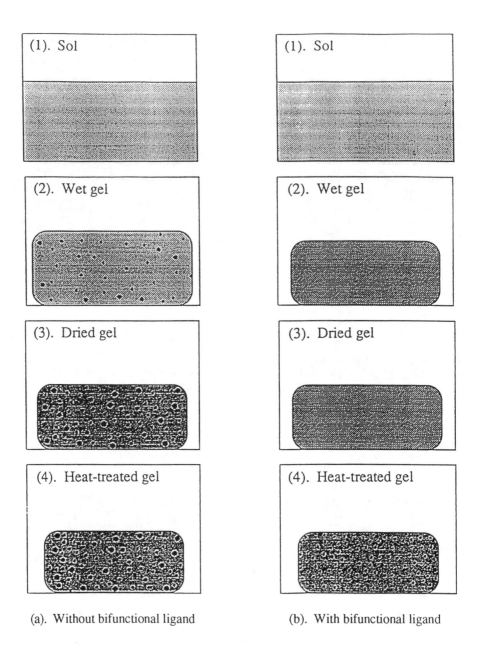

Fig. 8. Schematic comparison of the formation of quantum dots with and without the use of a bifunctional ligand in the sol-gel process for controlling particle size and size distribution[44].
(a). Without bifunctional ligand
(b). With bifunctional ligand

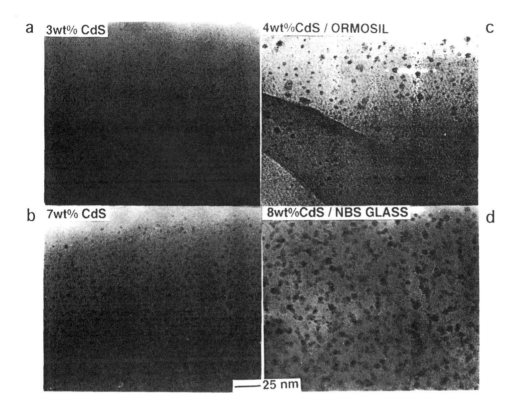

Fig. 9. Transmssion electron micrographs of CdS quantum dot NLO materials fabricated by the sol-gel process[44].
(a). 3 wt% CdS in APTES-modified silica gel matrix.
(b). 7 wt% CdS in APTES-modified silica gel matrix.
(c). 4 wt% CdS in Ormosil matrix.
(d). 8 wt% in sodium borosilicate glass matrix.

3.2(C). Metal cluster-doped materials

In the last few years, metal clusers dispersed in glasses have been discovered to have the potential of matching the high nonlinearities exhibited by the semiconductor-doped glasses. Ricard et. al. observed that Au cluster-doped glasses exhibit a $\chi^{(3)}$ of 10^{-8} esu and a response time of < 50 picosecond by DFWM at 532 nm[58,59]. They theorized that the high nonlinearities of metal cluster-doped glasses arise from local-field enhancement inside the metal particle at the plasmon resonance[59]. This model revealed a strong $1/r^3$ dependence of $\chi^{(3)}$ on metal cluster size r. Thus, the control of small particle size is important to enhance the nonlinearity of metal-cluster-doped materials. The glass-melting method only allows very low metal cluster concentration (<< 1%) to be incorporated without the undesirable segregation of large metal particles from the glass[60,61]. However, by ion implantation, Fukumi et. al. were able to incorporate a large concentration of Au clusters (7.6 vol%) in silica glass[62]. The high value of $\chi^{(3)}$ at about 1.2×10^{-7} esu by DFWM was attributed to the high metal concentration.

By the sol-gel technique, nano-sized metal clusters have been prepared in glassy[64-67] and Ormosils materials[68,69]. Kozuka et. al. fabricated 1 vol% Au-doped silica film[66]. Matsuoka et. al. fabricated 0.37 vol% Au doped-silica film and obtained a $\chi^{(3)}$ of 10^{-7} esu by DFWM[67]. This is of the same order of magnitude as those of the glasses prepared by ion implantation in spite of a lower Au concentration. In our laboratory, we have prepared metal cluster-doped Ormosils (Au, Pt, Ag, and Cu) in both film and bulk forms[68,69]. The materials were prepared by dissolving metal salts into the Ormosil precursor solution. After gelation, the metal ions were thermally or chemically reduced. A 5wt% Au film heat-treated at 200°C for 24 hrs had an average particle size of 2.5 nm. The Au particles were verified by X-ray diffraction to have the fcc structure. The response time of the Au-doped film may be faster than nanosecond, since by a nanosecond pump-and-probe technique, the third-order nonlinear response could not be detected[69]. In order to improve the nonlinearity of metal cluster-doped materials, bifunctional ligands have also been used to control the size and size distribution of metal clusters in SiO_2 gels[43].

The sol-gel process has enabled the synthesis of high-content, nano-sized CdS particles in glasses and Ormosils (<20 wt%) with high optical nonlinearity ($\chi^{(3)} \sim 10^{-6}$ to 10^{-7} esu). In addition, both the NBS and Ormosils samples exhibits reduced photodarkening effect compared to the melt-quenched filter glass. Through the gel process, we have demonstrated the advantage of using a bifunctional ligand in the starting sol to control dot size and narrow its size distribution. However, further research is necessary to improve the FOM [$\chi^{(3)}/(\tau\alpha)$] of the material for device applications. The nonlinearity is expected to increase with smaller size[34] and narrower size distribution[44]. The response time can be shortened through a modification of the surface of the quantum dot or by changing its local environment. This is because the response time has been found to be greatly affected by the charge carrier dynamics at the particle/matrix interface where surface defects (as well as deep traps) can cause electrons and holes to recombine[70,71]. Also, optical loss of the material can be reduced by minimizing the chemical impurities introduced during processing and scattering due to matrix inhomogeneity.

3.3. Solid-state lasers

For integrated optics, the laser material must be a solid. Research on solid-state lasers applicable to the sol-gel process has been on organic dyes and rare-earth ions (Re^{3+}). Presently, solvent-dispersed organic dyes are used as tunable lasers in the visible region. A closed dye/solvent system lacks photostability and exhibits reduced fluorescence yield due to easy dimerization of the monomers in the liquid environment[72]. Therefore, the presently-used organic dye laser system requires large, complex equipment which is obviously unfit for use in integrated optics. Rare-earth-doped lasers are of interest because the wavelengths of fluorescence match those presently used in telecommunications[73]. The sol-gel route allows the possibility of making both dye-doped and Re^{3+}-doped lasers in waveguide or fiber configurations for device integration.

3.3(A). Organic dye-doped lasers

Organic laser dyes have low threshold power for lasing, and large absorption and emission cross-sections[74]. They are tunable in the visible spectrum and have favorable gain characteristics for use in devices[74]. However, as solid-state laser material, organic dye-doped polymers exhibit relatively low efficiency for lasing, lack of photostability, and low thermal and mechanial durability due to the polymer matrix[74]. On the other hand, organic dye-doped inorganic and Ormosils matrices of improved optical, thermal, and mechanical properties has been prepared by the low-temperature sol-gel process without degradation of the dye.

To-date, numerous organic laser dyes have been incorporated and shown to lase in various sol-gel-derived matrices such as silica, aluminosilicates and Ormosils[75]. In particular, Ormosil matrices are excellent, since laser dyes doped in Ormosils have exhibited improved photostability over polymers hosts[76]. Coumarin doped in Ormosil exhibits one to two orders of magnitude improvement in photostability over that in poly(methylmethacrylate) (PMMA)[77]. We have also successfully incorporated Rh6G, Rh610, Rh620, and Rh640 dyes in Ormosils[78]. Table 3 shows that the photostability of these dyes is consistently better in Ormosils than in PMMA[79].

Table 3. Photostability (J/mm^3) of various Rhodamine dyes in Ormosil and PMMA matrices[79].

Matrix	Rh6G	Rh610	Rh620	Rh640
Ormosil	0.11	0.34	0.19	0.35
PMMA	0.05	0.04	0.09	0.08

* Photostability = NE_p/V, where V = Pumped volume, Ep = Energy/pulse, and N = Number of pump pulses needed for 50% gain reduction.

3.3(B). Rare-earth-doped lasers

Rare-earth ions are of interest because their wavelengths of fluorescence match those used in telecommunications, i.e. 1.3 µm and 1.5 µm[73]. These wavelengths are where dispersion and absorption are respectively the lowest in silica glass fibers[73,80]. Thus, Nd^{3+} and Er^{3+} ions, which have emissions around 1.3 µm and 1.5 µm, respectively, are of particular interest in optical communications. Signals from a glass fiber can be coupled directly into a Re^{3+}-doped solid-state laser. Lasing and light amplification can be accomplished by pumping with a diode laser[73].

Amongst the various types of glasses, silica glass is most advantageous for the propagation of high-intensity pulses under high average power[81]. It has high chemical durability, extremely low thermal expansion coefficient, low nonlinear refractive index[82], and high surface damage threshold at 1.064 µm[83]. Lasing has been observed in < 0.1 mol% Nd^{3+}-doped SiO_2 glasses and fibers made by the melt-quench method[84]. Only very low concentrations of Nd^{3+} ions can be incorporated (e.g. < 0.1 mol% Nd maximum) in silica glass[81,85]. At higher concentrations, concentration quenching causes a reduction in the quantum efficiency of the luminescence, and phase separation of the Nd_2O_3 additive causes scattering.

Through careful control of the chemistry of the sol, Thomas, et. al. successfully fabricated the first sol-gel-derived Re^{3+}-doped silica glasses of nominal compositions $98SiO_2$-$2Nd_2O_3$, and $98SiO_2$-$1.5Al_2O_3$-$0.5Nd_2O_3$ (in wt%). Glasses of $98SiO_2$-$1.5Al_2O_3$-$0.5Nd_2O_3$ exhibited lasing action with high quantum yield of 55%, long emission lifetime of 425µs, and optical scattering loss of 4.5%/cm[81]. The small addition of Al_2O_3 was found to assist the uniform dispersion of the Nd^{3+} ions[86].

Much can still be improved in order to realize an adequate material for laser systems. Higher Re^{3+} doping levels while maintaining a uniform dispersion of Re^{3+} ions is necessary to permit more efficient pumping. In addition to concentration quenching, the presence of OH^- groups in the glass also reduces the luminescence efficiency. Thomas et. al. have also observed reduction in quantum yields and fluorescent lifetimes with increasing OH^- concentration in their glasses[81]. Further research is needed to reduce the amount of OH^- groups to increase quantum yield to ~ 90% for device application.

In an effort to reduce the OH^- content, we have fabricated Er^{3+}-doped low-hydroxyl Ormosils[87]. This was accomplished through a non-hydrolytic reaction of methyl-modified silicon halides with tertiary butyl alcohol under controlled atmosphere. The material has been shown to have greatly reduced OH^- content compared to conventionally-prepared Ormosils. Luminescence of the Er^{3+} ion in the gels has been observed at 570 nm under an excitation wavelength of 488 nm.

3.4. Optical fibers

The conventional method of fabricating optical fibers for telecommunications is done by drawing fibers from GeO_2-doped SiO_2 glass preforms made by flame hydrolysis. These GeO_2-SiO_2 fibers exhibit losses <1.0 dB/km from ~1.1 nm to ~1.7 nm[80], and have a residual OH⁻ content of <1 ppb[88]. However, the flame hydrolysis method involves the use of corrosive and toxic metal chlorides which necessitates the use of pollution abatement equipment. Furthermore, large amounts of the GeO_2-SiO_2 particles generated at high temperatures (1200-1500°C) are lost in the carrier gas stream[89].

The sol-gel process offers a low-temperature non-corrosive route to make optical-quality preforms and glass fibers with high conversion efficiency. Glass fibers can be synthesized either directly from gel fibers[90] or by melt-drawing of gel-drived preforms. From gel-derived preforms, Matsuyama et. al made pure SiO_2 fibers with losses as low as 4 dB/km at 800 nm[91]. Susa et. al. made GeO_2-SiO_2 fibers (<5 mol% GeO_2) which exhibited losses of 20-30 dB/km at 630-830 nm[92,93]. During fiber drawing at temperatures higher than >2000°C, GeO_2 was found to cause severe bubbling in the glass, which increased with increasing GeO_2 content[92]. Kirkbir et. al. have succcessfully fabricated GeO_2-SiO_2 glass fibers with compositions >5 mol% GeO_2 without bubbles. These fibers exhibited optical losses of about 10-12 dB/km at 660 nm with OH⁻ content >10 ppm[89]. Fluorine-doped silica glass fibers with relatively low loss have been made by Shibata et. al[94]. The preform was made of a silica core and a F-doped silica clad. The core and the clad were individually synthesized by the sol-gel process and were subsequently bonded together. The fiber drawn from this preform was found to exhibit an optical loss of 1.6 dB/km at 1.69 μm and was found to have a low OH⁻ content of <1 ppm[94].

Gel-derived fibers are still inferior to vapor-deposited fibers in performance. In order to compete with vapor-deposited fibers, gel-derived preforms must have a lower OH⁻ content of <30 ppb, as well as defect-free core-clad interfaces[95]. Nevertheless, the sol-gel process offers the possibility of making active fibers. High-index, third-order nonlinear glass fibers (e.g. PbO-TiO_2[90,96] and chalcogenide[97] glasses) and Re^{3+}-doped laser fibers can have important usage in future integrated optics.

3.5. Gradient-index lenses

A gradient-index (GRIN) lens has a distribution of refractive index profile that varies in a controlled manner. The GRIN lenses can have radially, axially, or spherically varying profiles. The focusing action of a flat-surface radial-GRIN lens and that of a homogeneous lens is schematically shown in Fig. 10[98]. For integrated optics, GRIN lenses can be used as optical couplers, switches, and collimators[98].

Today, small radial-GRIN lenses are made commercially by the ion exchange process. However, the process consumes much energy and time. Also, it is limited to those monovalent cations which have sufficiently high diffusivity at the ion exchange temperatures to create the desired index gradient[99]. The key advantage of the sol-gel process is compositional flexibility, since precursors for almost all kinds of index-modifying dopants exist for designing the desired index profile. Also, the sol-gel process takes less energy, involves much higher diffusion coefficients of the ions in the gel, and can be used to form different size and shape GRIN lenses.

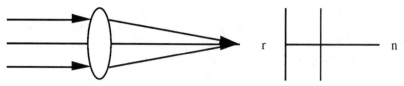

(a). Homogeneous lens and its refractive index profile.

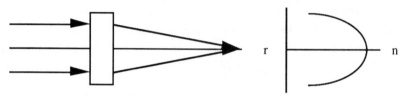

(b). Radial GRIN lens and its refractive index profile.

Fig. 10. Comparison of (a) homogeneous and (b) radial GRIN lenses and their refractive index profiles[98].

Radial GRIN's have been made by the sol-gel process by two methods: leaching of gels containing metal alkoxide dopants, and interdiffusion of index-modifying ions (see Fig. 11[99])[99-101]. In the leaching method, an index-modifying dopant such as Ge, Ti, Ta or Zr, is incorporated into the gel by using a metal alkoxide. After gelation, the gel is placed into an acidic solution to leach out the index-modifying ion. This creates a decreasing concentration profile and, therefore, a decreasing refractive index profile from the inside to the outside of the gel. This method typically allows good control over the index gradient. Similarly, the interdiffusion method involves the mixing of an alkoxide with an index-modifying metal salt. After gelation, the gel is placed into a second metal salt solution, resulting in the interdiffusion of both ions. Thus, compared to the leaching method, the interdiffusion method allows the use of various kinds of ions, which can create a higher index gradient. Table 4 shows some of the GRIN elements prepared by these two variations of the sol-gel method[99]. The interdiffusion method should be more suitable for making GRIN elements for integrated optics, since higher Δn could be obtained, and better control over the index profile is possible.

Table 4. Radial gradient-index lenses prepared by the sol-gel method[99].

Sol-Gel	Gradient-Index System	Properties	Refs.
Leaching	Ge, Ti, Ta out; H in SiO$_2$ gel matrix	2 - 3.4 mm dia. rods* $\Delta n = 0.015 - 0.025$	102-106
	Zr, Ti, Ge out; H in SiO$_2$, B$_2$O$_3$-SiO$_2$, Al$_2$O$_3$ matrices	3 - 8 mm dia. rods $\Delta n = 0.013 - 0.028$	107-108
Interdiffusion	Tl, Cs, Rb out; H, NH$_4$, K in SiO$_2$ gel matrix	10 - 13 mm dia. rods good focusing ability	109-113
	Pb out; K in B$_2$O$_3$-SiO$_2$ glass matrix	6 - 13 mm dia. rods $\Delta n = 0.02 - 0.05$	114-115

* For GeO$_2$-SiO$_2$ and TiO$_2$-SiO$_2$ systems only.

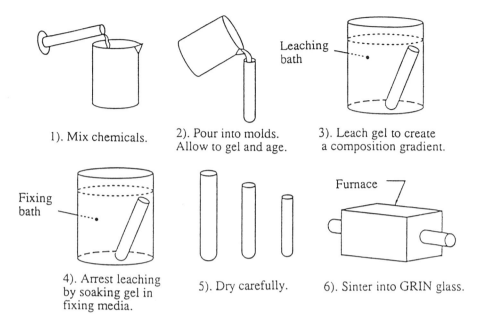

Fig. 11. Procedure for the fabrication of gradient-index lenses by the sol-gel method[99].

4. OTHER APPLICATIONS

4.1. Photochromic materials

When light is absorbed in a photochromic material, it can exhibit a reversible color change. The rate of change of the absorbance will determine the kind of application it can be used for, i.e. switching or optical memory[74,116]. For very fast transformations, photochromic materials can be used as optical switches; light of one wavelength causes a change in the absorption spectrum which either blocks or transmits light of another wavelength. On the other hand, for extremely slow transformations, photochromic materials can be used for optical data storage; light of one wavelength is used to encode the information, then light of another wavelength is used to read the encoded information.

In photochromism using AgCl, the absorption of light reduces Ag^+ to $Ag^°$. After the agglomeration of $Ag^°$, the material changes color. Mennig et. al. made photochromic coatings of AgCl in Na_2O-Al_2O_3-B_2O_3-SiO_2 gels on microscope slides[117]. The transformation time for this material is on the order of several ten minutes. Holographic experiments showed that this material in principle can be used for optical data storage. Also, photchromic organic molecules (e.g. derivatives of the molecule1', 3', 3'-trimethylspiro [2H-1-benzopyran-2, 2'-indoline], BIPS) have been incorporated in aluminosilicate[118] and silica[119] gels by the sol-gel process and has been shown to exhibit photochromism. In contrast to AgCl, the transformation time is on the order of a few seconds.

In principle, photochromism can be used for optical data storage in integrated optics, but awaits for an appropriate compound. The sol-gel process has the ability to incorporate this compound and allows the fabrication of photochromic films.

4.2. Photochemical hole-burning materials

Photochemical hole-burning (PHB) has been proposed for use in high-density optical data storage. Figure 12 shows schematically how a PHB material can be used for optical data storage, and how high-density data storage would be realized[120]. Photochemical hole-burning is a photobleaching phenomenon whereby irradiation with laser light on the material would cause a "hole" to be formed in the inhomogeneously broadened absorption spectrum.

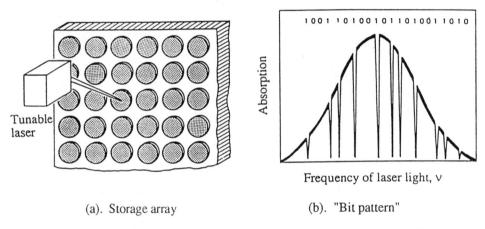

(a). Storage array   (b). "Bit pattern"

Fig. 12. Schematic description of a three-dimensional x-y-ν storage scheme[120].
(a). Typical x-y configuration with circles representing spatial data bits.
(b). A "bit pattern" in the frequency domain that could be inscribed within each spatial bit.

The low-temperature sol-gel process prevents thermal degradation of the organic, and provides stable gel matrices in which the optical properties of the molecules are preserved. Organic compounds such as 1,4-dihydroxy-9,10-anthraquinone (DAQ)[121], tetraphenylporphine tetrasulfonic acid (TPPS)[122], and tetrakis(1-methylpyridinium-4-yl)-porphine p-toluenesulfonate (TMPyP)[123] have been studied in sol-gel-derived silica matrices. Bulk solids and thin films were synthesized with or without final heat-treatment temperatures of <200°C. The PHB phenomenon has been observed in these materials at temperatures <20 K. However, the most important drawback of these PHB materials is that the encoding of information is limited to cryogenic temperatures. As the temperature increases, the "hole" has been found to gradually disappear[120].

A quantum dot-doped material may also be used for PHB[124]. This material would have a wide distribution of dots in order to create a broad spectral range of absorption upon excitation with light. Because the exciton resonance energy is a function of particle size, high-intensity light of a certain wavelength will be absorbed by a certain size particle. Subsequently, a photochemical change occurs at that particle, and a hole is burned in the absorption spectrum. Spectral-hole burning studies have been done on semiconductor quantum dot-doped glasses[45,125]. The sol-gel process allows the modification of the surface of the quantum dot (e.g. different matrix and the use of stabilizers) in order to provide information on how to improve the PHB response.

4.3. Protective coatings

Optical device components can be protected chemically, thermally, or mechanically from the environment through the use of coatings. Thick Ormosil coatings (several microns) have been shown to enhance the thermomechanical property of coated soda-lime silicate glasses[9]. Ormosils have also been applied as hard coatings for CR-39 lenses[126]. Hard Ormosils with high silica content have improved thermal properties over polymer coatings. Iwamoto et. al. have studied the mechanical properties of hard Ormosils prepared by the sol-gel process. Figure 13 and Table 5 show that hard transparent Ormosils have Vickers hardness numbers larger than that of polycarbonate polymer but smaller than that of borosilicate glass[127]. In addition, the porosity of these Ormosils is sufficiently small to prevent the penetration of isopropanol[127]. Thus, hard Ormosils may be used as protective coatings on components such as glass fibers and ridge waveguides.

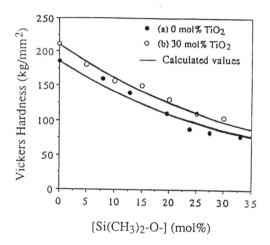

Fig. 13. Vickers hardness of hard Ormosils: (a). 0 mol% TiO_2, (b). 30 mol% TiO_2[127].

Table 5. Vickers hardness of some glasses and the hardest transparent plastics[127].

Material	Vickers Hardness (kg/mm^2)
Polyethyleneterephthalate (PET)	24
Polymethylmethacrylate (PMMA)	19
Polycarbonate (PC)	14 - 16
Borosilicate glass	220 - 350
Window glass	480 - 620

4.4. Ferroelectric materials

Ferroelectric thin films have wide-ranging applications as nonlinear materials. These films have been easily fabricated by the sol-gel process. Ferroelectric-like behaviors have been observed in transparent amorphous films with the $LiNbO_3$[128] and $Pb(Zr_{0.52}Ti_{0.48})O_3$ compositions[129]. After poling, these films exhibited ferroelectric-like behaviors, thus indicating the existence of dipoles. Through a simple heat-treatment of an amorphous thin film on an appropriate substrate, this can lead to single crystal-like thin films. Cheng, et. al. have prepared transparent single crystal-like $KNbO_3$ thin films via the sol-gel process[130].

5. CONCLUSIONS

The sol-gel process has demonstrated that it is suitable for the fabrication of a variety of amorphous materials for integrated optics. For many components, it has overcome numerous hurdles and added specific benefits to the property of each material. To realize actual device applications, further modifications of the sol-gel process is necessary. The sol-gel process has much potential to improve material performance because of its processing flexibility.

6. ACKNOWLEDGEMENTS

This research was funded by the Air Force Office of Scientific Research under Grant No. F49620-94-1-0071, and the Strategic Defense Initiative Organization under Grant No. AFOSR-91-0317. We gratefully acknowledge the collaboration with Professor Nassar Peyghambarian at the University of Arizona, Optical Sciences Center. We would also like to thank K. Hayashi for his assistance.

7. REFERENCES

1. B. Wherrett, "Introduction", in Optical Computing, Chap. 1, B. Wherrett, F. Tooley, Eds., IOP Publishing, Bristol, 1989.
2. E. Vogel, "Glass as nonlinear photonic materials," J. Am. Ceram. Soc., Vol. 72, No. 5, pp. 719-724, May, 1989.
3. S. Najafi, "Introduction", Chap. 1, in Introduction to Glass Integrated Optics, S. Najafi, Ed., Artech House, Norwood, 1992.
4. C. Brinker and G. Scherer, Sol-Gel Science, Academic Press, New York, 1990.
5. B. Zelinski, C. Brinker, D. Clark, and D. Ulrich, Better Ceramics Through Chemistry IV, Mat. Res. Soc. Symp. Proc., Vol. 180, 1990.
6. Y. Hu, and J. Mackenzie, "Rubber-like elasticity of organically modified silicates," J. Mat. Sci., Vol. 27, No. 16, pp. 4415-4420, Aug., 1992.
7. B. Lintner, N. Arfsten, H. Dislich, H. Schmidt, G. Philipp, and B. Seiferling, "A first look at the optical properties of Ormosils," J. Non-Cryst. Solids, Vol. 100, No. 1-3, pp. 378-387, 1988.
8. C. Li, J. Tseng, K Morita, C. Lechner, Y. Hu, and J. Mackenzie, "ORMOSILs as matrices in inorganic-organic nanocomposites for various optical applications," pp. 410-419, in Sol-Gel Optics II, J. Mackenzie, Ed., Proc. SPIE, Vol. 1758, 1992.
9. M. Mennig, G. Jonschker, and H. Schmidt, "Sol-gel derived thick coatings and their thermomechanical and optical properties," pp. 125-134, in Sol-Gel Optics II, J. Mackenzie, Ed., Proc. SPIE, Vol. 1758, 1992.

10. L. Hench and J. West, "The sol-gel process," Chem. Rev., Vol. 90, No. 1, pp. 33-72, 1990.

11. L. Hench and J. Nogues, "Sol-gel processing of net shape silica optics," Chap. 3, in <u>Sol-Gel Optics Processing and Applications</u>, L. Klein, Ed., Kluwer Academic, Norwell, 1994.

12. L. Hench and M. Wilson, "Processing of gel-silica monoliths for optics: Drying behavior of small pores," J. Non-Cryst. Solids, Vol. 121, No. 1-3, pp. 234-243, May, 1990.

13. L. Hench, S. Wang, and J. Nogues, "Gel-silica optics," pp. 76-85, in <u>Multifunctional Materials</u>, R. Gunshor, Ed., Proc. SPIE, Vol. 878, 1988.

14. S. Wang, C. Campell and L. Hench, p. 145, in: <u>Ultrastructure Processing of Advanced Ceramics</u>, J. Mackenzie and D. Ulrich, Eds., John Wiley & Sons, New York, 1988.

15. S. Najafi, <u>Introduction to Glass Integrated Optics</u>, Artech House, Norwood, 1992.

16. D. Uhlmann, J. Boulton, and G. Teowee, "Sol-gel synthesis of optical thin films and coatings," pp. 270-295, in <u>Sol-Gel Optics</u>, J. Mackenzie and D. Ulrich, Eds., Proc. SPIE, Vol. 1328, 1990.

17. I. Thomas, "Optical coating fabrication," Chap. 6, in <u>Sol-Gel Optics Processing and Applications</u>, L. Klein, Ed., Kluwer Academic, Norwell, 1994.

18. Y. Mitsuhashi, A. Matsuda, and Y. Matsuno, "Sol-gel technology for optical disk application," pp.105-112, in <u>Sol-Gel Optics II</u>, J. Mackenzie, Ed., Proc. SPIE, Vol. 1758, 1992.

19. M. Guglielmi, P. Colombo, L. Esposti, G. Righini, and S. Pelli, "Planar and strip optical waveguides by sol-gel method and laser densification," pp. 44-49, in <u>Glasses for Optoelectronics II</u>, G. Righini, Ed., Proc. SPIE, Vol. 1513.

20. B. Fabes, "Laser processing of sol-gel coatings," Chap. 21, in <u>Sol-Gel Optics Processing and Applications</u>, L. Klein, Ed., Kluwer Academic, Norwell, 1994.

21. B. Fabes, B. Zelinski, D. Taylor, L. Weisenbach, S. Boggavarapu, and D. Dent," Laser densification of optical films," pp. 227-234, in <u>Sol-Gel Optics II</u>, J. Mackenzie, Ed., Proc. SPIE, Vol. 1758, 1992.

22. D. Shaw and T. King, "Densification of sol-gel silica by laser irradiation," pp. 474-481, in <u>Sol-Gel Optics</u>, J. Mackenzie and D. Ulrich, Eds., Proc. SPIE, Vol. 1328, 1990.

23. D. Chen, B. Potter, and J. Simmons, "GeO_2-SiO_2 thin films for planar waveguide applications," J. Non-Cryst. Solids, 1994 (accepted).

24. M. Mennig, G. Jonschker, and H. Schmidt, "Sol-gel derived thick coatings and their thermomechanical and optical properties," pp. 125-134, in <u>Sol-Gel Optics II</u>, J. Mackenzie, Ed., Proc. SPIE, Vol. 1758, 1992.

25. P. Prasad, "Third-order nonlinear optical effects in molecular and polymeric materials," Chap. 3, in <u>Materials for Nonlinear Optics: Chemical Perspectives</u>, S. Marder, J. Solm, and G. Stucky, Eds., Am. Ceram. Soc. Symp. Series, Vol. 455, Wash., D.C., 1991.

26. P. Prasad, "Sol-gel processed inorganic and organically modified composites for nonlinear optics and photonics," pp. 741-746, in <u>Better Ceramics Through Chemistry IV</u>, B. Zelinski, C. Brinker, D. Clark, and D. Ulrich, Eds., Mat. Res. Soc. Symp. Proc., Vol. 180, 1990.

27. F. Nishida, B. Dunn, E. Knobbe, P. Fuqua, and B. Kaner, and B. Mattes, "Incorporation of polyaniline into a silica gel via the sol-gel technique," pp. 747-752, in Better Ceramics Through Chemistry IV, B. Zelinski, C. Brinker, D. Clark, and D. Ulrich, Eds., Mat. Res. Soc. Symp. Proc. Vol. 180, 1990.

28. R. Jain and R. Lind, "Degenerate four-wave mixing in semiconductor-doped glasses," J. Opt. Soc. Am., Vol. 73, No. 5, pp. 647-653, May, 1983.

29. A. Efros and A. Efros, "Interband absorption of light in a semiconductor sphere," Sov. Phys. Semicond. Vol. 16, No. 7, pp. 772-775, July, 1982.

30. L. Brus, J. Quant. Elect., Vol. 22, p. 1909, 1986.

31. D. Thomas and J. Hopfield, "Optical properties of bound exciton complexes in cadmium sulfide," Phys. Rev., Vol. 128, No. 5, pp. 2135-2148, Dec., 1962.

32. M. Dagenais, "Low power optical saturation of bound excitons with giant oscillator strength," Appl. Phys. Lett., Vol. 43, No. 8, pp. 742-744, Oct. 15, 1983.

33. M. Dagenais and W. Sharfin, "Picojoule, subnanosecond, all-optical switching using bound excitons in CdS," App. Phys. Lett., Vol. 46, No. 3, pp. 230-232, 1985.

34. L. Banyai, Y. Hu, M. Lindberg, and S. Koch, "Third-order optical nonlinearities in semiconductor microstructures," Phys. Rev. B, Vol. 38, No. 12, pp. 8142-8153, 1988.

35. M. Nogami, Y. Zhu, Y. Tohyama, and K. Nagasaka, "Preparation and nonlinear optical properties of quantum-sized CuCl-doped silica glass by the sol-gel process," J. Am. Ceram. Soc. Vol. 74, No. 1, pp. 238-240, 1991.

36. M. Nogami, M. Watabe, and K. Nagasaka, "Preparation of semiconducting sulfides microcrystalline-doped silica glasses by the sol-gel process," pp. 119-124, in Sol-Gel Optics, J. Mackenzie and D. Ulrich, Eds., Proc. SPIE, Vol. 1328, 1990.

37. T. Takada, T. Yano, A. Yasumori, M. Yamane, and J. Mackenzie, "Preparation of quantum-size CdS-doped Na_2O-B_2O_3-SiO_2 glasses with high non-linearity," J. Non-Cryst. Solids, Vol. 147 & 148, pp. 631-635, 1992.

38. C. Li, "Preparation and characterization of quantum dot doped Ormosils," Ph. D. Dissertation, University of California at Los Angeles, 1993.

39. S. Omi, H. Hiraga, K. Uchida, C. Hata, Y. Asahara, and A. Ikushima, "CdTe microcrystallite-doped glasses: Preparation and optical properties," pp. 181-186, in Science and Technology of New Glasses, S. Sakka and N. Soga, Eds., Tokyo, Oct. 16-17, 1991.

40. J. Yumoto, H. Shinojima, N. Uesugi, K. Tsunetomo, H. Nasu, and Y. Osaka, "Optical nonlinearity of CdS microcrystallites in a sputtered SiO_2 film," Appl. Phys. Lett., Vol. 57, No. 23, pp. 2393-2395, Dec., 1990.

41. S. Ohtsuka, T. Koyama, K. Tsunetomo, H. Nagata, and S. Tanaka, "Nonlinear optical property of CdTe microcrystallites doped glasses fabricated by laser evaporation method," Appl. Phys. Lett. Vol. 61, No. 25, pp. 2953-2954, 1992.

42. C. Li, M. Wilson, N. Haegel, J. Mackenzie, E. Knobbe, C. Porter, and R. Reeves, "Preparation of quantum-size semiconductor-doped Ormosils and their optical properties," pp. 41-46, in Chemical Processes in Inorganic Materials: Metal and Semiconductor Clusters and Colloids, Mat. Res. Soc. Symp. Proc., Vol. 272, 1992.

43. B. Breitscheidel, J. Zieder, and U. Schubert, "Metal complexes in inorganic matrices. 7. Nanometer-sized, uniform metal particles in a SiO_2 matrix by sol-gel processing of metal complexes," Chem. Mat. Vol. 3, No. 3, pp. 559-566, 1991.

44. T. Takada, C. Li, J. Tseng, and J. Mackenzie, "Control of particle size distribution of CdS quantum dots in gel matrix," J. Sol-Gel Sci. & Tech. Vol. 1, No. 2, pp. 123-132, 1994.

45. K. Kang, A. Kepner, Y. Hu, S. Koch, N. Peyghambarian, C. Li, T. Takada, Y. Kao and J. Mackenzie, "Room temperature spectral hole-burning and elimination of photodarkening in sol-gel derived CdS quantum dots," Appl. Phys. Lett., Vol. 64, No. 12, pp. 1487-1489, March 21, 1994.

46. M. Steigerwald and L. Brus, "Synthesis, stabilization, and electronic structure of quantum semiconductor nanoclusters," Annu. Rev. Mat. Sci., Vol. 19, pp. 471-495, 1989.

47. N. Herron, "Small semiconductor particles: Preparation and characterization," Chap. 38, in Materials for Nonlinear Optics: Chemical Perspectives, S. Marder, J. Sohn, and G. Stucky, Eds., Am. Ceram. Soc. Symp. Series Vol. 455, Wash., D.C., 1991.

48. M. Nogami, M. Watabe, and K. Nagasaka, "Preparation of semiconducting sulfides microcrystalline-doped silica glasses by the sol-gel process," pp. 119-124, in Sol-Gel Optics, J. Mackenzie and D. Ulrich, Eds., Proc. SPIE, Vol. 1328, 1990.

49. M. Nogami, S. Suzuki, and K. Nagasaka, "Sol-gel processing of small-sized CdSe crystal-doped silica glasses," J. Non-Cryst. Solids, Vol. 135, No. 2-3, pp. 182-188, 1991.

50. Y. Kobayashi, S. Yamazaki, and Y. Kurokawa, "Preparation of a transparent alumina film doped with CdS and its nonlinear optical properties," J. Mat. Sci.: Mat. Electron., Vol. 2, No. 1, pp. 20-25, Mar.,1992.

51. R. Reisfeld, "Nonlinear optical properties of semiconductor quantum dots and organic molecules in glasses prepared by the sol-gel method," pp. 546-556, in Sol-Gel Optics II, J. Mackenzie, Ed., Proc. SPIE, Vol. 1758, 1992.

52. N. Tohge and T. Minami, "Formation process of Cd and Zn chalcogenide-doped glasses via gels containing thiourea or selenourea complexes," in Sol-Gel Optics II, J. Mackenzie, Ed., Proc. SPIE, Vol. 1758, pp. 587-595, 1992.

53. M. Yamane, T. Takada, J. Mackenzie, and C. Li, "Preparation of quantum dots by the sol-gel process," pp. 577-586, in Sol-Gel Optics II, J. Mackenzie, Ed., Proc. SPIE, Vol. 1758, 1992.

54. T. Takada, J. Mackenzie, M. Yamane, K. Kang, N. Peyghambarian, R. Reeves, E. Knobbe, and R. Powell, "Preparation and nonlinear optical properties of CdS quantum dots in Na_2O-B_2O_3-SiO_2 glasses by the sol-gel process," J. Opt. Soc. Am. B, 1994 (accepted).

55. C. Li, J. Tseng, K. Morita, C. Lechner, Y. Hu, and J. Mackenzie, "Ormosils as matrices in inorganic-organic nanocomposites for various optical applications," pp. 410-419, in Sol-Gel Optics II, J. Mackenzie, SPIE Proc., Vol. 1758, 1992.

56. M. Wilson, C. Li, J. Mackenzie, and N. Haegel, "Photoluminescence excitation spectroscopy study of CdS nanocrystals in Ormosils," Nanostructured Materials, Vol. 2, No. 4, pp. 391-398, Jul.-Aug., 1993.

57. C. Li, "Preparation and characterization of CdS quantum dot doped ORMOSILs," Ph.D. Dissertation, University of California at Los Angeles, 1993.

58. D. Recard, P. Roussignol, and C. Flytzanis, "Surface-mediated enhancement of optical phase conjugation in metal colloids," Opt. Lett., Vol. 10, No. 10, p. 511-513, 1985.

59. F. Hache, D. Ricard, and C. Flytzanis, "Optical nonlinearities of small metal particles: Surface-mediated resonance and quantum size effects," J. Opt. Soc. Am. B, Vol. 3, No. 12, p. 1647-1655, Dec., 1986.

60. P. McMillan, Glass-Ceramics, Academic Press, London, 1964.

61. W. Weyle, Coloured Glasses, Soc. Glass Tech., Sheffield, 1951.

62. K. Fukumi, A. Chayahara, K. Kadono, T. Sakaguchi, Y. Horino, M. Miya, J. Hayakawa, and M. Satou, Jap. J. Appl. Phys., Vol. 30, No. 4B, pp. L742-L744, April, 1991.

64. J. Matsuoka, R. Mitzutani, H. Nasu, and K. Kamiya, "Preparation of Au-doped silica glass by sol-gel method," J. Ceram. Soc. Jap., Vol. 100, No. 4, pp. 599-601, 1992.

65. K. Kadono, T. Sakaguchi, M. Miya, J. Matsuoka, et. al., "Optical non-linear property of Au colloid-doped glass and the laser irradiation stability," J. Mat. Sci: Mat. Electron., Vol. 4, No. 1, pp. 59-61, Mar., 1993.

66. H. Kozuka and S. Sakka, "Preparation of gold colloid-dispersed silica coating films by the sol-gel method," Chem. Mat., Vol. 5, No. 2, pp. 222-228, 1993.

67. J. Matuoka, R. Mizutani, S. Kaneko, H. Nasu, K. Kamiya, K. Kadono, T. Sakaguchi, and M. Miya, "Sol-gel processing and optical nonlinearity of gold colloid doped silica glass," J. Ceram. Soc. Jap., Vol. 101, No. 1, pp. 53-58, 1993.

68. C. Li, J. Tseng, C. Lechner, and J. Mackenzie, "Preparation of metal-cluster-Ormosil nanocomposites," pp. 133-138, in Chemical Processes in Inorganic Materials: Metal and Semiconductor Clusters and Colloids, Mat. Res. Soc. Symp. Proc., Vol. 272, 1992.

69. J. Tseng, C. Li, T. Takada, C. Lechner, and J. Mackenzie, "Optical properties of metal-cluster-doped Ormosils nanocomposites," in Sol-Gel Optics II, J. Mackenzie, Ed., SPIE Proc., Vol. 1758, pp, 612-621, 1992.

70. T. Rajh, O. Micic, D. Lawless, and N. Serpone, "Semiconductor photophysics. 7. Photoluminescence and picosecond charge carrier dynamics in CdS quantum dots confined in a silicate glass," J. Phys. Chem., Vol. 96, No. 11, pp. 4633-4641, 1992.

71. X. Zhao, J. Schroeder, P. Persans, and T. Bilodeau, "Resonant-Raman-scattering and photoluminescence studies in glass-composite and colloidal CdS," Phys. Rev. B, Vol. 43, No. 15, pp. 12,580-12,589, 1991.

72. J. Altman, R. Stone, F. Nishida, and B. Dunn, "Dye activated Ormosil's for lasers and optical amplifiers," pp. 507-518, in Sol-Gel Optics II, J. Mackenzie, Ed., Proc. SPIE, Vol. 1758, 1992.

73. S. Najafi, "Waveguides and devices", Chap. 6, in Introduction to Glass Integrated Optics, S. Najafi, Ed., Artech House, Norwood, 1992.

74. J. Zink and B. Dunn, "Photonic materials by the sol-gel process," J. Ceram. Soc. Jap. Int. Edition, Vol. 99, pp. 858-871.

75. B. Dunn and J. Zink, "Optical properties of sol-gel glasses doped with organic molecules," J. Mater. Chem., Vol. 1, No. 6, pp. 903-913, 1991.

76. E. Knobbe, B. Dunn, P. Fuqua, and F. Nishida, "Laser behavior and photostability characteristics of organic dye doped silicate gel materials," Appl. Opt., Vol. 29, No. 18, pp. 2729-2733, June, 1990.

77. U. Itoh, M. Takakusa, T. Moriya, and S. Saito, "Optical gain of coumarin dye-doped thin film lasers," J. Appl. Phys., Vol. 16, pp. 1059-1060, 1977.

78. H. Lin, E. Bescher, J. Mackenzie, H. Dai, and O. Stafsudd, "Preparation and properties of laser dye-Ormosil composites," J. Mat. Sci., Vol. 27, No. 20, pp. 5523-5528, Oct., 1992.

79. H. Dai, H. Lin, and O. Stafsudd, "Optical gain and laser action in Rhodamine-doped solid-state PT-ORMOSIL composites," pp. 50-56, in Soid State Lasers IV, G. Quarles, M. Woodall II, Eds., Proc. SPIE, Vol. 1864, 1993.

80. H. Osanai, T. Shioda, T. Moriyama, S. Araki, M. Horiguchi, I. Izawa, H. Takata, "Effect of dopants on transmission loss of low OH^- content optical fibers," Electron. Lett., Vol. 12, No. 21, pp. 549-550, Oct., 1976.

81. I. Thomas, S. Payne, and G. Wilke, "Optical properties and laser demonstration of Nd-doped sol-gel silica glasses," J. Non-Cryst. Solids, Vol. 151, No. 3, pp. 183-194, Dec., 1992.

82. R. Adair, L. Chase, and S. Payne, "Nonlinear refractive index of optical crystals," Phys. Rev. B, Vol. 39, No. 5, pp. 3337-3350, Feb., 1989.

83. J. Campbell, F. Rainer, M. Kozlowski, C. Wolfe, I. Thomas, and F. Milanovich, "Damage resistant optics for a mega-joule solid-state laser," pp. 444-456, in <u>Laser Induced Damage in Optical Materials</u>, Proc. SPIE, Vol. 1441, 1990.

84. J. Stone and C. Burrus, "Neodymium-doped silica lasers in end-pumped fiber geometry," Appl. Phys. Lett., Vol. 23, No. 7, pp. 388-389, 1973.

85. E. Galant, B. Gorovaya, E. Demskaya, Y. Kondrat'ev, M. Golubovskaya, A. Przhevuskii, T. Prokhorova, and M. Tolstoi, "The spectroscopic and luminescent properties of high-silica glasses containing Neodymium," Sov. J. Glass Chem. Phys., Vol. 2, No. 5, pp. 429-433, 1976.

86. L. Ageeva, V. Arbuzov, E. Galant, E. Demskaya, A. Przhevuskii and T. Prokhorova, "Spectroscopic manifestations of the reorganization of the structure of Neodymium optical centers in high-silica glasses," Sov. J. Glass Phys. Chem. Vol. 13, No. 3, pp. 221-225, 1988.

87. S. Yuh, E. Bescher, and J. Mackenzie, "Rare-earth doped, low hydroxyl organically modified silicates," Mat. Res. Soc. Symp. Proc., San Francisco, California, April, 1994 (to be published).

88. F. Hanawa, S. Sudo, M. Kawachi, and M. Nakahara, "Fabrication of completely OH-free V.A.D. fibre," Electron. Lett., Vol. 16, No. 18, pp. 699-700, Aug. 28, 1980.

89. F. Kirkbir and S. Chaudhuri, "Optical fibers from sol-gel derived germania-silica glasses," pp. 160-172, in <u>Sol-Gel Optics II</u>, J. Mackenzie, Ed., Proc. SPIE, Vol. 1758, 1992.

90. K. Kamiya, "Sol-gel fabrication of glass fibers for optics," Chap. 5, in <u>Sol-Gel Optics Processing and Applications</u>, L. Klein, Ed., Kluwer Academic, Norwell, 1994.

91. I. Matsuyama, K. Susa, S. Satoh, and T. Suganuma, "Synthesis of high-purity silica glass by the sol-gel method," Am. Ceram. Bull., Vol. 63, pp. 1408-1411, 1984.

92. K. Susa, I. Matsuyama, S. Satoh, and T. Suganuma, "Sol-gel derived Ge-doped silica glass for optical fiber application. I. Preparation of gel and glass and their characterization," J. Non-Cryst. Solids, Vol. 119, No. 1, pp. 21-28, Mar., 1990.

93. K. Susa, I. Matsuyama, and S. Satoh, "Sol-gel derived Ge-doped silica glass for optical fiber application. II. Excess optical loss," J. Non-Cryst. Solids, Vol. 128, No. 2, pp. 118-125, April, 1991.

94. S. Shibata, T. Kitagawa, and M. Horiguchi, "Fabrication of fluorine-doped silica glasses by the sol-gel method," J. Non-Cryst. Solids, Vol. 100, No. 1-3, pp. 269-273, 1988.

95. A. Sarkar, F. Kirkbir, and S. Raychaudhuri, "Sol-gel optical fiber preforms," First International Conference on Applications and Commercialization of Sol-Gel Processing, Saarbrucken, Germany, Oct. 10-13, 1993.

96. Y. Dimitriev, V. Mihailova, V. Dimitrov, and Y. Ivanova, "Effect of the mode formation on the structure of amorphous materials in the TiO_2-PbO system," J. Mat. Sci. Lett., Vol. 10, No. 21, pp. 1249-1252, Nov., 1991.

97. H. Nasu, K. Kubodera, M. Kobayashi, M. Nakamura, and K. Kamiya, "Third-harmonic generation for some chalcogenide glasses," J. Am. Ceram. Soc, Vol. 73, pp.1794-1796, 1990.

98. T. Che, M. Banash, P. Soskey, and P. Dorain, "Gel derived gradient index optics--Aspects of leaching and diffusion," Chap. 17, in <u>Sol-Gel Optics Processing and Applications</u>, L. Klein, Ed., Kluwer Academic, Norwell, 1994.

99. T. Che, J. Caldwell, and R. Mininni, "Sol-gel derived gradient index optical materials," pp. 245-258, in <u>Papers on Gradient-Index Optics</u>, D. Moore and B. Thompson, Eds., SPIE Milestone Ser., Vol. MS 67, 1993.

100. M. Yamane, "Gradient-index (GRIN) elements by sol-gel interdiffusion," Chap. 18, in Sol-Gel Optics Processing and Applications, L. Klein, Ed., Kluwer Academic, Norwell, 1994.

101. M. Yamane, A. Yasumori, M. Iwasaki, and K. Hayashi, "Graded index materials by the sol-gel proess," pp. 237-244, in Selected Papers on Gradient-Index Optics, D. Moore and B. Thompson, Eds., SPIE Milestone Ser., Vol. MS 67, 1993.

102. K. Shingyouchi, S. Konishi, K. Susa, and I. Matsuyama, "Radial gradient refractive-index glass rods prepared by a sol-gel method," Electron. Lett., Vol. 22, No. 2, pp. 99-100, 1986.

103. K. Shingyouchi, S. Konishi, K. Susa, and I. Matsuyama, "r-GRIN TiO_2-SiO_2 glass rods prepared by a sol-gel method," Electron. Lett., Vol. 22, No. 21, pp. 1108-1110, 1986.

104. S. Konishi, K. Shingyouchi and A. Maskishima,"r-GRIN rods prepared by a sol-gel method," J. Non-Cryst. Solids, Vol. 100, No. 1-3, pp. 511-513, 1988.

105. K. Shingyouchi, A. Makishima, and S. Konishi, "Determination of diffusion coefficients of dopants in wet gels during leaching," J. Am. Ceram. Soc., Vol. 71, No. 2, pp. C-82 to C-84, 1988.

106. A. Makishima, K. Shingyouchi, Y. Kitami and M. Tsutsumi, "Microstructural studies of leached TiO_2-SiO_2 gel," J. Non-Cryst. Solid, Vol. 102, No. 1-3, pp. 275-279, 1988.

107. J. Caldwell, "Sol-gel methods for making radial gradient-index glass," Ph.D. thesis, University of Rochester, 1989.

108. J. Caldwell and D. Moore, "Sol-gel method for making gradient-index glass," U.S. Patent 4,797,376, Jan. 10, 1989.

109. S. Kurosaki, "Producing graded index optical glass articles from doped silica gel bodies," UK Pat. Appl. GB 2,084,990A, Sept. 15, 1981.

110. S. Kurosaki and M. Watanabe, "Producing graded index optical glass articles from silica gel bodies," UK Pat. Appl. GB 2,086,877, Sept. 15, 1981.

111. S. Kurosaki and M. Watanabe, "Process for the production of an optical glass article," U.S. Pat. 4,389,233, Jun. 21, 1983.

112. S. Kurosaki, "Process for the production of an optical glass article," U.S. Patent 4,436,542, Mar. 13, 1984.

113. T. Edahiro, N. Inagaki and S. Kurosaki, "Process for producing optical glass product," U.S. Pat. 4,528,010, July 9, 1985.

114. M. Yamane, H. Kawazoe, A. Yasumori, and T. Takahashi, "Gradient-index glass rods of PbO-K_2O-B_2O_3-SiO_2 system prepared by the sol-gel process," J. Non-Cryst. Solids, Vol. 100, No. 1-3, pp. 506-510, 1988.

115. M. Yamane, A. Yasumori, M. Iwasaki, and K. Hayashi,"GRIN rods of large diameter and large delta-N," pp. 717-725, in Better Ceramics Through Chemistry IV, B. Zelinski, C. Brinker, D. Clark, and D. Ulrich, Eds., Mat. Res. Soc. Symp. Proc., Vol. 180, 1990.

116. R. Bertelson, "Applications of photochromism," Chap. 10, in Photochromism, G. Brown, Ed., Techniques of Chemistry Vol. III, John Wiley & Sons, 1971.

117. M. Mennig, H. Krug, C. Fink-Straube, P. Oliveira, and H. Schmidt, "A sol-gel derived AgCl photochromic coating on glass for holographic application," pp. 387-394, in Sol-Gel Optics II, J. Mackenzie, Ed., Proc. SPIE, Vol. 1758, 1992.

118. D. Preston, J. Pouxviel, T. Novinson, W. Kaska, B. Dunn, and J. Zink, "Photochromism of spiropyrans in aluminosilicate gels," J. Phys. Chem., Vol. 94, No. 10, p. 4167, 1990.

119. D. Levy and D. Avnir, "Effects of the changes in the properties of silica cage along the gel/xerogel transition on the photochromic behavior of trapped spiropyrans," J. Phys. Chem. Vol. 92, p. 4734, 1988.

120. A. Gutierrez, J. Friedrich, D. Haarer, and H. Wolfrum, "Multiple photochemical hole burning in organic glasses and polymers: Spectroscopy and storage aspects," IBM J. Res. Develop., Vol. 26, No. 2, pp. 198-208, 1982.

121. T. Tani, A. Makishima, H. Namikawa, and K. Arai, "Photochemical hole burning study of 1,4-dihydroxyanthraquinone in amorphous silica," J. Appl. Phys., Vol. 58, No. 9, pp. 3559-3565, 1985

122. K. Kamitani, M. Uo, H. Inoue, and A. Makishima, "Synthesis and spectroscopy of TPPS-doped silica gels by the sol-gel process," J. Sol-Gel Sci. & Tech., Vol. 1, pp. 85-92, 1993.

123. A. Makishima, K, Morita, H. Inoue, M. Uo, T. Hayakawa, M. Ikemoto, K. Horie, T. Tani, and Y. Sakakibara, "Preparation and optical properties of amorphous silica doped with porphines and quinizarin by the sol-gel process," pp. 492-498, in Sol-Gel Optics II, J. Mackenzie, Ed., Proc. SPIE, Vol. 1758, 1992.

124. N. Peyghambarian, private communication, University of Arizona, Optical Sciences Center, 1993.

125. C. Spiegelberg, F. Henneberger, and J. Puls, "Spectral hole-burning of strongly confined CdSe quantum dots," Proc. SPIE, Vol. 1362, Pt. 2, pp. 951-958, 1991.

126. H. Schmidt, B. Seiferling, G. Philipp, and K. Deichmann, p. 651, in: Ultrastructure Processing of Advanced Ceramics, J. Mackenzie and D. Ulrich, Eds., John Wiley & Sons, New York, 1988.

127. T. Iwamoto and J. Mackenzie, "Ormosils of high hardness," Mat. Res. Soc. Symp. Proc., San Francisco, California, April, 1994 (to be published).

128. R. Xu, Y. Xu, and J.D. Mackenzie, "Amorphous thin films of ferroelectric oxides," pp. 261-273, in Sol-Gel Optics II, J. Mackenzie, Ed., Proc. SPIE, Vol. 1758, 1992.

129. Y. Xu, C. Cheng, R. Xu, and J. Mackenzie, "Electrical properties of crystalline and amorphous $Pb(Zr_xTi_{1-x})O_3$ thin films prepared by the sol-gel technique", pp. 359-364, in Better Ceramics Through Chemistry V, M. Hampden-Smith, W. Klemperer, and C. Brinker, Mat. Res. Soc. Symp. Proc., Vol. 271, 1992.

130. C. Cheng, Y. Xu, and J.D. Mackenzie, "The growth of single crystal-like and polycrystal $KNbO_3$ films via sol-gel process," pp. 383-388, in Better Ceramics Through Chemistry V, M. Hampden-Smith, W. Klemperer, and C. Brinker, Mat. Res. Soc. Symp. Proc., Vol. 271, 1992.

Fibers from gels and their application

Sumio Sakka

Institute for Chemical Research,
Kyoto University,* Uji, Kyoto-Fu,
611 Japan

ABSTRACT

Discussion has been made on two kinds of sol-gel derived silica-based fibers: fibers drawn from sol-gel derived silica preform and fibers directly drawn from sols. The former fibers can be used as optical fibers and have been developed in that direction. The optical loss of the sol-gel derived fiber has been reduced to 1 dB/km level for pure silica and 10 dB/km level for Ge-doped silica. The latter directly drawn fibers have been developed as heat resistant fibers.

1. INTRODUCTION

Silica-based optical fibers for communication are now produced by drawing fibers from preforms prepared by chemical vapor deposition of various types[1] such as outside vapor deposition, modified chemical vapor deposition and vapor phase axial deposition[2] methods. The optical loss of ultra-low loss SiO_2 fibers may be 0.25 dB/km.

The sol-gel method has been applied to formation of larger silica rods as preform for fiber drawing. As a result of overcoming the problem of crack formation during drying of wet gels, large, transparent silica preforms can be provided for formation of optical fibers. One refers to this method when one talks about preparation of optical fibers by sol-gel method. In this article, an emphasis is laid on the discussion of this method.

An alternative method of sol-gel fiber formation is based on the fiber drawing from the viscous sol and subsequent heating to 850-900°C for silica glass fibers. This method is referred to as the direct method. The fibers

----

* Present address for communication:
  Sumio Sakka, Kuzuha-Asahi 2-7-30, Hirakata City,
  Osaka-Fu, 611, Japan

prepared by this method are directed to the use for mechanical strengthening of plastic I.C. substrate and heat resistant supports for catalysts. At present, no research is made towards the application of fibers of this type as optical fibers. The discussion is also made on these fibers in the present paper.

2. SOL-GEL PREPARATION OF OPTICAL FIBERS

2.1. Preparation of large silica monolith as preform

At present sol-gel preparation of optical fibers goes through the preparation of silica preforms which may be transparent silica glass rods. Fig.1 shows the route for preparing optical fibers in this technique.

It is well known that to give low loss fibers, preforms must be free from metallic impurities as well as water or hydroxyl groups. Also, the source of gases has to be removed from the preform, so that no seeds might be formed

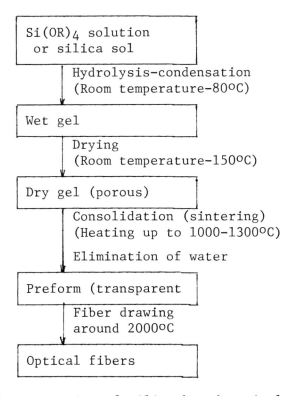

Fig.1 Preparation of silica-based optical fibers by the sol-gel method.

in the subsequent fiber drawing process. These problems have been resolved in the CVD preparation of silica-based optical fibers which are now practically used. Similar precautions and procedures may be applicable to the sol-gel preparation.

The most important problem encountered in the sol-gel formation of bulk material is the crack formation. Bulk bodies such as rods larger than 10 mm in diameter and plates larger than 10 cm x 10 cm in area and thicker than 5 mm are not easy. Cracks are often generated when a bulk wet gel body containing solvents is dried[3-5]. The seven techniques[6-18] of making dried bulk silica gel bodies so far proposed are listed in Table 1. The common principle underlying the techniques listed in Table 1 is to suppress the capillary force acting on the silica backbones consisting of silica particles in a wet gel body. This can be achieved when the sol particles are granular and large and the solvent remaining at the final stage of drying has a low

Table 1 Methods for making large-size, dried silica gels without cracks

Method	Reason for no cracks on drying	Reference
(a) Supercritical drying	No capillary force due to supercritical condition	6
(b) Hydrolysis-condensation at neutral or weakly basic condition.	Small capillary force due to formation of larger pores.	7,8
(c) Use os silica fine powder as starting material.	Small capillary force due large pores.	9,10
(d) Incorporation of silica fine powder.	Small capillary force due to large pores.	11
(e) Use of solvent with higher boiling point and lower surface tension than water.	Low surface tension of the solvent at the final stage of drying	12,13,14
(f) Use of tetramethoxy-silane solution of high HCl contents.	Small capillary force due to large pores.	15,16
(g) Use of phase separable solution containing polymers or organic compounds.	Small capillary force due to large pores.	17,18

surface tension. If the continuous fine pores are regarded as capillaries of diameter D, the solvent with surface tension γ causes the capillary force ΔP at the menisci

$$\Delta P = \frac{4\gamma \cos \theta}{D} \tag{3}$$

where θ is the contact angle between the capillary walls and the solvent. In other words, the surface dried layer of the gel body tends to shrink. This may induce the tensile stress in the surface layer, resulting in the crack formation or fracture when the stress exceeds the strength of the gel[6].

Using method (e) in Table 1, we could make a dried silica gel cylinder[13,14] having continuous fine pores of 16 mm in average diameter, as shown in (c) of Fig.2. The starting solution contained dimethylformamide (DMF) besides $Si(OCH_3)_4$, H_2O, CH_3OH and NH_3. No cracks were seen during drying at 40-150°C. This may be attributed to the presence of DMF in the starting solution. DMF remains at the final stage of drying because of its high boiling point of 153°C, suppressing the capillary force due to its lower surface tension of 38.6 dyn/cm compared with that of water (72 dyn/cm) The dried gel cylinder is converted to a transparent silica glass shown in (d) of Fig.2 by heating to 1050°C.

(a)   (b)   (c)   (d)

Fig.2 Photographs of gel and glass prepared from starting solution $1Si(OCH_3)_4 \cdot 1(CH_3)_2NCHO \cdot 2.2CH_3OH \cdot 10H_2O \cdot 3.7 \times 10^{-4} NH_4OH$. (a) Starting solution. (b) Wet gel. (c) Dry gel. (d) Glass.

For other works, refer to the references.

2.2. Comparison of the sol-gel method with conventional CVD method

It has been shown in the previous section that large-size, crackless, optically transparent silica glasses can be prepared by the sol-gel method. This gives us the idea that optical fibers may be prepared through this method. The next step is to solve the specific problems arising when one aims at optical fibers which exhibit very low optical losses. Before going into detailed discussion on the problems encountered in the sol-gel preparation, the rough comparison between the sol-gel method and the well-established CVD method (VAD method as example) will be made.

The comparisons are shown in Table 2. It is seen that the most significant difference is found in the process of formation of the porous gel mass (in sol-gel method), corresponding to the soot (in the VAD method). In the sol-

Table 2 Comparison of the sol-gel method with the CVD (VAD) method in preparing silica and silica-based optical fibers.

	Sol-gel	CVD (VAD)
Precursor	Solution (pure)	Liquid (pure)
Soot (porous gel) formation	Gelation	Deposition through gaseous phase
	Low temp. (<150°C)	High temp. (1400°C)
	Beaker, flask. High conversion efficiency.	Atomizer, burner.
Consolidation of the soot (porous gel)	1000-1300°C	1300-1500°C
Treatment with Cl_2 or Cl-containing chemicals during consolidation	Necessary	Necessary
Fiber drawing	$\geq 2000°C$	$\geq 2000°C$
SiO_2 fiber	Low loss	Low loss
SiO_2:F fiber	Low loss	Low loss
SiO_2:Ge fiber	Higher loss	Low loss

gel method, a solution containing pure raw materials is gelled and dried at low temperatures ($\leq 150^\circ C$) with very high conversion efficiency reaching almost 100 %. This is a low temperature process where the glass and plastic containers can be used as reaction vessel. All the source material for SiO_2 is recovered as gel. In the CVD method, the liquid is atomized and converted to silica particles in the gaseous atmosphere in a high temperauure furnace. The resultant silica particles are deposited on the silica rod or already formed soot agglomerate of rod shape. Since the deposition occurs in the gaseous state, a considerable portion of silica component is carried away with the vehicle gas, which makes the conversion efficiency lower.

The subsequent treatments, that is, the consolidation of porous silica into a transparent preform, the removal of water (hydroxyls), prevention of bubbling and swelling during heating at higher temperatures and fiber drawing at still higher temperatures are common to the sol-gel and CVD methods.

As shown in the bottom part of Table 2, very low optical loss fibers have been made by the sol-gel method for the non-doped SiO_2 composition, but the optical loss is still high for Ge-doped silica fiber.

2.3 Works aiming at low optical loss fibers

Considerable efforts have been made to produce low optical loss silica-based fibers by utilizing the sol-gel method. Table 3 lists the requirements which should be taken to reduce the optical loss in optical fibers.

Table 4 summarizes the values of optical losses in optical fibers prepared by the sol-gel method[8,19-26]. This table confirms the fact that the optical loss is sufficiently low for pure silica fiber, but that of Ge-doped glasses

Table 3 Minimization of optical losses in fibers
(1) Removal of transition metal impurities from starting materials.
(2) Removal of OH or water in heating of gels.
(3) Removal of voids in consolidation (sintering) of gels.
(4) Prevention of swelling or bubble formation in consolidation.
(5) Prevention of swelling or bubble formation in heating at higher temperatures and in fiber drawing.

Table 4 Performance(optical loss) of sol-gel prepared SiO_2 and silica-based optical fibers.

Author (Affil.)	Year [Ref.]	Optical loss	Kind of fiber	Method of fabrication
Matsuyama (Hitachi)	1982[8] 1984[19]	4dB/km at 0.80 μm	Pure SiO_2 core filter	Preform:control of reaction condition
Shibata (NTT)	1985[20]	1.8dB/km at 1.6μm	Pure SiO_2 fiber	Preform:control of reaction condition
Papanikolau (Philips)	1988[21]	2.5dB/km at 850nm	Pure SiO_2 fiber (Silica-silicone)	Dried in autoclave
Kitagawa (NTT)	1987[22]	0.43dB/km at 1.3 μm	F-doped SiO_2 fiber (Single fiber	Preform:control of reaction condition Graded index by F concentrat.
Puyane (Battel)	1982[23]	22dB/km	GeO_2 doped SiO_2 fiber	Layer formation inside a SiO_2 tube
Shibata (NTT)	1986[24]	9dB/km at 1.07 μm	GeO_2 doped SiO_2 core fiber	Preform:control of reaction condition
Mori (Seiko-epson)	1988[25]	16.7dB/km at 850nm	GeO_2 doped fiber	Preform:control of reaction condition
Susa (Hitachi)	1991[26]	20-30dB/km at 0.63-0.83 μm	GeO_2 doped SiO_2 fiber	Preform:control of reaction condition

is still high for their use as fiber core glass, as mentioned in the previous section. Some discussions will be made on these points in the followings.

The first task in lowering optical loss of optical fibers is to reduce the transition metal content to one ppb level ((1) in Table 3). When the source for SiO_2 is silicon alkoxide like tetramethoxysilane, as in the method employed by Susa et al.[8,19,27] (method (b) in Table 1), this problem is solved by using the purified raw materials.

When colloidal silica sols are used as starting material[9,28-30] (method (b) in Table 1), the transition metal impurities in the starting colloidal silica transfer into the fibers. Toki et al.[11] indicate that the transition metal impurities of the fibers correspond to those in the colloidal particles used in their method (method (d) in Table 1), and the purity of the fiber is much improved when the colloidal silica is synthesized from pure tetraethoxysilane by the sol-gel method. They use the starting solutions consisting of 50% tetraethoxysilane and 50 % colloidal silica.

Papanikolau et al.[21] discuss that in autoclave process of drying silica gel, impurities incorporated in the dried gel are removed by the treatment including second drying run in an oxidizing atmosphere at about 800°C and densification at about 1300°C in a mixture of He, O_2 and Cl_2.

The hydroxyls (OH^-) and water are the other cause for the optical absorptions around the wavelengths (0.6 - 1.6μm) corresponding to the wavelengths of the lights used in optical communication. As in chemical vapor deposition method, the reduction of the hydroxyl content of sol-gel derived gels is achieved by heating in chlorine-containing atmosphere at elevated temperatures[31]. 5-mm diameter rod-type gel samples of low density gels[31] were exposed to a pure chlorine atmosphere for 30 min at a given temperature followed by consolidation at 1300°C in helium. Up to 600°C the residual OH content was the same as in a nontreated sample, while at 700°C the hydroxyl content was drastically reduced, and above 800°C it was practically zero. Fig.3[31] shows typical transmission loss spectra for sol-gel core optical fibers. It is seen that the transmission loss of the OH-free optical fibers is as low as 4 dB/km at 0.80μm.

When OH^- is eliminated from sol-gel silica by treating the gel with Cl_2-containing atmosphere at 1300 - 1450°C, the further heating to 1600°C and higher temperatures may sometimes cause swelling or bloating[32]. Rabinovich et al.[32] suggest that the use of fluorine in place of Cl_2 is effective in both reducing the OH content and preventing possible swelling or bloating on heating at higher temperatures.

The most important cause for optical loss in optical fibers is the light scattering due to the presence of voids or bubbles in the fiber. In order to avoid the light scattering, a complete densification (No(3) of Table3) and

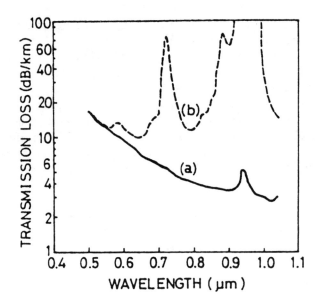

Fig.3 Typical transmission loss spectra for sol-gel core fibers with core materials of (a)chlorine-treated OH-free glass rod and (b)untreated OH-containing glass rod. After Matsuyama et al.[31]

prevention of bubble formation in the processes of consolidation (No(4) of Table3) and heating at higher temperatures (No(5) of Table 3) are compulsory.

Sintering of the dry gels and porous low density glass containing open pores consists of two steps, (1) densification of the low-density glass due to shrinkage of open pores and (2) shrinkage and collapsing of the closed pores. Step (1) takes places around 1000°C after organic impurities are removed on heating the gel in an oxidizing atmosphere at 250-600°C, hydroxyl groups are removed on treating the gel with a chlorine atmosphere and the chlorine content is reduced on heating the gel in an oxygen atmosphere[31]. Step (2) takes place at 1000-1300°C when the gel is heated in a helium atmosphere. When open pores are converted to closed pores and H_2O and CO_2 gases produced as a result of decomposition of organic matters are left in the closed pres, the gel may swell or bloat around 1000°C. A slow heating is necessary to avoid this. Closed pores disappear as a result of out diffusion of He gas, for 1000-1300°C. If much chlorine gas is left in the glass, swelling or bubbling may occur at higher temperatures

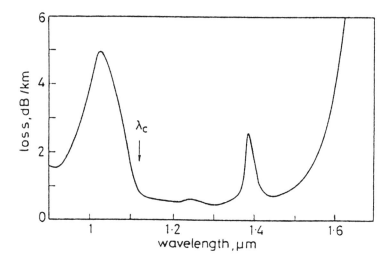

Fig.4 Transmission loss spectrum of single-mode optical fiber made from gel-derived, fluorine-doped silica glass. After Kitagawa et al.[22]

($\gtrsim 1600°C$) as a result of Cl_2 gas formation. Therefore, the elinination of chlorine gas important[19,32].

2.4. Results on sol-gel optical fibers

As seen from optical loss data summarized in Table 4, the sol-gel method can produce fibers of sufficiently low optical loss of 1 dB/km level for pure silica composition[8,19-21]. Fluorine-doped SiO_2 fibers have a very low loss (0.43 dB/km for single mode fiber[20], as seen from Fig. 4. It is assumed that the optical loss of B_2O_3-doped SiO_2 fibers is also low, although no data is shown. On the other hand, the optical loss is still high at 9-20 dB/km for GeO_2-doped SiO_2 fibers[23-26]. The loss spectrum for a GeO_2-doped SiO_2 fiber obtained by Shibata et al.[24] is shown in Fig.5 as example.

The reason for higher loss is being examined at present[26,33,34] and the substitute for Ge is sought for[35]. It is assumed that the reduction of Ge ions may be responsible for higher optical loss in GeO_2-doped silica fibers.

The above mentioned discussion suggests that at present the combination of Ge-doped SiO_2 (higher index) with non-doped SiO_2 (lower index) is not possible, and pure SiO_2 compositions have to be used as core, when one uses the sol-gel method for preparation of optical fibers. Fluorine

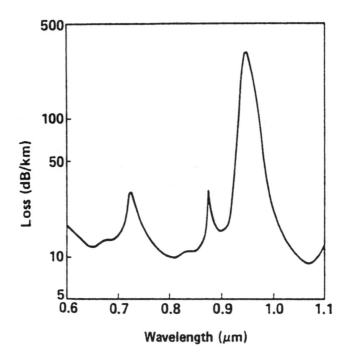

Fig.5 Optical loss spectrum of GeO$_2$-doped silica core optical fibers.   After Shibata et al.[24]

doped fibers can also be used at present.

3. DIRECT SOL-GEL PREPARATION OF SILICA FIBERS

3.1. Preparation of silica fibers

In the direct sol-gel preparation, silica fibers are drawn from a viscous solution at room temperature and the resultant gel fiber is converted by heating to 850ºC to silica glass fibers[36], as shown in Fig.6.

It is assumed that the silica glass fibers prepared by this method are applied as heat resistant fibers, and so far application as optical fibers has not been aimed at.

Fig.7 shows the time change of the viscosity of a drawable $Si(OC_2H_5)_4$-H_2O-C_2H_2OH-HCl solution with an [H_2O]/[$Si(OC_2H_5)_4$] ratio of 2 for three different temperatures[37].

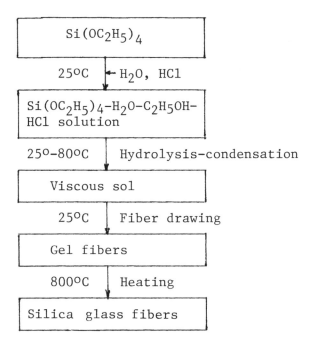

Fig.6 Direct preparation of silica fibers by the sol-gel method.

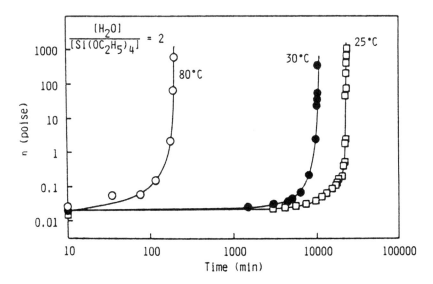

Fig.7 Variation of the viscosity of a drawable $Si(OC_2H_5)_4$ solution with the molar ratio $[H_2O]/[Si(OC_2H_5)_4]=2$ at 25, 30 and 80 °C as a function of time[37].

The viscosity of the solution increases with time as the hydrolysis-condensation reaction proceeds. When the viscosity reaches about 10 poise, the solution becomes sticky and spinnable, which makes fiber drawing possible. Fibers can be drawn by immersing a glass rod in the solution and pulling it up. It can be seen from Fig.7 that the time required for the solution to reach the drawable state is shortened for higher reaction temperatures, and at 80°C only a couple of hours is needed before the solution becomes drawable.

3.2. Spinnability of the solution

All the solutions which become gels finally show a continuous increase in viscosity with time until gelation. When a metal alkoxide solution is catalyzed with an acid and its water content is small at less than 4 or 5 in the water to alkoxide mole ratio and the reaction takes place in the open system, the solution (solutions 1 and 2 in Table 5) exhibits spinnability at viscosities above 10 poises and becomes drawable into gel fibers. No spinnability appears, however, when a solution contains much water or is catalyzed with an alkali like ammonia (solutions 3, 4 and 5 in Table 5).

It is also noted that the appearance of spinnability depends on the condition for hydrolysis-condensation reaction in the solution of the composition $Si(OC_2H_5)_4:H_2O:C_2H_5OH.HCl = 1:2:1:0.01$ in mole ratio exhibits spinnability,

Table 5 Gelling characteristics of $Si(OC_2H_5)_4$ solutions of various compositions[3,4]

Solution	Composition of solution $Si(OC_2H_5)_4$ (g)	H_2O (g)	C_2H_5OH (g)	Catalyst*	$H_2O/Si(OC_2H_5)_4$ (mole ratio)	Spinnability
1	169.5	14.5	239.7	HCl	1	Good
2	382.0	33.0	83.4	HCl	1	Good
3	169.5	292.8	37.5	HCl	20	No
4	50	3.8	47.6	NH_4OH	1	No
5	50	7.6	47.6	NH_4OH	2	No

* The mole ratio $[HCl]/[Si(OC_2H_5)_4$ or $[NH_4OH]/[Si(OC_2H_5)_4]$ is 0.01.

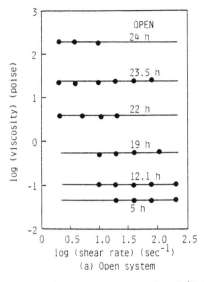 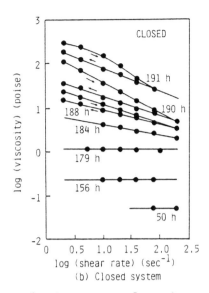

Fig.8 Viscosity of $Si(OC_2H_5)_4$ solutions as a function of the shear rate in the open (spinnable) and closed (non-spinnable) systems. Numbers attached to the lines denote the reaction time of the solution.

giving fibers on drawing, at the viscosities ranging from 10 to 100 poise in the open system, where the solvents can vaporize in air, but the same starting solution does not show spinnability in the closed system, where no vaporization is possible.

It is interesting to see that the flow characteristics of this solution is different with the reaction condition (open system or closed system ?). Fig.8 shows the viscosities as a function of shear rate. It can be seen that the solution which reacted in the open system retains Newtonian flow behavior at high viscosities ranging from 10 to more than 100 poises, where the solution exhibits spinnability. On the other hand, the viscosity behavior of the solution which is reacted in the closed system and exhibits no spinnability becomes non-Newtonian at high viscosities. That is, in the high viscosity range the viscosity decreases with increasing shear rate, indicating that the solution shows structural viscosity and then thixotropic flow.

This relation between the spinnability and flow behavior of the solution can be generalized to solutions of

Fig.9 Fibers drawn from the $Si(OC_2H_5)_4$-C_2H_5OH-H_2O-HCl solution.

other starting solutions of different compositions. That is, only Newtonian (or nearly Newtonian) solutions show the spinnability at higher viscosities, giving fibers on drawing.

The picture of gel fibers drawn from the solution are shown in Fig.9. The diameter of these fibers are about 20 μm. It can be said that thinner and thicker fibers can be made by controlling the viscosity of the solution at fiber drawing.

3.3. Direct preparation of optical fibers

As stated in the beginning of this section, the application of these derectly prepared fibers to the use as optical fibers is not attempted. As far as I know, no studies have been made along this line. In order to apply the direct method of preparing fibers to fabrication of optical fibers, the following improvements have to be made.

(1) The optical loss of the fiber has to be evaluated and reduced.

(2) The cross-sections of the fibers have to be controlled to a perfect circle. The cross-section of the fibers

shown in Fig.9 is not circular but of dumb-bell shape. In order to make the circular cross-section, the composition of the starting solution has to be selected.

(3) The mechanical strength of the fibers has to be improved.

4. CONCLUSION

The hopeful sol-gel method of fabricating optical fibers is to prepare transparent low loss preforms and then draw fibers from the preform. It has been shown that for non-doped SiO_2 and fluorine doped SiO_2 fibers very low losses comparable to optical fibers which are prepared by the conventional vapor phase method are attained, whereas the optical loss of sol-gel prepared GeO_2-doped SiO_2 fibers is still high at 10 dB/km level. The improvement of GeO_2-doped SiO_2 fibers are necessary.

Before applying the silica-based fibers directly fabricated be the sol-gel method, the evaluation of optical loss, improvements of the shape of cross-section and the increase in mechanical strength are required.

5. REFERENCES

1. P. C. Schultz, Fiber Optics, ed. B. Bendow and S. S. Mitra, 3, Plenum Press, New York and London (1979).
2. S. Sudo, M. Kochi, T. Edahiro and N. Inagaki, Reports of Practical Application of Research, Ibaragi Electric Communication Laboratory, 29, 1719 (1980).
3. S.Sakka, Science of Sol-Gel Method, Agne-Shofusha. (1988) pp. 221.
4. S.Sakka, Treatise on Materials Science and Technology, ed. M.Tomozawa and R.H.Doremus,Academic Press, 22, 129-167, (1982).
5. C. J. Brinker and G. W. Scherer, Sol-Gel Science, Academic Press, San Diego, 1990, pp.908.
6. J. Zarzycki, M. Prassas and J. Phalippou, J. Mater. Sci., 17, 1371 (1982).
7. M. Yamane and S. Okano, J. Ceram. Soc. Japan, 87, 434 (1979).
8. K. Susa, I. Matsuyama, S. Satoh and T. Suganuma, Electron. Lett., 449 (1982).
9. E. M. Rabinovich, D. W. Johnson, J. B. MacChesney and E. M. Vogel, J. Non-Cryst. Solids, 47, 435 (1982).
10. G. W. Scherer and J. C. Luong, J. Non-Cryst. Solids, 63, 103 (1984).

11. M. Toki, S. Miyashita, T. Takeuchi, S. Kanbe and A. Kochi, J. Non-Cryst. Solids, 100, 479 (1988).
12. S. Wallace and L. L. Hench, Mater. Res. Soc. Symp. Proc., 32, 47 (1984).
13. T. Asachi and S. Sakka, J. Mater. Sci. Lett., 22, 4407 (1987).
14. T. Adachi, S. Sakka and M. Okada, J. Ceram. Soc. Japan, 95, 970 (1987).
15. H. Kozuka and S. Sakka, Chem. Lett., 1791 (1987).
16. H. Kozuka and S. Sakka, Chem. Mater., 1. 398 (1989).
17. K. Nakanishi and N. Soga, J. Non-Cryst. Solids, 139, 14 (1992).
18. H. Kaji, K. Nakanishi and N. Soga, J. Sol-Gel Sci. Tech., 1, 35 (1993).
19. S. Satoh, K. Susa, J. Matsuyama and T. Suganuma, J. Am. Ceram. Soc., 68, 399 (1985).
20. S. Shibata and M. Nakahara, 11th European Conf. on Optical Communication, Venezia (1-4 Oct. 1985), Tech. Dig. I,3. ; cf. S. Shibata, T. Kitagawa, T. Hanawa and M. Horiguchi, J. Non-Cryst. Solids, 88, 345 (1986).
21. E. Papanikolau, W. C. P. M. Meerman, R. Aerts, T. L. Van Rooy, J. G. Van Lierop and T. P. M. Meeuwsen, J. Non-Cryst. Solids, 100, 247 (1988).
22. T. Kitagawa, S. Shibata and M. Horiguchi, Elect. Lett., 23, 1295 (1987).
23. R. Puyane, A. L. Harmer and C. Gonzalez-Oliver, 8th European Conf. on Optical Communication, Canne (21-24 Sep. 1982) p.623.
24. S. Shibata, T. Kitagawa, F. Hanawa and M. Horiguchi, J. Non-Cryst. Solids, 88, 345 (1986).
25. T. Mori, M. Toki, M. Ikejiri, M. Takei, M. Aoki, S. Uchiyama and S. Kanbe, J. Non-Cryst. Solids, 100, 523 (1988) ; M. Toki, Doctors Thesis, Kyoto University (1993) p. 87.
26. K. Susa, I. Matsuyama and S. Satoh, J. Non-Cryst. Solids, 128, 118 (1991).
27. S. Satoh, K. Susa, I. Matsuyama and T. Suganuma, J. Non-Cryst. Solids, 55, 455 (1983).
28. E. M. Rabinovich, D. W. Johnson, J. B. MacChesney and E. M. Vogel, J. Amer. Ceram. Soc. 66, 683 (1983).
29. D. W. Johnson, E. M. Rabinovich, J. B. MacChesney and E. M. Vogel, J. Amer. Ceram. Soc., 66, 688 (1988).
30. E. M. Rabinovich, J.B. MacChesney, D. W. Johnson, J. R. Simpson, B. M. Meagher, F. V. Di Marcello, D. L. Wood and E. A. Sigety, J. Non-Cryst. Solids, 63, 155 (1984).
31. I. Matsuyama, K. Susa, S. Satoh and T. Suganuma, Am. Ceram. Soc. Bull. 63, 1408 (1984).
32. E. M. Rabinovich, D. L. Wood, D. W. Johnson, D. A.

D. A. Fleming, S. M. Vincent and J. B. MacChesney, J. Non-Cryst. Solids, 82, 42 (1980).

33. K. Susa, I. Matsuyama, S. Satoh and T. Suganuma, J. Non-Cryst. Solids, 119, 21 (1990).

34. K. Susa, I. Matsuyama and S. Satoh, J. Non-Cryst. Solids, 146, 81 (1992).

35. S. Satoh, K. Susa and I. Matsuyama, J. Non-Cryst. Solids, 146, 121 (1992).

36. S. Sakka and H. Kozuka, J. Non-Cryst. Solids, 100, 142 (1988).

36. S. Sakka and K. Kamiya, Emergent Process Methods for High Technology Ceramics, Mater. Sci. Res. Vol. 17, ed. R. F. Davis, H. Palmour III and R. L. Porter (Plenum, New York, 1984) p.83.

Integrated-Optical Devices in Rare-Earth-Doped Glass

Kevin J. Malone

National Institute of Standards and Technology
Boulder, Colorado, 80303

ABSTRACT

In this paper, I will give a short overview of integrated-optical devices in rare-earth-doped glasses. I will discuss achievements in device performance. I will also describe fabrication of these components as well as analytical and diagnostic techniques that can improve performance. I will conclude with a discussion of some current topics in this field.

1. INTRODUCTION

Integrated-optical devices in rare-earth-doped glasses have emerged as an attractive new technology on the threshold of wide-scale manufacturing and commercial insertion. These devices can be used both as laser oscillators and optical amplifiers. They have been formed by a number of fabrication methods including ion exchange and thin-film deposition. Active integrated-optical devices are expected to be important elements in future optical fiber networks. Rare-earth-doped optical fiber devices provide nearly perfect amplification of signals in optical fibers. The performance of these rare-earth-doped optical fibers is so promising that researchers started investigating whether similar performance could be achieved in planar waveguides. The combination of passive integrated-optical components and rare-earth ions has produced many devices with impressive performance.

2. REVIEW

2.1 Rare-earth-doped glass integrated-optical devices

A rare-earth-doped glass integrated-optical device is a glass waveguide doped with one or more rare-earth ions. These devices are known by other names such as integrated-optic lasers or waveguide lasers. Rare-earth ions have many optical transitions in the visible and near infrared. Some of these transitions occur at wavelengths that are used for optical fiber telecommunications. Rare-earth-doped integrated-optical devices can provide optical gain on many of these transitions if the ions are promoted into excited states by absorbing light.

Figure 1 depicts such a planar device. The lengths of the these devices are on the order of centimeters, as opposed to meters for rare-earth-doped optical fiber devices.

Contribution of the U.S. Government, not subject to copyright.

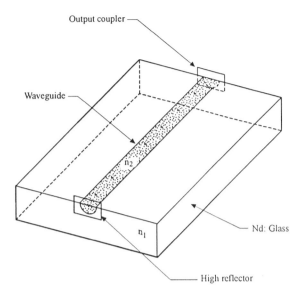

Figure 1. Ion-exchanged integrated-optical laser.

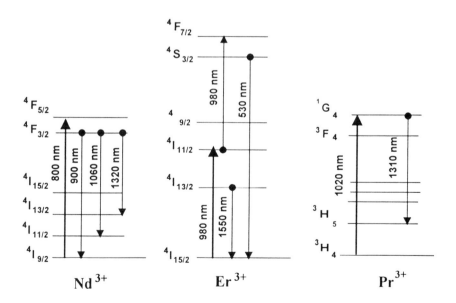

Figure 2. Energy-level diagrams of neodymium, erbium, and praseodymium.

The device or chip may have one or several waveguides embedded in the top surface. The pump light, which excites the rare-earth ions, is normally coupled into the device through the polished endfaces. If there are no mirrors attached to the endfaces and an optical signal at the laser transition is coupled into the waveguide along with the pump light, the device operates as an optically pumped, optical amplifier. If mirrors are attached to the endfaces, the device operates as an optically pumped laser oscillator.

The majority of these devices have used the $^4F_{3/2} \rightarrow {}^4I_{11/2}$ neodymium transition at 1.06 μm and the $^4I_{13/2} \rightarrow {}^4I_{15/2}$ erbium transition at 1.55 μm, which is in the third telecommunications window. Figure 2 depicts the energy levels of the erbium, neodymium, and praseodymium ions. Integrated-optical lasers have also been made in neodymium- and erbium-doped ferroelectric crystals, such as lithium niobate and lithium tantalate [1].

There are many ways to categorize these devices: according to which rare-earth ion they employ is one way. Another is according to their method of fabrication. In the second scheme, one class of devices uses a glass that is uniformly doped with rare-earth ions. The glass is then cut into plates, and optical waveguides are formed in these plates, usually by ion exchange. A second class of devices is made by depositing thin films of rare-earth-doped glass onto undoped substrates such as glass or silicon. Both techniques have been used to fabricate laser oscillators and optical amplifiers.

Integrated-optical lasers and amplifiers made from rare-earth-doped glasses are an attractive technology for several reasons. By borrowing methods originally developed for integrated-circuit fabrication, such as high-resolution lithography, we can fabricate many of these devices on a single, planar chip. The planar form of these devices allows novel integration of different functions on a single chip. Once production of these devices is scaled up, fabrication costs should be low compared to compound semiconductor devices with similar functions. It is important to remember, though, that this field is still in its infancy. With continuing efforts in academic, commercial, and government sectors, this technology should make the transition from laboratory demonstrations to commercial viability.

2.2 Achievements in rare-earth-doped glass integrated-optical devices

In this section, I will describe some achievements in rare-earth-doped-glass integrated-optical devices. It is interesting to present these results chronologically to show the improvements in device performance as this technology has evolved. As mentioned previously, most of these devices have used neodymium and erbium as the active ion. There are several reasons why the neodymium ion has been widely used in these devices. Lasers that use the $^4F_{3/2} \rightarrow {}^4I_{11/2}$ neodymium transition at 1.06 μm operate as four-level lasers. This means that population inversions can be established with vanishingly small pump powers. Lasers that operate as three-level lasers require greater pump powers because they must be bleached before a population inversion can be established. Neodymium ions can be easily pumped around 800 nm with inexpensive AlGaAs laser diodes. Further, laser fusion experiments at our national laboratories have used neodymium-doped glasses. Because of their importance in this field, a tremendous volume of research has been collected on the fabrication and analysis of neodymium-doped glasses.

The $^4I_{13/2} \rightarrow {}^4I_{15/2}$ erbium transition at 1.55 µm is also a popular candidate for these devices. The successes of erbium-doped fiber amplifiers that operate on this transition are well-known, and a number of lasers and amplifiers have been demonstrated using erbium-doped glasses. As the number of optical fiber networks that operate at this wavelength grow, there will be increasing demand for inexpensive glass components that provide modest optical gain. Integrated-optical devices made from erbium-doped glasses may provide a solution for this application.

The following discussion of results is not exhaustive, but merely mentions some highlights in device performance. In the past, many groups have reported luminescence, but not laser oscillation in several different host glasses. In general, I will cite groups that have demonstrated laser oscillation in these hosts. I would like to apologize, in advance, to any researchers whose work I may inadvertently omit.

The first report of an integrated-optical device using a rare-earth-doped glass was published in 1972 by Yajima and colleagues [2]. This device was formed by sputtering a thin layer of neodymium-doped barium crown glass onto a substrate of lower refractive index borosilicate glass. The planar waveguide was optically pumped from the back with a xenon flashtube. The device did not oscillate, but amplification of 1.06 µm radiation was demonstrated by injecting signal light into the planar film with prism couplers.

Saruwatari and Izawa reported a neodymium-doped glass laser with a three-dimensional optical waveguide in early 1974 [3]. Borosilicate glass doped with 4 mass percent Nd_2O_3 was the substrate. Ion migration was used to define a three-dimensional waveguide. The lateral dimensions of the waveguide were 600 by 300 µm, which is large compared to current devices. Dielectric mirrors were coated onto both sides of the device to provide optical feedback. While these waveguides were highly multimode, the transverse confinement provided by the waveguiding region reduced the lasing threshold to one half of that of lasers formed from the substrate material alone.

It is difficult to find published work on integrated-optical lasers between 1975 and 1987. In 1988, several groups were working in this field, but efforts concentrated on incorporating rare-earth ions into ferroelectric crystals. The year 1989 was a turning point for integrated-optical lasers. Hibino and colleagues reported a laser late in the year [4]. Their neodymium-doped-glass integrated-optical laser was fabricated by flame hydrolysis deposition (FHD) and reactive-ion etching (RIE) on a silicon substrate. This was the first modern report of a glass integrated-optical laser as well as the first buried-core device. The rare-earth ions were added to the device by soaking the waveguiding layer in a neodymium alcohol. The laser was optically pumped at 0.8 µm and had a threshold of 150 mW. The laser emission occurred at 1.05 µm, but they did not report output power.

Our group at NIST reported a neodymium-doped integrated-optical laser in late 1989 [5]. This was the first integrated-optical laser formed by electric-field-assisted ion exchange. The laser used a mixed-alkali-silicate glass, and the ion exchange was performed using a molten salt.

By 1990, several groups around the world were working on new integrated-optical lasers as well as improving the performance of previously demonstrated devices. In early 1990, Aoki and colleagues at Hoya Corp. reported a novel integrated-optical laser [6]. This device was also fabricated with electric-field-assisted ion exchange, but using a neodymium-doped phosphate glass and ion exchange from a solid silver film. This laser had large waveguide dimensions (the width of the waveguide was 200 μm) but gave very efficient performance. The threshold reported was 6.9 mW, and the maximum output power was 150 mW at 1.06 μm. Later that year, the same group demonstrated the first 1.3 μm emission from a neodymium-doped integrated-optical laser [7]. This laser used the same neodymium-doped phosphate glass as their earlier device, but used the $^4F_{3/2} \rightarrow {}^4I_{13/2}$ neodymium transition at 1.36 μm.

Mwarania and colleagues at the University of Southampton, UK, demonstrated a single-transverse-mode neodymium-doped waveguide laser using a remelted glass [8]. BK-7 glass has been widely used for passive ion-exchanged devices due to its superior ion-exchange properties. The Southampton group melted BK-7 and then added 2 mass percent Nd_2O_3 before recasting the glass and performing potassium ion exchange. This laser had output powers as high as 1.5 mW at 1.057 μm for a launched pump power of 40 mW. The lowest lasing threshold was 7.5 mW of launched pump power.

All of the integrated-optical lasers demonstrated up until this point used neodymium as the lasing ion. In early 1991, Kitagawa and his colleagues at NTT demonstrated a guided-wave laser using FHD, but with erbium as the lasing ion [9]. The fluorescence emission peak was at 1.53 μm, but the device oscillated at 1.6 μm. Output powers were as high as 1.5 mW.

The lasers I have mentioned all used on-chip Fabry-Perot cavities. In 1991, researchers began to investigate different configurations, as well as other properties of rare-earth integrated-optical lasers. In early 1991, our group at NIST demonstrated extended-cavity operation of a neodymium-doped-glass waveguide laser [10]. We reported mode-locked pulses of 80 ps and Q-switched pulses with peak powers of 1.2 W with pulsewidths of 75 ns. Linewidth narrowing and wavelength tuning were demonstrated with a grating placed in the extended cavity. Single-frequency operation was also achieved. Later that year, we successfully combined a neodymium-doped integrated-optical gain medium with a power divider [11]. The device operated as both a dual-output laser oscillator and a lossless splitter. The gain produced in the device compensated for the 3 dB splitting loss when the device was optically pumped.

The group at NTT meanwhile demonstrated that rare-earth-doped integrated-optic lasers could be optically pumped by laser diodes [12]. This demonstration was important, although it had been predicted for several years.

During 1992, several important results were reported. Kitagawa and co-workers demonstrated the first erbium-doped integrated-optic amplifier [13]. This device was fabricated by FHD/RIE. The waveguide length was increased by having several S-bends on the substrate. They achieved 13.7 dB of small-signal gain at 1.55 μm. This result was a tremendous breakthrough, since it demonstrated that high gain could be achieved in an active guided-wave device. This device is depicted in Fig. 3.

Shmulovich and colleagues at ATT demonstrated optical amplification at 1.55 μm in an erbium glass device. Their device was made by sputter deposition [14]. Feuchter and colleagues at Southampton demonstrated an erbium-doped ion-exchanged laser [15]. Once again, they used remelted BK-7 glass, but this time they added 0.5 mass percent of erbium oxide to the melt before recasting. The waveguides were formed by potassium ion exchange and operated in a single transverse mode. The threshold for lasing was 150 mW of launched pump power at 980 nm, with the emission occurring at 1538 and 1544 nm. The maximum output power was 0.4 mW, and the slope efficiency was 0.55 percent.

Later in 1992, Roman and Winick at the University of Michigan demonstrated the first integrated-optical laser containing a distributed Bragg reflector [16]. Distributed feedback (DFB) and distributed Bragg reflector (DBR) gratings have been used to produce very narrow linewidths in laser diodes and fiber lasers. Their laser was formed by silver ion exchange in a neodymium-doped glass. The grating was formed by holographic exposure and ion milling. The linewidth of the emission at 1057 nm was less than 37 MHz.

In early 1993, our group at NIST reported significant linewidth narrowing of a neodymium-doped integrated-optical laser [17]. The device used coupled cavities and was the integrated-optical equivalent of a Michelson laser. The linewidth of this laser was reduced by a factor of ten compared to a similar Fabry-Perot laser. The device is depicted in Fig. 4. That month we also reported an integrated-optical laser fabricated from a new neodymium-doped phosphate glass using ion exchange from solid silver films. This device emitted at 905, 1057, and 1356 nm. We achieved a slope efficiency of 57 percent at 1.06 μm and over 200 mW of output power [18].

It remains to be seen whether the improvements in device performance in the coming years will continue at this rapid pace. System designers are now interested in applying this new technology in future optical fiber networks. A great deal of work remains in high-volume manufacturing of rare-earth-doped devices before they will be cost and performance competitive with other technological options for future optical networks.

3. FABRICATION METHODS

3.1 General fabrication

In this section I will briefly discuss fabrication techniques for making both ion-exchanged and thin-film-based glass devices. Fabrication of these active devices is

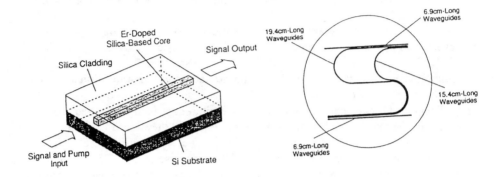

Figure 3.   Integrated-optical amplifier with 13.7 dB of gain at 1.535 μm [After Ref. 13].

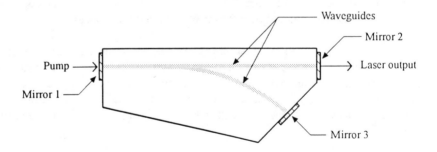

Figure 4.   Integrated-optical Michelson laser.

similar to fabrication of passive integrated-optical devices. Other speakers will address manufacturing of these devices in detail, so I will keep this section brief.

Figure 5 shows the complete process sheet for a waveguide laser made in a neodymium-doped phosphate glass. As the figure shows, there are quite a number of process steps. Moreover, this fabrication sequence was performed in a research laboratory. Specific process steps may need to be added or deleted in a production environment. Many of the process steps for fabricating a thin-film device are similar.

Steps like solvent and plasma cleaning are required to remove impurities from glass and silicon substrates. Electron beam evaporation is used to deposit metal films, typically, aluminum, silver, and titanium. These metal films are used as diffusion apertures for ion-exchanged devices and masking layers for thin-film devices. For the thin-film devices, plasma or thermal deposition is often used to grow the silica layer, which is doped with the rare-earth ions. Additional steps such as plasma etching and cladding deposition are required for thin-film devices.

High-resolution photolithography is performed on both device classes to define the waveguides. The glass wafer is first coated with a photosensitive dye or photoresist. A waveguide pattern on a chromium/glass photomask is transferred onto the photoresist with a semiconductor-grade mask aligner. The photoresist pattern can then be transferred to the metal film by acid or plasma etching.

A dicing saw is used to cut the wafer or plate into small chips. The endfaces of these chips are polished next. Several steps with different polishing compounds are usually used to prepare the endfaces. The polishing steps use progressively smaller particles, typically starting with several aluminum oxide slurries and finishing with a cerium oxide slurry.

3.2 Fabrication of ion-exchanged integrated-optical lasers

Ion-exchanged integrated-optical lasers use glasses that are uniformly doped with rare-earth ions. There are two primary tasks in the fabrication of an ion-exchanged integrated-optical laser. The first is the formation of waveguides by ion exchange and subsequent processing such as dicing and polishing. Most efforts in the past have concentrated on this area. Second, and equally important, however, is the preparation of the glass itself. I will discuss this in detail in a subsequent section.

Many different rare-earth-doped glasses have been developed. Most of these glasses have used neodymium as the active species and were developed for laser fusion experiments. Many of these glasses, however, are unsuitable for ion exchange. Other papers presented at this conference deal exclusively with ion-exchanged optical waveguides. However, a brief discussion of ion exchange will be helpful.

The most common forms of ion exchange substitute either potassium (K^+) or silver (Ag^+) ions for sodium (Na^+) ions in the glass. The presence of the new ions in the surface of the glass changes the refractive index of the glass according to the Gladstone-

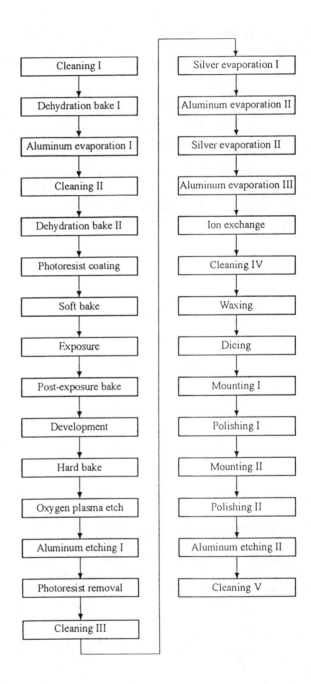

Figure 5. Process sheet for an integrated-optical laser formed by solid-silver ion exchange in a neodymium-doped phosphate glass.

Dale relations [19]. If the refractive index of this surface layer is greater than the unexchanged glass, a planar optical waveguide has been formed. If the surface of the glass is masked, the ion exchange takes place only in the unmasked region, producing a channel waveguide. Silver ion exchange typically produces a refractive index difference of about 0.05. Potassium, on the other hand, gives a much smaller refractive index difference, typically 0.008.

There are two common methods used to fabricate ion-exchanged waveguides. The first method uses molten salts as the ion source. This method often uses molten silver nitrate ($AgNO_3$) or molten potassium nitrate (KNO_3) as the ion source. The ion exchange is performed in the following manner. A small amount of the silver nitrate, for example, is placed in a crucible. The crucible and the glass substrate are then placed in an oven. The temperature of the oven is raised above the melting point (350°C) of the silver nitrate. After the melt has heated, the glass is placed in the melt, and the Ag^+ ions in the melt substitute for the Na^+ ions in the glass.

Another common ion exchange method also exchanges Ag^+ ions for Na^+ ions in the glass. This method uses a thin film of silver, and not molten silver nitrate, as the ion source. The ion exchange is performed in the following way. A thin silver film is deposited on the top surface of a polished glass plate. Another metal film is deposited on the back surface of the plate. An electric field is applied across the sample, and the ion exchange is performed around 350°C. The difference between ion exchange from a molten source and a solid film is that the solid film does not contain silver ions, but rather silver atoms. Silver ions are released from the metallic silver film as the result of an electrochemical reaction where the glass acts as a solid electrolyte and the silver film acts as the anode.

3.3 Fabrication of integrated-optical lasers by thin-film deposition

Several thin-film deposition techniques have been used to fabricate integrated-optical lasers. RF sputtering has been used to deposit rare-earth-doped glass films. A rare-earth-doped glass is machined into a target usually 75 cm or so in diameter. The target is used in a vacuum chamber equipped with an RF planar magnetron. Argon gas is fed into the chamber. RF energy at the magnetron creates a plasma containing energetic argon ions. These heavy ions collide with the glass target and remove or sputter the glass particles. The particles are deposited onto a substrate several centimeters from the target.

Sputtering in pure argon usually produces oxygen-deficient films. Therefore a small amount of oxygen is usually added to the argon. Good quality, glassy films can be deposited by carefully controlling the deposition pressure, gas feed rates, RF power, and substrate temperature. This deposition process can be slow, however; It can take as long as 8 h to grow a 1 μm thick film. Another disadvantage of RF sputtering is that the different constituents of a multi-component glass sputter at different rates. The composition of the glass film will be different than the target due to the different sputter yields of the constituents. This makes it difficult to predict the composition of the glass film. Photolithography and reactive ion etching can then be used to define rib or channel waveguides in these sputtered films.

A more common deposition technique is flame hydrolysis deposition (FHD). FHD uses oxygen, hydrogen, and $SiCl_4$ to form SiO_2. The gases are combined in a torch which decomposes the $SiCl_4$. The SiO_2 is deposited onto a wafer as a fine white soot. The oxygen/hydrogen torch is then used to consolidate the soot into SiO_2 glass. For waveguide fabrication, the refractive index of the guiding layer must be higher than the cladding. This can be accomplished by delivering $AlCl_3$ with helium as a push gas to the torch.

The fabrication of an FHD device usually starts with a silicon wafer, on which a thick (10 μm or so) film of SiO_2 has been grown by high-pressure thermal oxidation (HIPOX). An aluminum oxide-rich core layer is then deposited. Photolithography is performed after the core is deposited. The patterned photoresist may be used as the mask, or the pattern in the resist may be transferred onto a thin metal film, by acid or plasma etching. Reactive ion etching (RIE) is then used to transfer this pattern onto the high refractive index core. CF_4 or C_2F_6 are the most common RIE gases for silica-based devices. Once the core has been patterned, a low refractive index SiO_2 cladding is deposited. The wafer can then be diced and polished.

The rare-earth ions have been introduced into the core of these devices by two fashions. The first method is solution doping. After the high-refractive index core is deposited, but before it is consolidated, the wafer is soaked in a rare-earth alcohol, such as $NdCl_3$. It is difficult to incorporate large amounts of rare-earth ions using this technique. A better method uses vapor phase transport of chelates of rare-earth ions. The most successful to date has used the vapor transport of the neodymium chelate 2,2,6,6-tetramethyl-3,5-heptanedione [Nd(thd)$_3$] and the similar erbium chelate [Er(thd)$_3$].

One problem with FHD is that it requires sintering of the glass soot. After the core has been patterned and defined by reactive ion etching, a cladding layer is deposited and consolidated in the same manner as the core. The core can soften and flow during this step, however. This may make it difficult to fabricate components such as directional couplers with this technique. If the spacing between two adjacent waveguides changes during the cladding deposition, the coupling coefficients of the structure can be adversely effected.

Another promising method for depositing thin films of rare-earth-doped glass is chemical vapor deposition (CVD). This technique is widely used in the semiconductor industry for depositing gate oxides and interlayer dielectrics. Silane (SiH_4) or a silane-related compound such as tetraethylorthosilane (TEOS) can be decomposed into silica in several ways. Atmospheric pressure thermal CVD requires heating the substrate to around 700°C. Plasma-enhanced CVD (PECVD) typically uses 13.56 MHz RF to decompose the silane. Excellent quality SiO_2 can be grown without the high substrate temperatures required for thermal CVD. Growth rates for PECVD vary, but it is possible to grow a 1 μm film in about 30 min. It is possible to incorporate rare-earth ions in silica films grown by PECVD, by using the same chelate delivery as FHD. An index-raising species such as aluminum or phosphorus can be added with push gases, such as helium.

A recent paper reported using CVD of ZBLAN glass [20]. ZBLAN is an acronym for the ZrF_4-BaF_2-LaF_3-AlF_3-NaF five-component glass system. This fluoride glass has been used in 1.3 μm optical fiber amplifiers that use praseodymium as the active ion. CVD of these glass particles used metal β-diketonates as the precursors and hydrogen fluoride and fluorine gas in the reaction chamber. This glass system may prove useful for rare-earth-doped integrated-optical components. It may be possible to incorporate rare-earth ions into the core of such a planar device using similar chelates. I will mention more about this glass system in Section 5.4.

A new method for depositing thin films of SiO_2 is electron cyclotron resonance based PECVD (ECR-PECVD). ECR-PECVD uses a microwave source at 2.4 GHz to generate an ultra-high-density oxygen plasma. The highly energetic plasma is directed to the reaction chamber, where it combines with either silane or TEOS to form SiO_2. ECR-PECVD has become common because it has two advantages over traditional PECVD. The first is that it is a low temperature process. The substrate temperature for ECR-PECVD is less than 200°C, as opposed to 300°C or greater for traditional PECVD. Low temperature deposition is required for certain integrated circuits. A greater advantage is that deposition rates for ECR silica are very high. A good quality, 1 μm thick SiO_2 film can be deposited in 1 min. It should be possible to incorporate rare-earth ions and refractive index raising species in these films using chelate delivery.

4. TOOLS FOR IMPROVING PERFORMANCE

4.1 Optimizing performance

In the last section I discussed methods for fabricating these devices. In this section, I will address analytical and diagnostic methods for improving the performance of these devices and discuss some of these in detail. The first question to ask is, "Why are these devices are not yet widely manufactured." A second question is, "Are the barriers preventing commercialization of these devices technical or are they related to manufacturing?" I will not address large-scale manufacturing of these components. It is clear, however, that cost-efficient manufacturing techniques are a prerequisite for wide-scale use of these components. Before I discuss specific performance improvements, let me start in more general terms. Which properties of these devices are relevant and need to be analyzed, and which analytic and diagnostic tools can be used to investigate these properties?

Many important properties of an ion-exchanged waveguide laser can be determined by analysis of the glass host itself. A glass host for a rare-earth-doped waveguide laser must possess most of the properties of an optical glass. These properties include excellent refractive index homogeneity, few bubbles, and good chemical durability. In addition, the glass must have low attenuation at the lasing wavelength, low birefringence, and resistance to optical damage. Attributes such as these are known as the optical properties of the glass.

The spectroscopic properties of a waveguide laser glass concern the absorption and emission of light by the rare-earth ions in the glass. The spectroscopic properties of a glass, such as the absorption and emission bands, radiative lifetimes, and cross sections, largely determine whether the host is suitable for a waveguide laser.

Compositional properties of a waveguide laser glass, such as alkali content, are important for designing a successful ion exchange. The chemical composition of the waveguide region is especially important. Mechanical properties of the glass, such as the density, specific heat, transformation point, and thermal conductivity, are important especially during melting and casting.

Optical properties of the waveguides are critical. Properties such as mode diameter and mode ellipticity affect coupling between these components and optical fibers. Ion-exchange artifacts such as scattering centers and excess absorption can dramatically increase the losses of these devices.

For example, consider a laser source fabricated using ion exchange in a neodymium-doped glass. A desirable property of such an optically pumped, glass device is a low threshold for lasing. The lasing threshold is a function of many properties including the upper-state lifetime, stimulated-emission cross section, waveguide losses, and waveguide dimensions.

Spectroscopic analysis can predict stimulated-emission cross sections and radiative lifetimes. This information can then be used to modify the composition of the glass so that these quantities are maximized. This will reduce the threshold for lasing Microprobe techniques such as secondary ion mass spectroscopy (SIMS) and atomic force microscopy (AFM) can provide important information about the waveguides, which may be useful for reducing the transverse waveguide dimensions. This would also lower the threshold of laser oscillation. Optical techniques such as photothermal deflection can be used to analyze waveguide losses. Chemical analysis can reveal whether the glass is compatible with standard cleanroom fabrication procedures. There are many tools at our disposal to analyze and improve the performance of these components. I will describe some techniques that I have found particularly useful.

4.2   Spectroscopic Analysis

The performance of a waveguide laser is critically dependent on the spectroscopic properties of the rare-earth-doped glass. Consider a neodymium-doped device again. The two main laser transitions take place between excited states of the neodymium ion. It can be difficult to measure transitions between excited states of ions. Measurements between excited states of rare-earth ions often use pump/probe techniques. These techniques use a laser to populate an excited state (pump) and a second laser, operating at a different wavelength, to promote the ions to a higher energy excited state (probe). The wavelength of the probe laser must correspond to the energy difference of the desired excited-state transition. A complete analysis of the excited-state transitions would require a great many laser sources, which few laboratories can afford.

Direct measurement of some spectroscopic properties may depend on absolute measurements of radiative quantities, such as fluorescence. While relative measurements of fluorescence are easy, absolute measurements can be difficult. Some spectroscopic properties such as stimulated-emission cross section can be inferred from other measurements, such as small-signal gain. These measurements themselves are difficult and prone to errors. Estimation of other radiative properties using these uncertain measurements is questionable. Absolute spectral absorbance, on the other hand, is relatively simple to measure.

A semiempirical approach for calculating the strength of the *4f* optical transitions of rare-earth ions was developed independently by Judd and Ofelt [21, 22]. The Judd-Ofelt method computes absolute radiative properties of rare-earth-doped glasses based on simple and accurate ground-state spectral absorbance measurements. Ground state absorption is used to determine three phenomenological parameters. Once these three host-dependent parameters are known, it is possible to estimate radiative properties such as spontaneous emission probabilities and radiative lifetimes for excited-state transitions. The accuracy of the Judd-Ofelt method is around 20% because of approximations in the method. It is a powerful analytic technique, however, because it can determine the strengths of transitions that are difficult or impossible to measure directly.

In practice, the method is performed by measuring the ground-state absorption of the rare-earth-doped glass with a computer-controlled spectrophotometer. These data are numerically integrated, and an equivalent oscillator strength is computed for each ground state absorption transition. The oscillator strengths are then used to compute the three phenomenological Judd-Ofelt parameters Ω_2, Ω_4, and Ω_6. Once these parameters have been computed, it is relatively simple to calculate spectroscopic properties such as spontaneous emission probabilities, stimulated-emission cross sections, branching ratios, and radiative lifetimes.

Other spectroscopic measurements are important as well. Direct measurements of upper-state lifetime can reveal excess water in rare-earth-doped glasses. Fluorescence measurements can be used to estimate stimulated-emission cross sections, as well. The location and width of certain ground state absorption transitions is important for laser diode pumping.

4.3 Waveguide Analysis

Many diagnostic and analytical tools can be applied to analyzing and improving the waveguides in both ion-exchanged and thin-film active integrated-optical devices. These include optical and microprobe techniques. Computer simulations can provide information for improving fabrication.

Optical properties such as the number of propagating modes, mode field distribution, and optical losses dramatically affect the performance of these devices. High optical gain is negated by high optical losses. Accurate measurement of optical properties will become increasingly important as these components find use in optical fiber networks.

As an example, consider the mode-field of a single-transverse-mode glass integrated-optical component. A component like this may be used in an optical fiber network as an optical amplifier. The mode diameter of the glass waveguide must match the mode diameter in the optical fiber for efficient coupling. The optical mode in the planar component should be circular and not elliptical. The propagation characteristics at the pump wavelength should be similar to those at the design wavelength. Many devices, in both fiber and planar forms, support a single transverse mode at their design wavelength but support two or more modes at their pump wavelength. This can reduce the gain efficiency of the amplifier.

Scattering losses of ion-exchanged devices can be reduced by burying the waveguide below the surface of the glass plate. Optical techniques such as near-field optical microscopy with dark-field illumination can reveal the degree of burial. Photothermal deflection and optimum coupling are two techniques for accurately measuring optical losses [23, 24]. Low-coherence reflectometry can reveal reflections in waveguides on a micrometer scale [25, 26]. Refractive index distributions of waveguide components can be estimated using the inverse-WKB method with measured mode indices [27].

Microprobe techniques can also be used to measure refractive index distributions. Secondary ion mass spectroscopy is a technique in which a beam of heavy ions is directed at a sample. The ion beam is approximately 10 μm in diameter and slowly sputters away a small amount of glass. The sputtered or secondary ions of the glass are then detected in a mass spectrometer. SIMS measurements give ion counts per second. The mass spectrometer must be calibrated for the desired species for accurate concentration-versus-depth measurements.

Figure 6 shows the refractive index versus depth of a silver ion-exchanged waveguide in a neodymium-doped phosphate glass. SIMS, refractive index, and x-ray fluorescence measurements were recorded for several similar glass samples uniformly doped with Ag_2O. The refractive index of the bulk samples was linear with silver concentration. This information was also used to determine a relative sensitivity factor for the waveguide measurements. This factor is then used to convert raw counts per cycle SIMS data into silver concentration versus time. The time data were converted into depth by measuring the depth of the ion crater with a stylus profilometer. The accuracies of the concentration and depth data are estimated to be 10%.

Energy dispersive spectroscopy (EDS) uses a scanning electron microscope (SEM) for single-species detection. The electron beam in the SEM is directed at the sample and generates secondary X-rays. The energy of these X-rays is related to the composition of the sample. The absolute accuracy and the spatial resolution of EDS measurements are both poor. Therefore, EDS measurements should be considered only qualitative. Figure 7 shows a backscattered image of the endface of a polished glass waveguide. A portion of a 600 μm wide planar waveguide region, as well as 6 waveguides exchanged through 10 μm apertures is visible.

We have successfully used atomic force microscopy (AFM) at our laboratory to

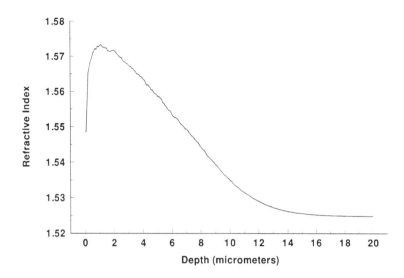

Figure 6. Refractive index versus depth of a silver ion-exchanged integrated optical device.

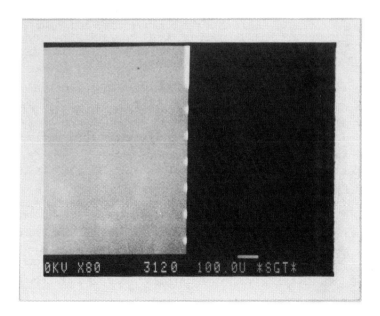

Figure 7. EDS image of a waveguide endface.

investigate stress in waveguide lasers. Figure 8 is an AFM trace of a potassium-ion-exchanged waveguide and shows a trench over the waveguide surface. This technique should find much greater use due to its high spatial resolution.

A tremendously useful tool for the design and analysis of these devices is simulation software. There are now commercial codes that can model the diffusion/migration equation for ion-exchanged waveguides. If material properties, such as diffusion coefficients and ion concentration have been accurately determined, these simulations can provide concentration profiles that agree well with measurements. Figure 9 depicts the predicted silver concentration profile for a silver-ion-exchanged waveguide in a neodymium-doped phosphate glass. In addition, there are several commercial products that can analyze the propagation of optical modes in integrated-optical devices using techniques such as the beam propagation method (BPM).

4.4 Glass Analysis

Analysis of the base glass of rare-earth-doped integrated-optical devices can also be used to improve their performance. Compositional studies have been very successful in determining how changes in glass composition affects spectroscopic and thermo-mechanical properties of rare-earth-doped glasses. These studies rely on accurate measurements of glass composition. X-ray fluorescence is a common technique for accurately determining composition.

Chemical durability is another important property of these glasses. Some chemicals used in cleanroom processing may attack certain glasses. It is important that glasses for these active devices be compatible with common fabrication chemicals. We have found that increased alumina content in one of our glasses dramatically improves its durability in one of our processing steps.

Nonradiative decay of excited rare-earth ions can prevent laser oscillation on certain transitions. Nonradiative decay rates of glasses with low-energy phonons, such as heavy-metal fluorides, can be substantially lower than those of oxide glasses such as silicates and phosphates. Low-phonon-energy glass hosts are important for praseodymium devices that operate at 1.3 μm. Raman spectroscopy is powerful technique for determining the phonon spectra of such glasses, and their suitability for praseodymium hosts.

Techniques have been developed for measuring the composition of compound semiconductor devices during growth. These in-situ methods, such as atomic absorption spectroscopy, can provide accurate real-time control of layer composition and thickness in III-V quantum-well devices. They may be used during CVD of rare-earth-doped glass devices to enable tight control of layer thickness, composition, and doping levels.

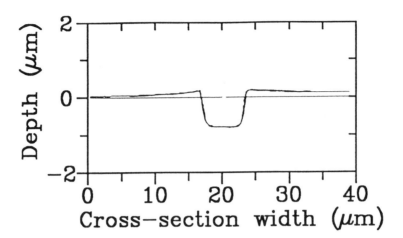

Figure 8.   AFM trace of a trench over an ion-exchanged waveguide.

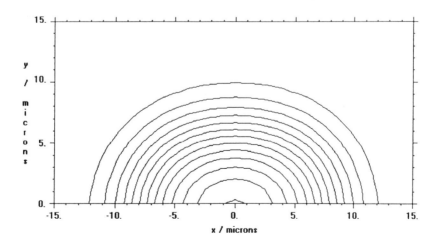

Figure 9.   Calculated silver concentration profile of an integrated-optical device formed by ion exchange from a solid-silver film.

5. COMMENTARY

5.1 General Comments

In this last section, I would like to comment on some current topics in rare-earth-doped glass integrated-optical devices and give some of my opinions on where I think this field is headed. I would like to give some general comments first and then discuss some of the particular rare-earth ions and their optical transitions.

I am firmly convinced that this technology will be embraced in future optical communication networks. These networks will require many moderate-gain optical amplifiers. I think that rare-earth-doped planar devices will fill this need. This technology may prove useful in other areas as well. It is not difficult to envision these compact laser sources in applications such as printing, data storage, or compact wavelength standards.

Mass production of these components may prove challenging. However, consider the length of time between the first demonstrations of laser diodes in the mid-1960's and large-scale manufacturing of these components in the last decade or so. Many innovations in integrated-circuit fabrication had to occur before laser diodes could be mass-produced. Many of these same innovations, such as deep UV lithography, high resolution photoresists, plasma etching, and plasma deposition, can be used for fabrication of rare-earth-doped glass integrated-optical devices. Most, if not all, of the manufacturing technology required to bring these components to the marketplace exists now. Many issues related to manufacturing and cost/performance issues remain unsolved, but I am confident that solutions can be found to these problems.

There are many new developments that we can use to improve performance, regardless of which rare-earth ion is used in the device or which fabrication method is used to make a device. Computer-aided design tools are essential. The optical properties of waveguide devices, both passive and active, can be significantly improved with novel cavity designs. Features such as modified cosine S-bends and improved tapers can reduce optical losses and improve performance. The sophistication of software for designing optical waveguide device photomasks is nowhere near that of layout tools for multiple-level integrated circuits. Rapid progress is being made in this area, and analytical tools such as the beam propagation method are being integrated with these layout tools.

Considerably more effort needs to be devoted to producing ion-exchangeable rare-earth-doped glasses. There are many excellent ion-exchangeable glasses. There are also many excellent rare-earth-doped glasses, having very desirable spectroscopic properties. There are not enough glasses, however, that combine superior spectroscopic properties and ion-exchangeability. Many glass hosts that are suitable for passive ion-exchanged devices have poor performance when they are doped with rare-earth ions. The contrary is also true. Many glasses that have desirable spectroscopic properties, such as large stimulated-emission cross sections, are poor candidates for ion exchange.

5.2 Neodymium devices for 1.06 μm

Significant performance has been achieved using the $^4F_{3/2} \rightarrow {}^4I_{11/2}$ neodymium transition at 1.06 μm. Many different types of devices ranging from simple, yet high-power Fabry-Perot lasers, to lasers using distributed feedback and coupled cavities have also been demonstrated. Both thin-film deposition and ion exchange have been used to make these devices. I have heard the comment that 1.06 μm is not an "interesting wavelength" since it is outside the 1.3 μm or 1.55 μm telecommunications windows.

I think that devices utilizing the $^4F_{3/2} \rightarrow {}^4I_{11/2}$ neodymium transition at 1.06 μm will continue to be important devices. They may find wide use in local area networks. Future networks deployed in aircraft, ships, and possibly automobiles will be highly tapped, requiring a number of modest-gain optical amplifiers. The cost of an optical fiber system must include the source and detector. If the system is not dispersion-limited, why not operate the network at 1.06 μm, where inexpensive silicon detectors can be used? Indeed, some recent cable television systems are using 1.06 μm for signal distribution.

5.3 Erbium devices for 1.55 μm

Impressive performance has been achieved in both laser sources and amplifiers fabricated by ion-exchange and thin-film deposition. Integrated-optical devices utilizing erbium-doped glasses will continue to be important due to optical fiber telecommunications at this wavelength. This wavelength is also an eye-safe region of the spectrum. Compact, erbium-doped glass waveguide devices may be useful for rangefinding.

One problem with erbium at 1.55 μm is that it is a three-level transition. This means that portions of the active waveguide that are not pumped will absorb the 1.55 μm signal. Thin-film deposition may be an especially useful solution to this problem. Optical pumping of these devices usually occurs through one or both endfaces. Regions of the device that are undoped will be transparent to both the pump and signal. It may be possible to integrate erbium-doped and undoped regions along with evanescent couplers on a single device. The undoped regions and the couplers would allow pump light to be injected into the active region at several locations along the device. This would allow complete population inversions along the length of the device and would increase the available gain.

Additional dopants added during thin-film deposition may increase the stimulated-emission cross sections of the erbium-doped glass layers. Saturation powers of these devices are steadily improving. Studies of the noise properties of these devices are lacking; however, their noise properties are not expected to be significantly different than erbium-doped fiber amplifiers.

5.4 Rare-earth-doped glass devices for 1.3 μm

The majority of optical fiber networks operate at 1.3 μm. Optical amplifiers that operate at this wavelength have been elusive. Recently, however, praseodymium-doped ZBLAN optical fibers have emerged as a viable technology [28]. I anticipate that integrated-optical components using this material or a similar glass with low-energy phonons will be demonstrated in the near future.

The $^1G_4 \rightarrow {}^3H_5$ praseodymium transition at 1.31 μm suffers from severe nonradiative decay. Ions in the $^1G_4$ level rapidly decay to the $^3F_4$ level due to their close spacing. Nonradiative decay is a multiple-phonon process. If the energy of the phonons of the glass host can be reduced, the nonradiative decay rates can be reduced. The phonon energies of heavy-metal fluoride glasses are much lower than those in oxide glasses such as silicates and phosphates. Indeed, it is only when praseodymium is doped into low-phonon energy glasses that high population inversions and significant gain can be achieved.

While the performance of praseodymium-doped ZBLAN optical amplifiers is impressive, the glass itself is difficult to work with. The precursors are anhydrous fluorides and must be kept in nitrogen atmospheres. Ion exchange in this glass may prove to be difficult, and the glass is readily attacked by water. It may be possible to deposit thin films of this glass by sputtering. CVD may also be useful. There are still several technical barriers, such as incorporating index-raising species, brittleness, passivation, and devitrification. With sufficient effort, I expect that these problems can be solved. Another promising low-phonon-energy glass for praseodymium is Ga_2S_3:La_2S_3 [29]. Nonradiative relaxation rates in this glass are lower than those in ZBLAN glasses. It may be possible to fabricate rare-earth-doped integrated-optical devices in this host.

The $^4F_{3/2} \rightarrow {}^4I_{13/2}$ neodymium transition at 1.35 μm is another candidate for 1.3 μm optical fiber communications. This transition suffers from excited-state absorption (ESA). Consider a neodymium ion that is in the $^4F_{3/2}$ level. A photon near 1310 nm that interacts with the excited ion can stimulate emission at that wavelength. This causes emission of two coherent 1310 nm photons. If excited state absorption occurs, however, the excited neodymium ion will absorb the 1310 nm photon. This will promote the neodymium ion into the $^4G_{7/2}$ level. The result of this process is that the signal photon is absorbed, and stimulated emission does not occur. In many neodymium-doped glasses, the ESA cross section is greater than the stimulated-emission cross section at the center of the telecommunications band. This precludes the possibility of optical gain in this region.

1.31 μm ESA is noticeably reduced in certain families of neodymium-doped glasses, such as phosphates, fluorophosphates, and fluorozirconates [30]. It may be possible to shift the wavelength range of the ESA out of the telecommunications band by modifying the composition of neodymium-doped glasses. The ESA oscillator strength may also be reduced by compositional changes. If the excited-state absorption can be reduced, neodymium-doped glass integrated-optical devices may be used in 1.3 μm optical fiber networks.

5.5 Integrated-optical lasers based on cooperative upconversion

Upconversion integrated-optical lasers using rare-earth-doped glasses are another area that merits further study. Upconversion lasers have been demonstrated in both rare-earth-doped fibers and rare-earth-doped crystals. Upconversion lasers use excited-state-absorption (or cooperative upconversion). They operate by absorbing two (or three) infrared photons and emitting a single blue, green, or red photon [31]. If the glass host has low nonradiative decay rates, significant population inversions can be established in these excited states. Erbium ions can absorb two 980 nm photons and emit one 530 nm green photon. This is shown in Fig. 2. Praseodymium-doped ZBLAN is a popular glass host for these lasers, but holmium-doped devices have been demonstrated as well. Integrated-optical lasers based on cooperative upconversion in rare-earth ions could be very useful as compact blue lasers for data storage applications.

6. CONCLUSION

I have attempted to present a concise overview of integrated-optical devices that use rare-earth-doped glasses. I think that, in the future, we will see continuing improvement in the performance of these devices. Many new components will also be demonstrated in the future. Some may use common rare-earth ions, familiar optical transitions, and current fabrication techniques. Some of these new devices may use novel cavity designs, new fabrication methods, and unfamiliar transitions of other rare-earth ions. This field has come a long way in a few years. In the years ahead, I expect integrated-optical devices that use rare-earth-doped glasses will make the difficult transition from the laboratory into the marketplace.

7. ACKNOWLEDGEMENTS

I thank my colleagues in the Dielectric Materials and Devices Project at NIST, Boulder for their comments and suggestions. Much of our research is collaborative and it is a great pleasure to acknowledge my co-workers Norman Sanford, David Veasey, Andy Aust, and Robert Hickernell. I also thank Joe Sauvageau of NIST for helpful discussions of plasma deposition.

8. REFERENCES

[1] E. Lallier, "Rare-earth-doped glass and $LiNbO_3$ waveguide lasers and optical amplifiers," Appl. Opt., Vol. 31, p. 5276 (1992).

[2] H. Yajima, S. Kawase, and Y. Sekimoto, "Amplification at 1.06 μm using a Nd:glass thin-film waveguide," Appl. Phys. Lett., Vol. 21, p. 407 (1972).

[3] N. Saruwatari and T. Izawa, "Nd-glass laser with three dimension optical waveguide," Appl. Phys. Lett., Vol. 24, p. 603 (1974).

[4] Y. Hibino, T. Kitagawa, M. Shimizu, F. Hanawa, and A. Sugita, "Neodymium-doped silica optical waveguide laser on silicon substrate," IEEE Photon. Tech. Lett., Vol. 1, p. 349 (1989).

[5] N. Sanford, K. Malone, and D. Larson, "Integrated-optic laser fabricated by field-assisted ion exchange in neodymium-doped soda-lime-silicate glass," Opt. Lett., Vol. 15, p. 366 (1990).

[6] H. Aoki, O. Maruyama, and Y. Asahara, "Glass waveguide laser," IEEE Photon. Tech. Lett., Vol. 2, p. 459 (1990).

[7] H. Aoki, O. Maruyama, and Y. Asahara, "Glass waveguide laser operated around 1.3 μm," Electron. Lett., Vol. 26, p. 1910 (1990).

[8] E. Mwarania, L. Reekie, J. Wang, and J. Wilkinson, "Low-threshold monomode ion-exchanged waveguide lasers in neodymium-doped BK-7 glass," Electron. Lett., Vol. 26, p. 1317 (1990).

[9] T. Kitagawa, K. Hattori, M. Shimizu, Y. Ohmori, and M. Kobayashi, "Guided-wave laser based on erbium-doped silica planar lightwave circuit," Electron. Lett., Vol. 27, p. 334 (1991).

[10] N. Sanford, K. Malone, and D. Larson, "Extended-cavity operation of rare-earth-doped glass waveguide lasers," Opt. Lett., Vol. 16, p. 1095 (1991).

[11] N. Sanford, K. Malone, and D. Larson, "Y-branch waveguide glass laser and amplifier," Opt. Lett., Vol. 16, p. 1168 (1991).

[12] K. Hattori, T. Kitagawa, Y. Ohmori, and M. Kobayashi, "Laser diode pumping of waveguide laser based on Nd-doped silica planar lightwave circuit," IEEE Photon. Tech. Lett., Vol. 3, p. 882 (1991).

[13] T. Kitagawa, K. Hattori, K. Shuto, M. Yasu, M. Kobayashi, and M. Horiguchi, "Amplification in erbium-doped silica-based planar lightwave circuits," Electron. Lett., Vol. 28, p. 1818 (1992).

[14] J. Shmulovich, A. Wong, Y. Wong, P. Becker, A. Bruce, and R. Adar, "Er^{3+} glass waveguide amplifier at 1.5 μm on silicon," Electron. Lett., Vol. 28, p. 1181 (1992).

[15] T. Feuchter, E. Mwarania, K. Wang, L. Reekie, and J. Wilkinson, "Erbium-doped ion-exchanged waveguide lasers in BK-7 glass," IEEE Photon. Tech. Lett., Vol. 4, p. 542 (1992).

[16] J. Roman and K. Winnick, "Neodymium-doped glass channel waveguide laser containing an integrated distributed Bragg reflector," Appl. Phys. Lett., Vol. 61, p. 2744 (1992).

[17] N. Sanford, J. Aust, K. Malone, and D. Larson, "Linewidth narrowing in an imbalanced Y-branch waveguide laser," Opt. Lett., Vol. 18, p. 281 (1993).

[18] K. Malone, N. Sanford, and J. Hayden, "Integrated optic laser emitting at 906, 1057 and 1358 nm," Electron. Lett., Vol. 29, p. 691 (1993).

[19] D. Ryan-Howard and D. Moore, "Model for the chromatic properties of gradient-index glass," Appl. Opt., Vol. 24, p. 4356 (1985).

[20] Y. Nishida, K. Fujiura, H. Sato, S. Sugawara, K. Kobayashi, and S. Takahashi, "Preparation of ZBLAN fluoride glass particles by chemical vapor deposition process," Jpn. J. Appl. Phys., Vol. 31, p. 1692 (1992).

[21] B. Judd, "Optical absorption intensities in rare earth ions," Phys. Rev., Vol. 127, p. 750 (1962).

[22] G. Ofelt, "Intensities of crystal spectra of rare earth ions," J. Chem. Phys., Vol. 37, p. 511 (1962).

[23] R. Hickernell, D. Larson, R. Phelan, and L. Larson, "Waveguide loss measurement using photothermal deflection," Appl. Opt., Vol. 27, p. 2636 (1988).

[24] M. Haruna, Y. Segawa, and H. Nishihara, "Nondestructive and simple method of optical waveguide loss measurement with optimisation of end-fire coupling," Electron. Lett., Vol. 28, p. 1612 (1992).

[25] B. Danielson and C. Wittenburg, "Guided-wave reflectometry with micrometer resolution," Appl. Opt., Vol. 26, p. 2836 (1987).

[26] K. Takada, M. Takato, and J. Noda, "Characterization of silica-based waveguides with an interferometric optical time-domain reflectometry system using a 1.3 μm superluminescent diode," Opt. Lett., Vol. 14, p. 706 (1989).

[27] K. Chiang, "Construction of refractive-index profiles of planar dielectric waveguides from the distribution of effective indexes," J. Lightwave Technol., LT-3, p. 385 (1985).

[28] Y. Miyajima, T. Sugawa, and Y. Fukasaku, "38.2 dB amplification at 1.31 μm and the possibility of 0.98 μm pumping in Pr^{3+}-doped fluoride fibre," Electron. Lett., Vol. 27, p. 1706 (1991).

[29] D. Hewak, et al., "Low phonon-energy glasses for efficient 1.3 μm optical fibre amplifiers," Electron Lett., Vol. 29, p. 237 (1993).

[30]   S. Zemon, et al., "Excited-state absorption cross sections and amplifier modelling in the 1300-nm region for Nd-doped glasses," IEEE Photon. Tech. Lett., Vol. 4, p. 244 (1992).

[31]   R. Smart, et al., "CW room temperature upconversion lasing at blue, green, and red wavelengths in infrared-pumped Pr^{3+}-doped fluoride fiber," Electron. Lett., Vol. 27, p. 1307 (1991).

Commercial Devices

Ion-exchanged glass waveguide devices for optical communications

Seppo Honkanen

Nokia Research Center, Helsinki, Finland
and
Helsinki University of Technology, Espoo, Finland

ABSTRACT

Recent developments of ion-exchanged glass waveguide devices for optical communications are reviewed. The focus is on waveguide configurations for passive devices and on their optical characteristics. First the ion-exchanged glass waveguide structures are briefly described. The most important ion-exchanged passive components, power splitters and dual-wavelength multi/demultiplexers (WDM), are discussed in detail. We also discuss the recent advancement on other components such as narrow-band WDM-devices, modefield transformers and ring resonators. A view to the important issue of fiber attachment and device packaging is briefly given, and finally the integration of several ion-exchanged components (e.g. wavelength multi/demultiplexers and splitters) in a single glass chip is briefly discussed.

1. INTRODUCTION

The application of optical fibers is increasing rapidly in telecommunication and CATV access networks, and the first mass deployment of fibre in the loop (FITL) has started in Germany.[1] One promising cost effective topology that has been widely considered and chosen is the Passive Optical Network (PON)[2], which is a tree type architecture requiring a very large number of passive splitting devices. When PONs are designed an important issue is the easy possibility for upgrading the network, for example, for future broadband services. Wavelength division multiplexing (WDM) and optical amplification are believed to have an important role in the evolution of the PONs. In the future, optical amplification together with dense-WDM is expected to revolutionize the fiber-optics communications leading to a new generation of lightwave systems[3]. Therefore, in addition to passive 1 x n and n x m splitting devices, variety of passive WDM-devices are needed. In fact, high performance dual-wavelength multi/demultiplexers for 1.30/1.55 µm wavelength windows are already presently widely demanded. For optical amplifiers (OFAs) wavelength combiners for pump and signal wavelengths of an amplifier are needed. In the case of the well established erbium-doped OFA the wavelength combinations required are 0.98/1.55 µm and 1.48/1.55 µm. In the longer timescale, dense WDMs with several densely spaced wavelengths within one wavelength window, and also narrow-band add-drop WDM-devices are required. Furthermore, in optical communications there are applications for several other types of passive components such as ring resonators, polarization splitters and modefield transformers.

High performance passive optical devices have been produced commercially for quite some time by using micro-optics components or by fused fibre couplers. However, before the passive components are employed in large volumes in optical communications the production costs of them have to be low enough, while the high performance of the components is obviously still demanded. It seems that all the requirements of passive components for optical communications can be met only by utilizing planar integrated optical technologies, which offer the inexpensive mass production possibilities.

Although some very interesting techniques and promising results utilizing planar polymer waveguides have recently been proposed[4], there are definitively two major, and already commercial, waveguide technologies for passive integrated optical components. These technologies are: ion-exchanged glass waveguides[5] and silica waveguides on silicon[6]. The purpose of this paper is to review the recent developments of the planar ion-exchanged glass integrated optical components for optical communications. The advancement of components realized by silica waveguides on silicon are reviewed elsewhere in this proceedings. The focus here is on waveguide configurations for passive components and on their optical characteristics. We start (chapter 2) by briefly describing the ion-exchanged glass waveguide structures for passive components. The most important ion-exchanged passive components, power splitters and dual-wavelength WDM-devices, are discussed respectively in chapters 3 and 4. Chapter 5 describes the recent advancement on narrow-band WDM-devices, modefield transformers and ring resonators. Chapter 6 gives a view to the very important issue of fiber attachment and device packaging. In chapter 7, we conclude and discuss briefly the more integrated ion-exchanged circuits (e.g. several wavelength multi/demultiplexers and splitters in a single chip).

2. ION-EXCHANGED GLASS WAVEGUIDE STRUCTURES

2.1 Surface waveguides

Ion-exchanged glass waveguides are fabricated by exchanging alkali ions (e.g. Na^+) originally in the glass substrate for other ions (e.g. Ag^+, Tl^+), which increases locally the refractive index of the glass[5]. The ion exchange (through a proper mask to get a channel waveguide) is carried out at an elevated temperature, typically > 300 °C, and usually by using molten salts as ion sources. An electric field is often used to enhance and get more flexibility to the process. In the case of the electric-field assisted technique a deposited silver film can also be used as an ion source for silver ion exchange. This electric-field assisted silver-film technique combined by post-annealing is utilized by Optonex, which is a small European producer of commercial passive glass waveguide components. This surface waveguide technique by Optonex together with an inexpensive multipurpose glass results in waveguide losses below 0.2 dB/cm and coupling losses below 0.2 dB with optical fibers. These values are low enough for most passive components for optical communications. Figure 1a shows an example of a calculated index profile of this type of surface waveguide and Fig. 1b a measured mode profile at 1.523 µm wavelength of a fabricated waveguide.[7] Similar profiles are obtained by surface waveguides realized utilizing molten salt ion sources cmbined with post-annealing. In many components higher index changes and smaller mode sizes are required. For these components a two-step process, potassium and silver double-ion exchange, is an interesting possibility. It results in low waveguide losses (below 0.2 dB/cm), even if an inexpensive multipurpose substrate glass and aluminum as masking material during silver ion exchange are utilized.[8]

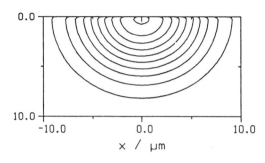 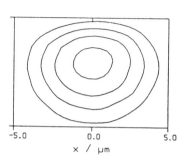

a) b)

Fig. 1. a) Calculated index distribution of a silver-film ion-exchanged and post-annealed surface waveguide in Corning 0211. The calculation was performed for following process parameters: a mask opening width 4 μm, amount of Ag in the glass 6.1 μg/m and post-annealing duration 3200 s at 343 °C. The contours are for index increase from substrate index at 0.001 intervals. b) The measured mode intensity distribution at 1.523 μm wavelength of a channel waveguide fabricated in a Corning 0211 glass with similar process parameters. The contours are for normalized intensities of 0.3, 0.5, 0.7, and 0.9.

2.2 Buried waveguides

After fabrication of an ion-exchanged surface waveguide a second unmasked ion exchange is often carried out, in which the ions originally in the glass (e.g. Na^+) are exchanged back to the glass surface. This results in burial of the waveguides below the glass surface. NSG, the major non-European producer of ion-exchanged components employes a purely thermal ion exchange process for burying the waveguides.[9] The mode profiles of these thermally buried waveguides are very similar to the mode profiles of the post-annealed surface waveguides.

By utilizing an electric field in the second process step the waveguides can be buried very deep below the glass surface. This type of ion exchange technique is used, for example, by IOT and Corning, which are the two major European producers of passive ion-exchanged integrated optical components.[10] The electric-field assisted burial in connection with glass substrates, which are specially developed for waveguide fabrication by ion exchange, results in very low waveguide losses (below 0.1 dB/cm) and low coupling losses with optical fibers (below 0.1 dB). Figures 2a and 2b illustrate respectively the calculated index profile and mode profile of a buried waveguide obtained by electric-field assisted burial.

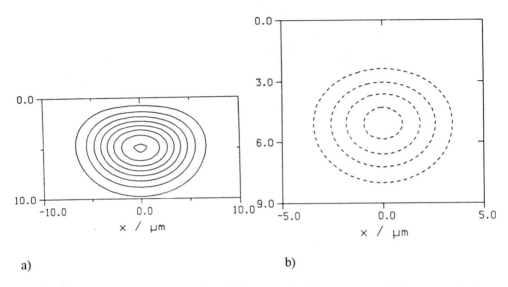

Fig. 2. a) Calculated index distribution of a buried silver ion-exchanged waveguide in Corning 0211 glass obtained by electric-field assisted burial. The calculation was performed for following process parameters: mask opening width 4 μm, amount of Ag in glass 6.1 μg/m (after 1st step) and duration of 1400 s of the electric-field assisted burial (40 V/mm at 343 °C). The contours are for index increase from substrate index at 0.001 intervals. b) The calculated mode intensity distribution at 1.523 μm wavelength of the waveguide of Fig. 2a. The contours are for normalized intensities of 0.3, 0.5, 0.7, and 0.9.

3. SPLITTING COMPONENTS

Power splitters, such as 1x4 and 1x16, are the most important passive components for the access networks. The performance of the splitters should be polarization insensitive and also independent of the wavelength used. This enables the future upgrading of the network by utilizing WDM. Power splitters can be constructed, for example, by cascading directional couplers or utilizing a so called multifunnel waveguide[11] at the output of a slab waveguide region. These waveguide configurations feature low losses, but their behaviour is not wavelength independent. A straightforward way to realize a 1xN splitter is to use cascaded symmetrical Y-branches, which are naturally polarization and wavelength insensitive. A conventional mask design of an ion-exchanged 1xN splitter is sketched in Fig. 3a. Here, the Y-branches are formed by two mirror imaged S-bends with constant radius of curvature. Improved mask designs have been reported[12] and are used to reduce the loss/branching point, the bend loss and the size of the device. For example, the Y-branch region can be easily optimized with the aid of design tools for ion-exchanged waveguides, and gradually bent S-bends can be utilized. One example of an improved mask design for 1x4 splitter is sketched in Fig. 3b. It has a special feature that one of the output ports has no lateral offset with respect to the input port.

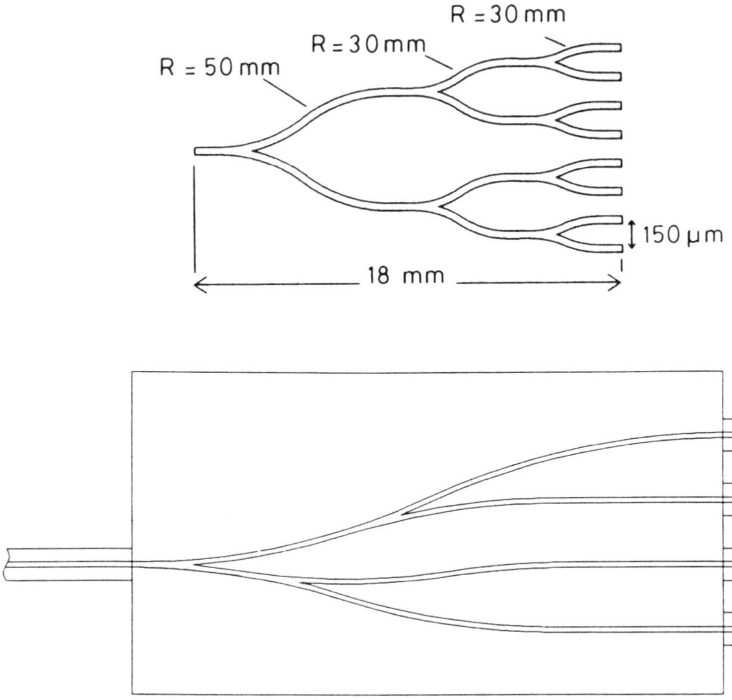

Fig. 3. Example of a) a conventional and b) an optimized mask design for 1xN splitters.

Cascaded Y-branch 1xN splitters made by ion exchange in glass have typically somewhat lower losses than splitters made by silica on silicon techniques. This is due to the slightly higher excess loss/branching point that is typical for silica on silicon waveguides[13]. A good demonstration of an ion-exchanged 1xN splitter is a fiber pigtailed 1x16 splitter by Corning[10]. It has an excess loss of less than 1 dB, including coupling losses, and a splitting uniformity of less than 1 dB both at the 1.3 μm and the 1.55 μm wavelength windows. Fiber pigtailed 1xN splitters with similar characteristics have been reported also by IOT[14] and NSG[12].

Achromatic 2xN splitting components are also useful components for access networks. They can be used, for example, to multiplex different signals or in controlling/securing the network. If a symmetrical Y-branch as a combiner would be used at the input of a 2xN splitter, it would result in an additional 3 dB loss. To avoid this loss a symmetrical directional coupler designed for 3 dB splitting at 1.30 μm and 1.55 μm wavelengths could be used. However, the behaviour of directional couplers is not achromatic and their reproducible fabrication requires tight tolerances for fabrication parameters. One possibility to realize a fully achromatic 2xN splitter is to use an adiabatic asymmetrical Y-branch as a combiner at the input of a 1xN splitter. Figure 4a shows schematically the operation of an adiabatic asymmetric Y-branch as a mode splitter. The waveguide withs, i.e. mask opening widths, and branching angle shown in the figure are suitable for ion-exchanged mode splitter. Fig. 4b sketches its use as a combiner in a 2xN splitter. When used as a combiner the adiabatic asymmetric Y-branch couples light (independent of the wavelength) to the fundamental mode of the junction, if the wider waveguide having higher effective mode index is used as an input port. If the narrower waveguide is used as an input port, light couples to the 1st order mode of the junction (two mode region). When this type of Y-branch combiner is followed by a symmetrical Y-branch, as in the 2x8 splitter in Fig. 4b, both the fundamental and the 1st order modes are split evenly to the two branches of the symmetrical Y-branch.

2xN splitters have been announced both by Corning and IOT. Corning has reported excellent achromaticity in a very wide wavelength region (1.20 μm - 1.64 μm).[15] Their spectral transmission measurements indicate that, instead of a directional coupler, a waveguide structure having a similar function with the adiabatic asymmetric Y-branch is used at the input.

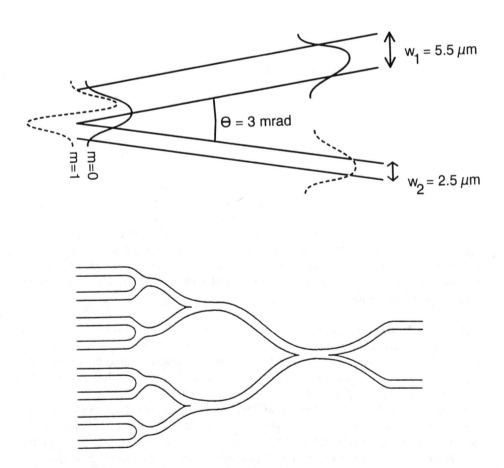

Fig. 4. a) Schematics of an adiabatic asymmetric Y-branch as a mode splitter and b) its use as a combiner in a 2x8 splitter.

4. DUAL-WAVELENGTH MULTI/DEMULTIPLEXERS

4.1 Symmetrical directional coupler WDM-device

The most common configuration for dual-wavelength multi/demultiplexer has been a symmetrical directional coupler[15,16] sketched schematically in Fig. 5. For WDM application the length of a directional is usually designed to be L_c for shorter wavelength and $2 \times L_c$ for longer wavelength. Here, L_c is the length with which total coupling from one waveguide to the other occurs. The WDM-component based on symmetrical directional coupler has the advantage of small size and low losses due to its simple structure. In fact, high performance 1.30/1.55 µm multi/demultiplexers with low losses and crosstalk has been demonstrated by several groups. The crosstalk in 1.30/1.55 µm WDM-component has to be low, not only at the designed center wavelength, but within a relatively wide bandwidth. Requirements for crosstalk are highly demanding, typically < -30 dB for about 60 nm bandwidth. This requirement cannot be met with only one symmetrical directional coupler, but it is possible to cascade them to achieve additional filtering. However, the wavelength response of a symmetrical directional coupler is very sensitive to fabrication parameters. For example, in order to maintain the crosstalk lower than -20 dB even at the designed center wavelength of the "stop-band", the process temperature has to be controlled better than ± 1.5 °C. This has been calculated for an ion-exchanged and post-annealed surface waveguide.[17] In the calculation the other process parameters, such as mask opening width, were fixed to give the perfect characteristics, and polarization sensitivity of the couplers were not taken into account. To conclude, it seems extremely difficult to produce 1.3/1.55 µm WDM-components based on symmetrical directional couplers with good reproducibility. However, they may be utilized as wavelength multiplexers, for example, in fiber amplifiers to combine the pump and signal wavelengths.

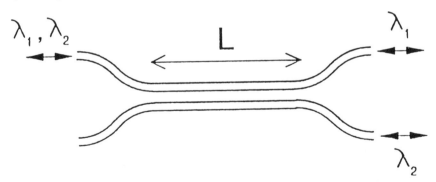

Fig. 5. Schematics of a symmetrical directional coupler as a dual WDM-device.

4.2 Asymmetrical Mach-Zehnder interferometer WDM-device

Dual-wavelength configurations having no symmetrical directional couplers have also been developed. One example is an asymmetrical Mach-Zehnder interferometer[18], which is sketched in figure 6a. Its operation principle as a demultiplexer is following: Light is first split evenly into the two arms of the Mach-Zehnder interferometer by a symmetrical Y-branch. Light then interferes in a so called 4-port hybrid coupler, which consists of a symmetrical Y-branch and an adiabatic asymmetric Y-branch (mode splitter), which was described already in Sec. 3. The wavelength selectivity of the device is based on the arm length difference of the Mach-Zehnder interferometer, which produces phase difference between light from the two interferometer arms. At wavelength λ_1 the phase difference is zero and the local fundamental mode at the 4-port hybrid coupler is excited, and light couples to the wider waveguide of the mode splitter. At wavelength λ_2 the phase difference is π and the 1st order mode is excited, and light couples to the narrower waveguide of the mode splitter. The major advantage of this dual WDM-device is a very low sensitivity to variations in fabrication parameters. It was first demonstrated for 1.3/1.55 μm wavelength demultiplexing[18], and later its suitability also for 1.48/1.55 μm[19] and 0.98/1.55 μm[20] wavelength combining has been demonstrated.

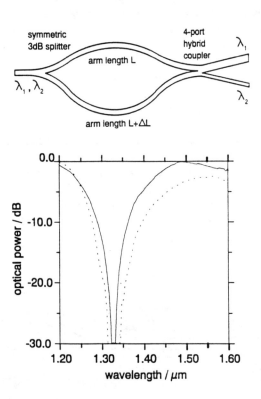

Fig. 6. a) Schematic diagram of an asymmetric Mach-Zehnder interferometer for WDM-application, and b) measured spectral transmission of a WDM-device before (solid line) and after an additional annealing (dashed line).

In Fig. 6b the low sensitivity to fabrication parameters of an asymmetric Mach-Zehnder interferometer dual WDM-device is illustrated[21]. The solid line shows the measured spectral transmission from the 1.55 μm output-port of a 1.3/1.55 μm demultiplexer. This measurement was performed after the initial fabrication process; a short field-assisted silver film ion exchange followed by a 20 min annealing at 340 °C. Afterwards, the device was further annealed for 60 min at 340 °C, and the spectral transmission measurement was repeated. The measurement result is shown with dashed line in the figure, and it is seen that the center wavelength of the "stop band" around the 1.3 μm wavelength remains still the same. It even appears that the "stop band" has become wider after the annealing. This widening, however, is due to additional excess loss especially in bends, which was caused by the long post-annealing.

In order to achieve the low crosstalk in a wide wavelength window, which is usually required, one has to cascade 1.3/1.55 μm asymmetric Mach-Zehnder interferometer WDM-devices. Figure 7 shows the layout and expected characteristics of the Mach-Zehnder type double-filtered fibre-pigtailed 1.3/1.55 μm WDM-device.[22]

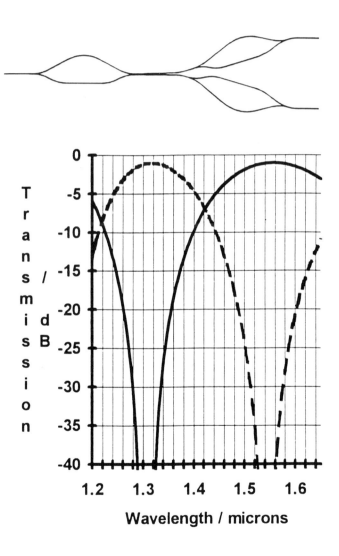

Fig. 7. The layout for 1.3/1.55 μm WDM-device and its expected transmission spectra.

Typically the path length difference in a Mach-Zehnder interferometer WDM-device is realized by using different lengths of curved waveguides in the two interferometer arms. If the path length difference is relatively long, as is the case when the two wavelengths are close to each other, the device becomes somewhat sensitive to fabrication parameters. This is due to the fact that modefields are shifted outwards in bends. This shift is dependent of the fabrication parameters. For example, post-annealing reduces the index difference and mode confinement, and modefield shift in bend is increased. Therefore, the optical path length becomes longer. This sensitivity can be totally eliminated by realizing the asymmetry by using only different lengths of straight waveguides in the two arms of the interferometer (as in the device in Fig. 7). This, however, results in a somewhat longer device. Also, in some cases it would be advantageous to have a possibility to tune the device afterwards. The possibility of tuning the spectral transmission of the asymmetric Mach-Zehnder interferometer WDM-devices by post-annealing was demonstrated in Ref. 23. The studied device was a 1.48/1.55 μm WDM-device having a geometrical path length difference of 10.75 μm, and it was fabricated by potassium and silver double-ion exchange in Corning 0211 glass. Fig. 8a shows the spectral transmission of the device after the initial ion exchange and Fig. 8b after an additional 68 min post-annealing at 300 °C. The tuning possibility is clearly seen.

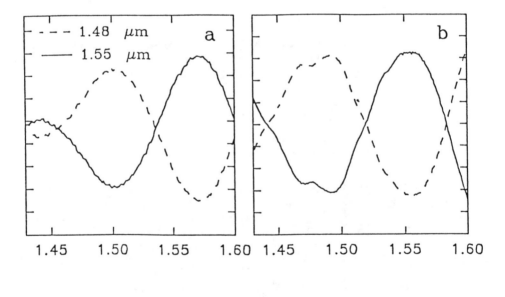

Fig 8. Measured spectral transmission of a 1.48/1.55 μm WDM-device a) before and b) after an additional annealing.

4.3 Asymmetrical adiabatic Y-branch WDM-device

An asymmetrical adiabatic Y-branch can also be employed as a dual wavelength WDM-device. Its fabrication has been proposed, for example, by potassium and silver double-ion exchange[24], and the device is sketched in Fig. 9a. Here, the narrower branch with higher Δn is silver ion-exchanged and the wider branch with lower Δn is potassium ion-exchanged. For proper operation as a dual wavelength WDM-device, the effective mode index at shorter wavelength in the silver ion-exchanged waveguide has to be larger than in the potassium ion-exchanged waveguide, and at the longer wavelength the effective mode index has to be larger at the potassium ion-exchanged waveguide. This situation can be easily achieved, if the two wavelengths to be multi/demultiplexed are far from each other. Although the device seems quite promising with relaxed tolerances for variations in fabrication conditions, no experimental demonstration, to our knowledge, has been reported. However, detailed device designs have been reported for 1.3/1.55 μm and 0.98/1.55 μm WDM-devices.[25,26] As an example, Fig. 9b shows a BPM-simulation for an 0.98/1.55 μm WDM-device[26]. The advantage of this particular wavelength multiplexer for fiber amplifier applications is that the coupling losses, i.e. mode mismatch losses, with different components can be minimized. The potassium ion-exchanged waveguide with large mode size is well-matched with the optical fiber and the narrow silver ion-exchanged waveguide with high Δn is compatible with the 0.98 μm laser diode.

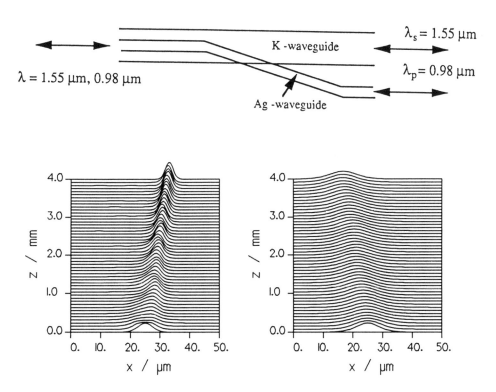

Fig. 9. a) Schematics of a double-ion-exchanged asymmetrical adiabatic Y-branch as a 0.98/1.55 μm WDM-device and b) its BPM-simulation at 0.98 μm (left) and at 1.55 μm (right) wavelength.

5. OTHER COMPONENTS

5.1 Narrow-band wavelength multi/demultiplexers

Passive devices for dense WDM-applications have received a lot of attention recently. They can be realized, for example, by utilizing arrayed-waveguide gratings or by different types of etched gratings in waveguides. The focus with ion-exchanged glass waveguides has been on devices based on etched gratings. One example of the devices recently proposed utilizes several linear gratings (with different grating period) in a slab waveguide. In the device, light is first coupled from the input channel waveguide to a guided plane wave utilizing an adiabatically tapered waveguide region (no lenses are required).[27] After reflection from the Bragg grating light is coupled to the output channel waveguide utilizing a similar taper. With the proposed rather simple waveguide configuration, it is possible to multi/demultiplex several densely spaced wavelengths.

As an example of an efficient Bragg grating in an ion-exchanged waveguide, Fig. 10 shows a measured transmission of a potassium and silver double-ion-exchanged channel waveguide with an etched grating.[28] The grating depth was 0.15 μm and its period was 0.42 μm.

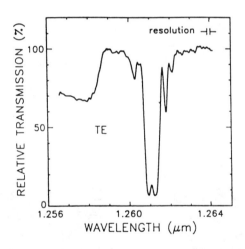

Fig. 10. Transmission spectrum of a double-ion-exchanged waveguide with grating.

In Ref. 5, an interesting narrow-band add-drop WDM-device based on a combination of a Bragg grating and a symmetrical Mach-Zehnder interferometer with 3 dB directional couplers at the input and output of the device was described. Very recently, a similar narrow-band add-drop WDM-device was proposed[29], in which the directional couplers are replaced by four-port hybrid couplers (see Sec. 4.2). In this new add-drop WDM-device the total length of the two arms of the Mach-Zehnder interferometer are the same, but they have a rather long path length difference before and after the Bragg grating as is sketched in Fig. 11a. The device operates as following: non-Bragg-resonant light (wavelengths λ_2, λ_3, ... λ_n) coupled to the input port is split equally, with no phase difference, in the first hybrid coupler to the two arms of the interferometer. Since the total optical path length in the two interferometer arms is exactly the same, light from the two arms arrive to the second hybrid coupler without any phase difference and is coupled to the wider output port of the second hybrid coupler; Bragg-resonant light (λ_1) is reflected by the Bragg grating and, for proper operation, the total phase difference between light from the two arms as they recombine at the first hybrid coupler is π, and light is coupled to the narrower port of the first hybrid coupler. The advantage of the device is that the phase difference can be tuned simply by thermally annealing the device, since the path length difference before the grating is realized by using a curved waveguide in one of the arms enabling thermal tuning (see Sec. 4.2). Fig. 11b shows the calculated phase difference accumulated before the grating as a function of post annealing time at 343 °C. The calculation was performed for silver ion-exchanged waveguide in Corning 0211 glass and for 1.55 µm wavelength. The geometrical path length difference before the grating was about 70 µm. The possibility for tuning is very important, since otherwise even very small errors in alignment of the grating would deteriorate the behaviour of the device. Here, the tuning is possible, since the behaviour of the hybrid couplers is not sensitive to thermal annealing as in the case of 3 dB directional couplers.

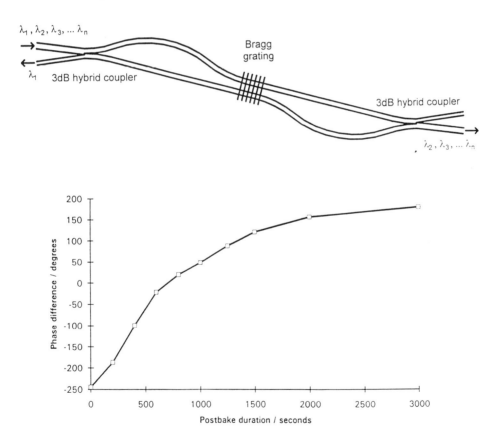

Fig. 11. a) Schematics of an ion-exchanged narrow-band add-drop WDM-device, and b) phase difference between light in the two arms of the Mach-Zehnder interferometer before the Bragg grating as a function of post-annealing time.

5.2 Modefield transformers

In many ion-exchanged devices it is advantageous to utilize waveguides with high Δn resulting in tight mode confinement. This is useful, for example, in ring resonators and other devices having small radius bends, and in waveguide Bragg gratings. However, the small mode size results in unacceptable mode mismatch loss with optical fibers. The mismatch loss can be avoided by tapering the waveguide end to adiabatically enlarge the mode size of the waveguide. An interesting and simple technique to produce modefield transformers in ion-exchanged waveguides was demonstrated in Ref. 30. Low loss modefield transformers were achieved by thermally postbaking silver ion-exchanged waveguides in a suitable temperature gradient. An alternative technique for modefield transformers utilizes waveguide couplers for potassium and silver ion-exchanged waveguides[31]. This type of double-ion-exchanged waveguide coupler is sketched schematically in Fig. 12a. In Fig. 12b the measured mode intensity distributions of the large mode size potassium ion-exchanged and the small mode size silver ion-exchanged waveguide of the coupler are presented. The device was fabricated in Corning 0211 substrate glass and the measurements were performed at 1.3 μm wavelength. The application of demonstrated waveguide coupler is not limited to loss reduction of coupling between an ion-exchanged waveguide and an optical fiber. The full benefit of the waveguide coupler is achieved in applications, in which the advantages of both potassium and silver ion-exchanged waveguide structures in the same device are desired.

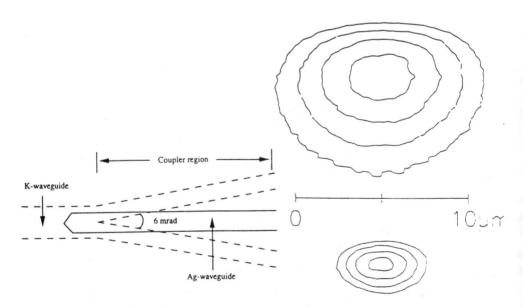

Fig. 12. a) Schematics of a waveguide coupler as a modefield transformer and b) the measured mode intensity distributions at 1.3 μm wavelength of (top) the potassium ion-exchannged and (bottom) the silver ion-exchanged waveguide of the fabricated coupler. Contours are for normalized intensities 0.3, 0.5, 0.7, and 0.9.

5.3 Ring resonators

Ring resonators can be applied in optical communications as optical frequency filters or as narrow-band WDM-devices. Most of the ion-exchanged ring resonators demonstrated utilize a directional coupler as an input/output coupler for the closed loop (ring) waveguide.[32,33] Here, we present a recently demonstrated configuration[34], in which the directional coupler is replaced by an asymmetrical Mach-Zehnder interferometer (discussed in Sec. 4.2) as an input/output coupler. The structure of the ring resonator and its coupler are shown in Fig. 13a. The device was fabricated by potassium and silver double-ion exchange in Corning 0211 glass and its optical behaviour was measured at 1.523 µm wavelength. The resonance curve shown in Fig. 13b was obtained by first heating the device and measuring the output while the device was cooling. The finesse of 5 was estimated from the resonance curve, which gave a loss of 0.17 dB/cm in the 12 cm long ring waveguide.

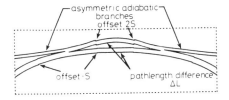

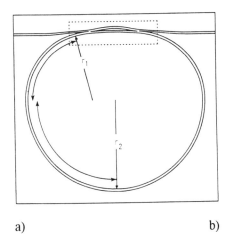
a)

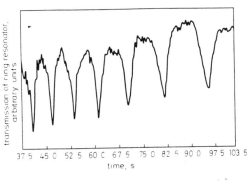
b)

Fig. 13. a) Schematic diagram of the ring resonator and an asymmetrical Mach-Zehnder interferometer as an input/output coupler and b) the measured resonance curve at 1.523 µm wavelegth (when the sample was gradually cooling after initial heating).

The advantage of the new ring resonator configuration is that the wavelength response of the coupler can be easily designed and realized. In the demonstrated device, the asymmetrical Mach-Zehnder interferometer was essentially an 0.98/1.55 µm wavelength multi/demultiplexer, which couples most of the power around 0.98 µm and small part of the power around 1.55 µm wavelength into the ring waveguide. If the device is combined with Er-doped glass overlays or composite waveguides, it can operate as a nearly lossless ring resonator or as a ring laser at 1.55 µm wavelength region when pumped at 0.98 µm wavelength.

6. FIBER ATTACHMENT AND PACKAGING

Attachment of the optical fibres and packaging has been regarded as the "bottle-neck" in low-cost production of passive integrated optical components. There are three approaches for fiber attachment that has been developed:

1; The fibres can be positioned individually by active power monitoring, and then fixed to the waveguides with a suitable bonding material. This technique ensures a very good alignment, but it tends to be time consuming, especially if the number of fibers to be attached is large.

2; The most widely adopted fiber attachment method is to use fiber arrays, and in the alignment to actively monitor the power of the two outer ports only, and then to fix the array to waveguides. Recently, this technique has become more and more popular as the fabrication of high quality fiber arrays has developed rapidly.

3; The most attractive fiber attachment concept is the passive alignment without any power monitoring by using precisely etched fiber guiding grooves on the substrate. This passive alignment is a special feature of silica on silicon technology, since accurate V-grooves can be formed on silicon by anisotropic etching. However, the progress on passive attachment technology has not been as rapid as was believed in the past. In fact, the most important producers of silica on silicon passive integrated optical components utilize the fiber array technique described.

An important issue in the fiber attachment is that the return loss remains high (> 55 dB) after the fiber pigtailing. This can be achieved by using tilted fiber and waveguide endfaces, which ensures that the reflected light is not coupled back to the fiber. The tilted endface technique enables the use of glass materials having a refractive index slightly different from that of the fiber. For example, commercially available glass substrates can be utilized in ion exchange. It is worth mentioning that it may be necessary to use tilted endfaces even if the waveguide index matches accurately the index of the fiber. This is due to the temperature dependence of refractive index of the bonding materials, which may reduce the back-reflection losses of the devices in the field, if tilted endfaces are not employed.

The packaging of passive ion-exchanged integrated optical components has advanced rapidly. For example, all the major manufacturers of ion-exchanged glass waveguide devices have reported reliable performane of their 1xN splitters during and after several mechanical and environmental tests and test sequences. In addition, at least Corning has reported good long term stability of the devices.

7. DISCUSSION

The past 5 years or so have shown that the fabrication processes of passive ion-exchanged integrated optical components are mature for large scale production of passive components for optical communications. Several 1xN splitters and dual-wavelength multi/demultiplexers with high optical performances have been demonstrated by utilizing surface waveguides and buried waveguides. Also, several promising new waveguide configurations for other applications, for example, for dense WDM-applications have been proposed. The progress in fiber attachment techniques has been significant and it seems that the passive ion-exchanged integrated optical components can meet the low-cost requirements of the optical communications networks.

One important issue in passive integrated optical components for optical communicatios is the birefringence of the waveguides. In most cases the device behaviour has to be independent of polarization. Fortunately, there is very little birefringence in the ion-exchanged waveguides and many of the device configurations, such as symmetrical Y-branches for splitters and asymmetrical Mach-Zehnder interferometer for WDM-devices, have low sensitivity to polarization. In some device configurations, however, the waveguide birefringence has to be totally eliminated. One example of such devices is a waveguide Bragg grating for narrow-band WDM-applications. The birefringence in ion-exchanged glass waveguide can be controlled and totally eliminated by utilizing potassium and silver double-ion exchange[35]. This behaviour, in fact, was one of the motivations to utilize these so called "dual-core" waveguides in the waveguide grating study of Ref. 28. On the other hand, there is also need for polarization splitters in optical communications, for example, in polarization-independent coherent optical receivers. It is somewhat difficult to realize polarization splitters by ion exchange in glass due to the low birefringence. However, polarization splitters based on potassium ion-exchanged symmetrical directional couplers have been demonstrated[36]. It is also possible to increase the birefringence by adding overlayers on the waveguides.

To fully utilize the benefits of integrated optics several components should be integrated in a single substrate. This increases the reliability and reduces the overall costs for fiber attachment and packaging. One integrated device, sketched in Fig. 14, was recently demonstrated[37]. The device utilizes asymmetrical Mach-Zehnder interferometers and was fabricated by double-ion exchange in Corning 0211 glass. It involves a 1.48/1.55 μm WDM-device, and a 1.30/1.55 μm WDM-device followed by a 1x8 splitter in a single glass substrate. This integrated optics circuit in connection with an Er-doped fiber may be applied in future WDM access networks. The measured facet-to-facet excess loss in the device was about 2.5 dB, and the splitting uniformity of the 1x8 splitter was about ± 0.5 dB both at 1.30 μm and 1.55 μm wavelengths.

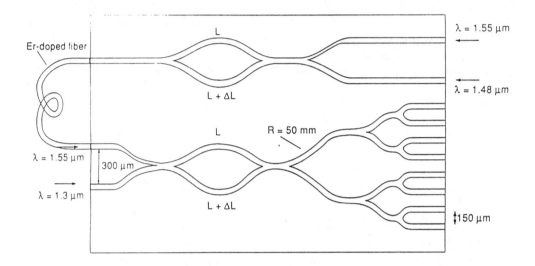

Fig. 14. Schematic diagram of an ion-exchanged integrated optical circuit (40 mm long) in connection with an Er-doped fiber.

10. REFERENCES

1. R. Green, "Status and future prospects of European fibre-in-the-loop deployment", in proc. EFOC&N '93, Optical Access Networks, The Hague, 1993, pp. 35-38.

2. D.W. Faulkner, P.J. Smith, C.A. Wade, J.R. Stern, "Evolution of PON systems", in proc. EFOC&N '93, Optical Access Networks, The Hague, 1993, pp. 69-72.

3. E. Desurvire, "The golden age of optical fiber amplifiers", Physics Today, January 1994, pp. 20-27.

4. A. Neyer, T. Knoche, L. Muller, P.C. Lee, J.H. Kim, M.A. Andrade, J. Carvalho, J.L. Figueiredo, A.P. Leite, "Design and fabrication of low loss passive polymeric waveguides based on mass replication techniques" in proc. ECIO'93, Neuchatel, 1993, pp. 9/10-9/11.

5. S.I. Najafi [ed.], **Introduction to Glass Integrated Optics**, Artech House, Boston, 1992.

6. R. Kazarinov, "Silica waveguides and waveguide devices on silicon: technology and applications", in proc. EFOC&N '93, Invited Plenary Papers, The Hague, 1993, pp. 21-24.

7. A. Tervonen, P. Pöyhönen, S. Honkanen, M. Tahkokorpi, S. Tammela, "Examination of two-step fabrication methods for single-mode fiber compatible ion-exchanged glass waveguides", Appl. Opt. **30**, pp. 338-343, 1991.

8. W.J. Wang, S. Honkanen, S.I. Najafi, A. Tervonen, "Loss characteristics of potassium and silver double-ion-exchanged glass waveguides", J. Appl. Phys. **74**, pp. 1529-1533, 1993.

9. M. Seki, H. Hashizume, R. Sugawara, "Two-step purely thermal ion-exchange technique for single-mode waveguide devices in glass", Electron. Lett. **24**, pp. 1258-1259, 1988.

10. M. Mc Court, "Status of glass and silicon-based technologies for passive components", in proc. ECIO'93, Neuchatel, 1993, pp. 9/1-9/4.

11. H. Takahashi, K. Okamoto, Y. Ohmori, "Integrated-optic 1x128 splitter with multifunnel waveguide", IEEE Photon. Technol. Lett. **5**, pp. 58-60, 1993.

12. S. Kobayashi, K. Nakama, S. Sato, H. Wada, H. Hashizume, I. Tanaka, M. Seki, "Novel waveguide y-branch for low-loss 1xN splitters", in proc. OFC'92, San Jose, 1992, p. 145.

13. H. Takahashi, Y. Ohmori, M. Kawachi, "Design and fabrication of silica-based integrated-optic 1x128 power splitter", Electron. Lett. **27**, pp. 2131-2132, 1991.

14. K. Grosskopf, N. Fabricius, R. Fuest, "Performance of integrated optical singlemode multiport splitters under environmental test conditions", in proc. ECIO'93, Neuchatel, 1993, pp. 9/6-9/7.

15. A. Beguin, P. Laborde, C. Lerminaux, "Ion-exchange in glass components for telecommunication applications: High isolation WDM and 2xN", presented at EFOC&LAN'92, Paris, 1992.

16. H.C. Cheng, R.V. Ramaswamy, "Symmetrical directional coupler as a wavelength multiplexer-demultiplexer: theory and experiment" IEEE J. Quantum Electron. **27**, pp. 567-574, 1991.

17. A. Tervonen, S. Honkanen, S.I. Najafi, "Analysis of symmetric directional couplers and asymmetric Mach-Zehnder interferometers as 1.30- and 1.55 µm dual-wavelength demultiplexers/multiplexers" Opt. Eng. **32**, pp. 2983-2091, 1993.

18. A. Tervonen, P. Pöyhönen, S. Honkanen, M. Tahkokorpi, "A guided-wave Mach-Zehnder interferometer structure for wavelength multiplexing, Photon. Technol. Lett. **3**, pp. 516-518, 1991.

19. A. Tervonen, P. Pöyhönen, S. Honkanen, M. Tahkokorpi, "Integrated optics 1.48/1.55 µm wavelength division multiplexer for optical amplifier application" J. Mod. Opt. **39**, pp. 1615-1618, 1992.

20. W.J. Wang, S. Honkanen, S.I. Najafi, A. Tervonen, "Four-port Mach-Zehnder interferometer in glass" Appl. Phys. Lett. **61**, pp. 150-152, 1992.

21. P. Pöyhönen, Ph.D. Thesis, Univ. Helsinki, Acta Polytechnica Scandinavica Ph 185, 1992.

22. Optonex, preliminary data sheets, 1994.

23. G. Zhang, S. Honkanen, A.Tervonen, S.I. Najafi, P. Katila, "Ion-exchanged glass waveguide Mach-Zehnder interferometers for wavelength multi/demultiplexing and the effect of thermal post-annealing on spectral transmission", Pure and Appl. Opt., in press.

24. S. Honkanen, M.J. Li, W.J. Wang, S.I. Najafi, A. Tervonen, P. Pöyhönem, "Ion exchange processes for advanced glass waveguides", in proc. 1st Intl. Workshop on Photonic Networks, Components and Applications (World Scientific, New Jersey,1991), pp. 428-432.

25. F. Xiang, G.L. Yip, "New Y-branch wavelength multi/demultiplexers by K^+ and Ag^+ ion exchange for $\lambda=1.31$ and 1.55 µm", Electron. Lett. **28**, pp. 2262-2264, 1992.

26. S. Honkanen, S.I. Najafi, W.J. Wang, P. Lefebvre, A. Tervonen, "Integrated optical devices in glass by ionic masking", in SPIE proc. Integrated Optical Circuits, OE/Fibers Symp., Boston, Sept. 1992.

27. S.I. Najafi, M. Kahverad, private communication.

28. M.J. Li, S. Honkanen, W.J. Wang, R. Leonelli, J. Albert, S.I. Najafi, "Potassium and silver ion-exchanged dual-core glass waveguides with gratings", Appl. Phys. Lett. **58**, pp. 2607-2609, 1991.

29. A. Tervonen, S. Honkanen, S.I. Najafi, "New ion-exchanged narrow-band add-drop WDM-device: design considerations, in proc. ICAPT, Toronto, June 1994.

30. A. Mahapatra, J.M. Connors, "Thermal tapering of ion-exchanged channel guides in glass", Opt. Lett. **13**, pp. 169-171, 1988.

31. S. Honkanen, P. Pöyhönen, A. Tervonen, S.I. Najafi, "Waveguide coupler for potassium- and silver-ion-exchanged waveguides in glass", Appl. Opt. **32**, pp. 2109-2111, 1993.

32. R.G. Walker, C.D.W. Wilkinson, "Integrated optical ring resonators made by silver ion exchange in glass", Appl. Opt. **22**, pp. 1029-1035, 1983.

33. J.M. Connors, A. Mahapatra, "High finesse ring resonators made by silver ion exchange in glass", J. Lightwave Technol. **LT-5**, pp. 1686-1689, 1987.

34. W.J. Wang, S. Honkanen, S.I. Najafi, A. Tervonen, "New integrated optical ring resonator in glass", Electron. Lett. **28**, pp. 1967-1968, 1992.

35. S. Honkanen, A. Tervonen, M. McCourt, "Control of birefringence in ion-exchanged glass waveguides", Appl. Opt. **26**, pp. 4710-4711, 1987.

36. A.N. Miliou, R. Srivastava, R.V. Ramaswamy, "A 1.3-µm directional coupler polarization splitter by ion exchange", J. Lightwave Technol. **11**, pp. 220-225, 1993.

37. G. Zhang, S. Honkanen, A. Tervonen, S.I. Najafi, "Glass integrated optics circuit for 1.48/1.55 µm and 1.30/1.55 µm wavelength division multiplexing, and 1/8 splitting, Appl. Opt., in press.

Ion-exchanged glass waveguide sensors

Ludwig Ross

IOT Integrierte Optik GmbH
D-68744 Waghäusel-Kirrlach - Germany

ABSTRACT

Integrated optics in glass are establishing as a reliable new technique with a wide range of applications. The most advanced applications are in the sensor and in the telecommunication field. The paper describes a number of different sensors based on interferometry, absorption effects, spectroscopy, Faraday rotation etc..

A very important property which ensures the usefulness in practice is the reliability and the environmental stability.

1. INTRODUCTION

The driving force for the development of integrated optics was the realization of low loss optical fibers for optical communication techniques. The first definition of integrated optics was introduced by Miller[1] in 1969: "Using photolithographic techniques ... making laser (light) circuits using combinations of elements such as" splitters, couplers, modulators, lasers etc. which can be combined in a planar substrate to give complex optical circuits in monolithic chips. One class of the most favoured substrate materials are special glasses. Their advantage is the optical isotropy so that no interfering polarisation effects are expected. In 1971 Koizumi et al.[2] published the first waveguides made by ion-exchange in a planar glass substrate.

In the following years a number of researcher groups started to develop and to improve special glasses and the ion-exchange techniques to achieve suitable waveguides for several applications. An overview is given for example in ref.[3,4].

While in the first period researchers concentrated on the development of the technological basis after 1980 a more applicational oriented period of development followed. The recent years are characterized by testing and introduction of new products based on integrated optics in glass into the market.

Allthough the technique of integrated optics in glass has the potential to realize nearly any sensor type there are only a few examples for commercial products while in telecommunication a multi million $ market grew for glass integrated optical devices in the last two years. In the sensor field the costs for an integrated optical element are too high if compared with a sensor element realized in a traditional technique. Therefore only a few niches have been identified in which a glass integrated optical device is much superior in technique or gives a unique solution.

In principle glass integrated optical elements can be used to detect nearly any physical or chemical measure. In the next chapters the basic technology to manufacture an integrated optical device in glass, basic waveguide elements, sensor devices for physical measurements and sensor devices for chemical measurements are described. It will be shown that there is a large potential for future applications of glass integrated optics in the sensor field.

2. BASIC TECHNOLOGY

Integrated optics in glass can be made by several different processes. The most intensely investigated process is the ion-exchange process[3,4]. But in recent years also deposition processes have been introduced successfully[5]. Ion-implantation has also been investigated but due to the high expenditure and the low flexibility of the process it became not important[6].

2.1 Glasses and materials for integrated optical devices

The type of substrate material for integrated optics in glass depends on the process used for waveguide fabrication. For the ion-exchange process special optical multi component glasses are used with a sufficient content of exchangeable ions. Typically these ions are Na^+ or K^+ which can be locally replaced by ions like Ag^+, Tl^+ or Cs^+, so that the ion-exchanged regions show an increased refractive index to form the waveguide. Higher electrical charged ions are not preferred because they have a very low mobility even at a temperature near the transition temperature of the glass which is the upper limit for the ion-exchange process.

In the literature standard optical glasses like BK 7 have been described as suitable substrate material[7-9] for the ion-exchange process but also special glasses like BGG 21[10] for Cs-ion-exchange or BGG 31[11] for Ag-ion-exchange have been described. The main components of the special glass BGG 31 are SiO_2, Al_2O_3, B_2O_3, and Na_2O with a partial substitution of oxygen by fluorine. The content of Na_2O is 12.5 mole-% which is fairly enough to achieve a high refractive index increase of about 0.06, when the sodium ions are replaced by silver ions.

In the case of planar waveguides formed by a deposition process it is not necessary that the substrate material has to be a glass because all glass layers (low refractive cladding and high refractive core) which are required to form a waveguide structure can be deposited separately onto the surface of such a wafer. Preferred substrates are single crystalline silicon wafers as they are used in semiconductor techniques. The electronic and micromechanical properties of silicon may allow a combination of the pure optical elements with e.g. electrooptical elements on one chip or may simplify the coupling technique with other optical or electrooptical elements.

But also silica, dielectric crystals like saphire or ceramics can be used as substrate material.

2.2 Ion-exchange process

The ion-exchange process in glass is a long known technique, which has been used to improve the mechanical surface stability of glasses like eye lenses. At elivated temperatures the mobility of one plus ions increases so that these ions can be exchanged during a diffusion process. The most frequently used process is the thermal ion-exchange in a salt melt:

$$A^+ \text{ glass}^- + B^+ NO_3^- ==> B^+ \text{ glass}^- + A^+ NO_3^- \qquad (1)$$
(for A = Na, K; for B = Ag, Tl, K, Cs)

According to the diffusion equation (2) and its solution (3) an error function concentration profile of ions B^+ is formed with decreasing B^+ concentration from the surface to the interior of the glass substrate.

$$dN_A/dt = d/dx \, (D \, dN_A/dx) \qquad (2)$$

where $N_A = C_A / (C_B + C_A)$ is the mole fraction of the ion A^+ in the glass (C = concentration) and D is the diffusion coefficient.

$$C_A(x,t) = C_{A0} \, \text{erfc} \, (x/2Dt) \qquad (3).$$

Covering the glass surface by a metallic or dielectric mask a local control of the diffusion process is achieved. A small slit with an opening of 1 to 2 µm allows to produce a single-mode strip waveguide. The process steps are shown in Fig. 1.

The waveguides formed by this thermal ion-exchange process are placed directly below the surface. These surface waveguides are preferred for sensors which have to measure an optical effect in a sensitive material. Especially chemical or biochemical sensors are based on such an interaction between a surface waveguide and a reaction layer on top of it.

On the other side surface waveguides are disadvantageous for sensor elements in which such surface interactions could disturb the measurement. In this case a second ion-exchange process follows to bury the waveguides below the surface. Such buried waveguides have very low losses and exhibit a very good matching of their nearfield to the optical fiber.

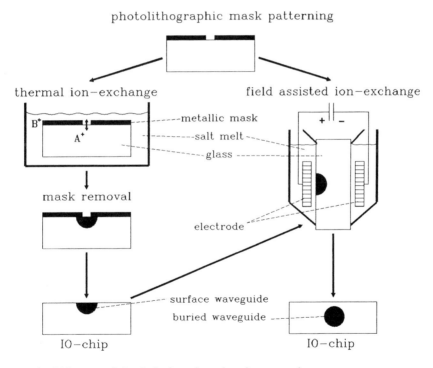

Fig. 1: Waveguide fabrication by ion-exchange

The burial process is a field assisted ion-exchange. After the first ion-exchange the mask is removed from the surface and the wafer is placed between two salt melts, which are electrical isolated. The type of cations of these salt melts is the same as in the original glass (A^+). Applying an electrical field between the two salt melts the ions A^+ enter the glass from the cathode side and leave it at the opposite side. The ions B^+, which are located direcly below the glass surface at the cathode side, are driven by this process into the interior of the glass. Additionally the shape of ion B^+ distribution changes from a semicircular to a more circular shape. The solution of the diffusion equation (4) for the case of a strong electrical field gives a more step index like ion profile.

$$C_A(x,t) = C_{A0}/2 \text{ erfc } ((x - E\mu t)/W_0) \qquad (4)$$

The buried waveguides show very low losses of < 0,1 dB/cm. Another advantage is that especially the silver ion-exchanged buried waveguides in the above mentioned special glass BGG 31 have nearly no mechanical stresses, which is a necessary condition for sensors measuring polarization effects.

2.3 Waveguides by deposition processes

In recent years deposition techniques have been introduced to produce glass waveguides on planar substrates. A number of different methods have been developed:
* flame hydrolysis[12]
* chemical vapour deposition (e.g. PECVD or LCVD)[13]
* sputtering[14]
* sol gel deposition[15]

These different methods have the same principle production steps (see Fig. 2). In the first step a low refracting layer of SiO_2 is deposited on the surface of a planar substrate. Then a layer of a higher refractive material follows. The refractive index increase of this layer is achieved by doping the SiO_2 with TiO_2 or GeO_2. A photolithographic process follows to form the waveguide structure by wet or dry etching the high refracting layer. The last optional step is then a further deposition of a low refracting layer to form a cladding. In the case of an evanescent field sensor this step may be abandonned.

The deposition process gives chemical pure layers. Nevertheless the homogenity of the deposited layer is a critical point to come to very low loss waveguides. Additionally mechanical stresses between the layers and between the lowest layer and the substrate produce polarization effects which could disturb in some applications.

2.4 Packaging techniques for sensors

Integrated optical chips for sensor applications can be used as (a) optical signal processing devices or as (b) sensor elements or as (c) a combination of both functions. In the case of (a) some standard coupling techniques have been developed to couple optical fibers or other optical elements to the endface of the chip. A very stable setup is shown in Fig. 3. It is an array coupling in which the optical fibers are fixed in a high precision V-groove array made of silicon and butt-coupled to the integrated optical chip.

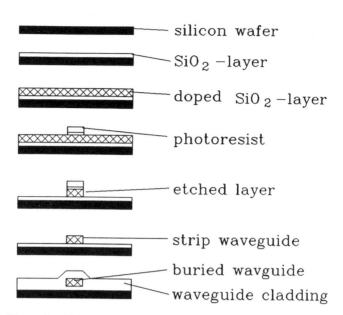

Fig. 2: Waveguide production by deposition processes

Tab. 1: Environmental testing conditions

Test	Temp. [°C]	R.H. [%]	Time[h]
10 temp. cycles	-40 to 85	no cond.	60
high temp.	85	no cond.	336
low temp.	-40	no cond.	336
10 temp. cycles	-40 to 85	no cond.	60
humidity	60	>90	336
56 temp. cycles	-40 to 85	no cond.	336

In a similar way other optical elements like prisms, Selfoc lenses or mirrors are coupled to the IO-chip. Such devices are very stable even in a very harsh environment. Table 1 shows a testing program to demonstrate the environmental stability and reliability. The tests show that an integrated optical device (e.g. a 1:4 splitter) works in a wide temperature range from -40°C to 85°C with only a very small effect of $< \pm 0.2$ dB on the optical transmission properties. Long term tests at an elevated temperature of 60°C and 90% relative humidity, drop tests or vibration tests demonstrate that integrated optics in glass is now a very reliable technique, which fulfils any specification even in a harsh industrial environment. Besides this general coupling technique

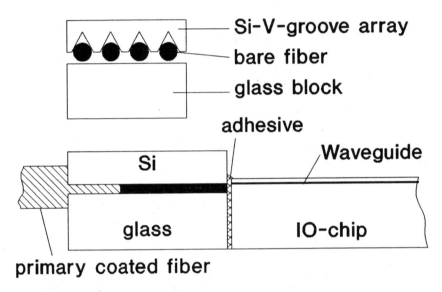

Fig. 3: Fiber chip coupling technique

described above the packaging technique for integrated optical sensors is very complex and depends on the individual application. Therefore it is not discussed here.

3. Integrated optical sensors for physical measurements

Nearly any physical dimension can be measured by an optical method. Miniaturization of the optical setups is now possible based on integration of waveguide structures in optical chips. In the following chapter some examples are described, which demonstrate the potential and the power of integrated optics for such sensor elements.

3.1 Integrated optical distance sensor

Optical interferometry is a well established method to measure distances or a movement along an axis. A setup for this measurement is the classical Michelson interferometer, which is shown in Fig. 4(a).

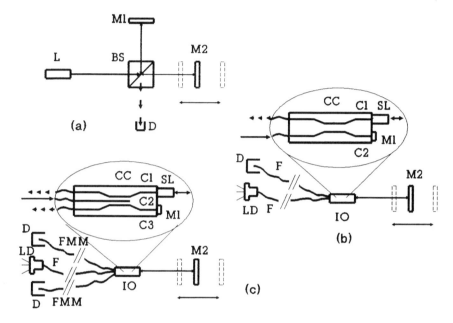

Fig. 4: Michelson interferometer (a) classical setup, (b) elementary integrated optical version (c) advanced integrated optical version.

(BS - beam splitter; C1/C2/C3 - channels 1, 2, 3; CC - coupler chip; D - detector; F - single-mode fiber; FMM - multimode fiber; IO - integrated optical component; L - laser; LD - laser diode; M1 - fixed mirror; M2 -retroreflector; SL - Selfoc lens)

It requires a very stable and precise arrangement of its optical elements. This makes it heavy and expensive and it excludes a number of important applications.

Some researcher groups developed monolithic optical chips on glass integrating the main elements of a Michelson interferometer. The most elementary arrangement is shown in Fig. 4(b). It has been demonstrated by Jestel et al.[16]. The laser light is guided by a single-mode fiber to the integrated optical chip which has a simple coupler structure. The coupler splits the light into two channels. At the end of channel 2 the light is reflected by a mirror which is fixed at the endface of the chip. Channel 1 ends at a Selfoc lens making the divergent beam parallel. The mobile retroreflector at the end of the distance to be measured reflects the beam back to the Selfoc lens which focuses it into the waveguide channel 1. The two back reflected beams in channel 1 and 2 come together in the coupling region to interact and give an interference signal into the output channel which is coupled through a fiber to a detector.

The interferometric signal gives the measure for a movement of the retroreflector. A uniform movement of the retroreflector gives a sine curve with maxima and minima. Unfortunately the singleness of direction is lost in the maxima and minima of the curve. Additionally the sensitivity is reduced in the flater parts of the sine curve near the maxima and minima. This problem can be solved by frequency modulation of the laser diode which is a very expensive and critical technique.

In the literature some waveguide structures are discussed to solve the problem in an all optical way. The proposals for a passive phase diversity detection are based (a) on a 3x3-coupler proposed by Fuest.[17] (see Fig. 4(b) and 5(a)), (b) on a double coupler structure with two reference channels of $1/8\lambda$ path length difference[18] (see Fig. 5(b)), (c) on a matrix of 4 cross connected 2x2-couplers and a 90° phase shifter between two of the couplers[19] (see Fig. 5(c)) and on an imaging film waveguide generating 4 quadrature signals[20] (see Fig. 5(d)).

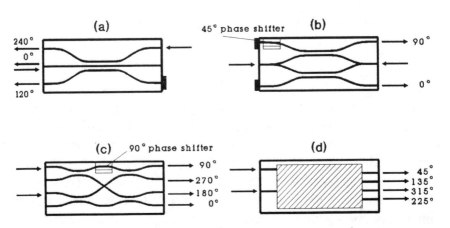

Fig. 5: Passive integrated optical chips for signal processing

Here the solution based on the 3x3-coupler chip is discussed in detail. This version of an integrated optical Michelson interferometer has a single-mode input fiber coming from a stabilized laser diode and 2 or 3 multimode output fibers guiding the output signals to the detectors and electronic signal processor.

The incoming light is split into the channels 1 and 3 while channel 2 is terminated. At the end of channel 3 is a fixed mirror, which reflects the light back into channel 3. Channel 1 ends also in a selfoc lens which makes the divergent beam parallel (see above). The backreflected beams come together in the coupling region to interfere. In the case of the 3x3-coupler the interference of the two beams produces interferometric signals in all 3 channels on the left side of the integrated optical chip shown in Fig. 4(c). The 3 signals are characterized by a phase shift of 120° (Fig. 6). Comparison of at least two of the three signals gives the singleness of the direction over the total measuring range of the interferometer.

This Michelson interferometer is a commercially product (IOT and EUCHNER, both Germany). The technical specifications for the interferometer are shown in Tab. 2.

In the standard version signal processing is done by counting the maxima, so that the resolution is $\lambda/2$. By interpolation of the interferometric signal a much higher resolution can be achieved.

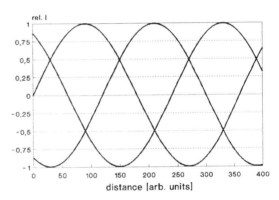

Fig. 6: Interferogram of a Michelson interferometer based on an integrated optical 3x3-coupler

Tab. 2: Technical properties of the integrated Michelson interferometer from Euchner based on a 3x3-coupler chip from IOT

measurement range	max. 500 mm
wavelength	780 nm
resolution	0.39 μm (standard) 0.016 μm (option)
relative accuracy	1 ppm
speed maximum	45 m/min

At any position of the retroreflector at least one of the three phase shifted signals is in the very sensitive and nearly linear region of the sine curve, so that the signal processing is very simple.

The 3x3-coupler chip is fabricated by Ag-ion-exchange in BGG 31 in a 2 step ion-exchange process. The second ion-exchange is also a thermal ion-exchange process in a pure $NaNO_3$ melt. This second ion-exchange is carried out to bury the waveguides slightly below the surface and to achieve a better coupling efficiency.

The integrated optical 3x3-coupler chip is coupled to the optical fiber by an array coupling technique. The Selfoc lens and the mirror are also mounted on a glass block and butt-coupled to the chip. The device is incapsulated in a metallic tube with a diameter of 17 mm and a length of about 90 mm. The optical connection to the electrooptical and signal processing unit is a stable optical cable which allows a flexible operation with nearly no restriction on the distance between sensor element and control unit. This miniaturized sensor element is very useful for high precision measurements in very compact setups like tools maschines, microscopes or semiconductor stepper maschines.

The integrated optical Michelson interferometer is a universal optical element for a wide range of sensor applications. In combination with appropriate setups like a membran for pressure detection, like a magnetostrictive layer on top of the measuring channel for magnetic fields, like a cuvette for refractive index measurement, like two different covering layers on top of the two interferometer channels for temperature or like scales for weighing.

3.2 Integrated optical current sensor

In dielectrical materials like optical glasses a linear polarized beam is rotated when a magnetic field is applied. This Faraday effect can be used to build up a current sensor. For a bulk optical setup the same disadvantageous are identified than for the classical optical interferometer described in chapter 3.1.

An integrated optical waveguide structure, which has been designed for high power current detection, is shown in Fig. 7.

An extremely low birefringence of the waveguide in the glass chip is a basic condition for this element. In BGG 31 buried waveguides have been fabricated with a stress induced part of birefringence of about $\Delta n \approx 2 \times 10^{-6}$ and a temperature dependance of $\delta \Delta n < 1 \times 10^{-7}$ in the temperature range -40 to 80 °C[21].

The very low birefringence of the buried BGG 31 waveguides is also used in a similar sensor system with a fiber coil surrounding the electric power cable. For the sensor system two single-mode splitters are required. The splitters work nearly neutral so that their influence on the polarization state of the guided wave can be neglected.

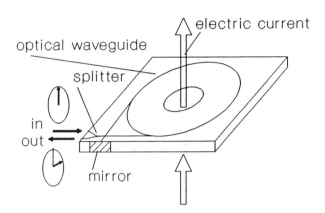

Fig. 7: Integrated optical current sensor

3.3 Integrated optical refractive index sensor

Refractive index sensors are very useful to control chemical processes in which the concentrations of two different chemicals in a liquid mixture has to be controlled.

Two different interferometer types have been proposed to detect changes in refractive index. The Mach-Zehnder interferometer and the Fabry-Perot interferometer[22] are shown in Fig. 8. Both sensors use the interaction between the guided light of a surface waveguide and a superstrate (liquid mixture).

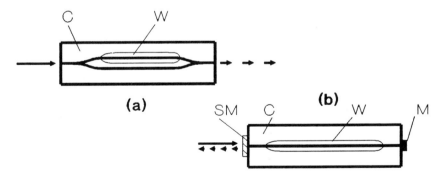

Fig. 8: Integrated optical (a) Mach-Zehnder interferometer and (b) Fabry-Perot interferometer - (C cladding; M mirror; SM semitransparent mirror; W window)

The penetration of the evanescent field of the guided wave depends on the refractive index of the superstrate. The effective mode index changes according to the refractive index change of the superstrate, so that the phase of the guided wave is shifted. The phase shift is given by equation (5).

$$\Delta\phi = k_0 L * \Delta n_{eff} \qquad (5)$$
$$\text{with } k_0 = 2\pi/\lambda_0$$

The Mach-Zehnder interferometer type refractive index sensor has been fabricated by Cs-ion-exchange in the glass BGG 21[23]. A single-mode surface waveguide splits into two arms which join by a second Y-branch to form the output channel. The chip is covered by a low refracting cladding layer of SiO_2 except a small window on one of the two interferometer arms. A cuvette is mounted on top of the chip so that a liquid can flow through and can come into contact with the surface waveguide in the window area.

The measuring range of the refractive index sensor is limited by the refractive index of the substrate glass BGG 21 (< 1.45). The sensitivity increases with increasing refractive index of the superstrate. At $n = 1.3$ the accuracy is 5×10^{-4} and increases to 5×10^{-5} at $n = 1,44$.

3.4 Rotational sensor by an integrated optical ring resonator

Optical gyroscopes are very sensitive rotational sensors which are mainly used in long distance navigation. These high precision laser gyros exhibit extremely low shifts and are very expensive. For applications with lower requirements on accuracy and stability (e.g. car navigation) an integrated optical ring resonator has been proposed[24,25].

Fig. 9 shows the principal structure of the integrated optical chip.

The Sagnac effect causes a difference in optical path length for the contrary rotating waves coupled into the ring so that a drift of the resonance signal can be observed. To come to an unambigious signal the two waves have to be modulated. This can be done by piezoelectric or thermooptical modulation of the guided wave.

4. CHEMICAL AND BIOCHEMICAL INTEGRATED OPTICAL SENSORS

Monitoring of environmental pollution becomes extremely important and is called in by a rapid increasing number of laws and regulations.

Optical methods exhibit a lot of advantages for such sensor applications especially when they can be miniaturized. Integrated optics in glass is one of the most powerful basis to develop such sensors. A number of researcher groups are working on this field and seem to be at the beginning of a technological breakthrough.

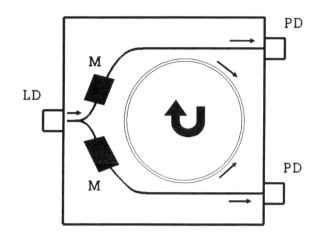

Fig. 9: Integrated optical Sagnac interferometer

Besides monitoring environmental pollution in air and water other applications for integrated optical chemical or biochemical sensors like process control, remote spectroscopy, biomedical diagnostics and biotechnology are of great interest.

There are only a few basic optical principles which are used for developing a specific sensor element. However the number of specific chemical reactions is nearly as high as the number of the different chemicals to be detected. Therefore the following text will concentrate on the different optical aspects but not on the numerous chemical reaction mechanisms.

4.1 Absortive sensors

The characteristic spectroscopic behaviour of a chemical compound can be used to detect this compound and to determin its concentration in a mixture with one or more other compounds. The glass integrated optical basic structures which can be used for such measurements are shown in Fig. 10.

Classical spectroscopic measurements are based on the transmission of light through the analyte, which has to be investigated. Surface waveguides allow another method, in which the light does not completely pass the analyte. The majority is guided in the surface waveguide, while a smaller portion of the guided wave penetrates the analyte on top of the

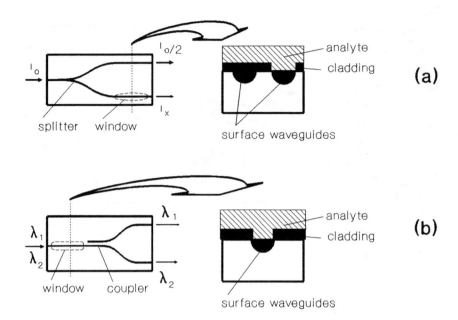

Fig. 10: Integratred optical chips for chemical sensing by absorption effects (a) based on a splitter (b) based on a directional coupler

waveguide (superstrate - see Fig. 11). This evanescent field effect has been described in chapter 3.3. In the case of sensors based on a spectral absorption the guided wave is attenuated according to this absorption.

The sensor shown in Fig. 10(a) is based on a 1:2 splitter. The light is split into two arms, a reference arm and a measuring arm. The constant intensity of the reference arm is compared with the intensity I_x of the measuring arm which is reduced by the absorbing analyte. The ratio $I_x/(I_o/2)$ is a measure for the spectral absorption effect of the analyte.

In Fig. 10(b) a second

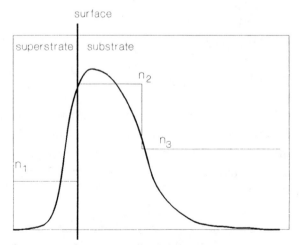

Fig. 11: Nearfield pattern showing the evanescent field effect

basic structure for spectral chemical sensing is shown. It can be used to detect the spectral change of an indicator dye. Light of two different wavelengths (λ_1 close to the absorption maximum of state 1 of the indicator dye and λ_2 close to the absorption maximum of state 2) is coupled into a waveguide, which has an interaction window followed by a wavelength devision coupler. The two wavelength are separated into the two output arms of the integrated optical chip and guided to photodiodes to measure the intensities $I(\lambda_1)$ and $I(\lambda_2)$. The ratio of both intensities gives the state of colour of the dye, which is a measure for the concentration of an analyte showing a specific reaction with the dye.

A more complex sensor is a spectrometer. The integrated optical version of this classical analytical instrument is shown in Fig. 12. The polychromatic light transmitted through a gas or a solution to be analysed is guided to the integrated optical chip via an optical fiber. The diverging beam is focused by a mirror on a grating which splits it into its spectral components. The separated parts are reflected and focused to N output channels. In the example shown in Fig. 12 thirteen multimode output fibers are coupled to the chip and guide the light to 13 photodetectors. This integrated optical spectrometer has been designed for the wavelength range from 1100 to 1500 nm. A dense arrangement of the

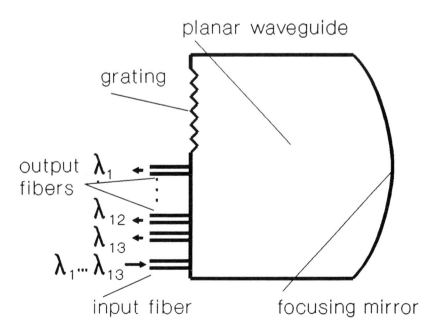

Fig. 12: Integrated optical spectrometer

multimode output fibers gives a high yield of light for each channel. The band width of each channel is 30 nm and the separation is better >20 dB. A much higher resolution can be achieved, when the output fiber array is substituted by a linear photodiode array.

4.2 Interferometric chemical and biochemical sensors

In chapter 3.3 a refractive index sensor based on a Mach-Zehnder interferometer has been described. Covering the window field with a solid reactive layer gives the basis for a class of chemical sensors with high sensitivity and high selectivity[26]. Depending on the reactive layer such a sensor can detect chemicals in gaseous or liquid mixtures.

Polysiloxane layers exhibit a very specific absorption behaviour when they come into contact with halogenated or non halogenated hydrocarbons. The absorption depends on the concentration of the compound in a gas mixture and changes the refractive index of the layer. Gauglitz and Ingenhoff[26,27] investigated the typical absorption of thin layers of the commercial available siloxanes PDMSH, PDMSV, VPMT and VMPS for an number of chemicals like n-pentane, n-hexane, n-heptane, n-octane, dichloremethane, trichloremethane, tetrachloremethane, toluene and xylene. Each siloxane layer shows a typical refractive index change for each of these chemicals depending on the concentration in a gas mixture. For a multicomponent sensor Ingenhoff[26] proposed a sensor array with four parallel Mach-Zehnder interferometers each covered with one of the above mentioned polysiloanes (Fig. 13). The complex signals can be processed using a method like the partial least squares or some similar mathematic methods to give a concentration value for each component of the analyte mixture.

For the detection of pesticides or other organic pollutants in water an enzymatic reaction layer covering the Mach-Zehnder window

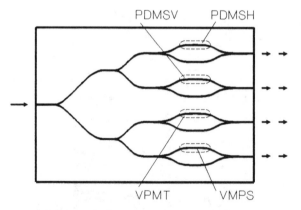

Fig. 13: Mach-Zehnder interferometer array for detection of halogenated and non halogenated hydrocarbons

can be used. Some prototypes of such a biochemical sensors have been demonstrated. The most critical part of such a sensor is the stability and reproducibility of the reactive layer. In several research institutes the preparation of these layers is under development.

The signal of a Mach-Zehnder interferometer has the disadvantage, that it is not unambiguous in the maxima and the minima of the interferogram as in the case of the Michelson interferometer described in chapter 3.1. Therefore an improved waveguide structure is proposed which com-

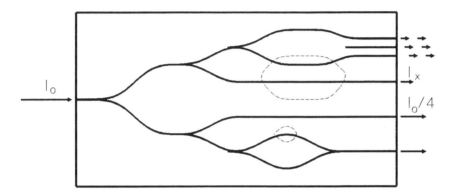

Fig. 14: Integrated optical chip for chemical sensors

bines a high resolution Mach-Zehnder interferometer with a 3x3 coupler to give the quadrature signals, a short Mach-Zehnder interferometer with only a 1/4 interferometric period over the total measurment range and two straight waveguides for absorption control.

For these elements only one laser is required. The light is transmitted through a single-mode fiber, which is coupled to the integrated optical chip and split into 4 waveguide channels guiding to the 4 sensor elements. The complete sensor structure is shown in Fig. 14.

The two examples demonstrate the potential of the technique of integrated optics in glass. A combination of several functions can be integrated on a single chip to give a high sophisticated sensor element.

5. REFERENCES

1. S. E. Miller. "Integrated optics - An introduction". Bell Syst. Techn. J. Vol. 48. pp. 2059-2069, 1969.
2. T. Sumitomo, S. Matsushita, K. Koizumi, M. Furukawa. "Dielectric miniaturized optical circuits made by ion-exchange methods". Jap. Soc. Appl. Phys., The Science Lecture Meeting 1971, p. 83.

3. L. Ross. "Integrated optical components in substrate glasses". Glastechn. Ber. Vol. 62 pp. 285-297, 1989.

4. S. I. Najafi. (Editor) "Introduction to glass integrated Optics" Artech House, Boston, London 1992.

5. T. Ikegami, M. Kawachi. "Passive paths for networks". Physics World, Sep. 1991, p. 50-54.

6. P. D. Townsend. "Effects on ion beam processing on optical properties". In C. J. McHargue et al. (ed.): "Structure properties in surface-modified Ceramics". Kluwer Academic Publ., pp. 355-370, 1989.

7. R. G. Eguchi, E. A. Maunders, J. K. Naik. "Fabrication of low-loss waveguides in BK 7 by ion-exchange". SPIE Proc. Vol. 408, pp. 21-26, 1983.

8. H. F. Schlaak, A. Brandenburg. "Integrated optical circuits with curved waveguides". SPIE Proc. Vol. 651, pp. 38-45, 1986.

9. O. G. Leminger, U. Rothamel, R. Zengerle. "Characterization of ion-exchanged glasses". SPIE Proc. 651, pp. 51-57, 1986.

10. L. Ross, H.-J. Lilienhof, H.-J. Hölscher et al.. "Improved substrate glass for planar waveguides by Cs-ion-exchange". Topical Meeting on Integrated and Guided-Wave Optics IGWO, Atlanta, Techn. Dig. THBB2, 1986.

11. N. Fabricius, L. Ross et al.. "BGG 31 - A new glass for multimode waveguide fabrication". EFOC/LAN, Amsterdam, Proc. pp. 59-62, 1988.

12. M. Kawachi. "Silica waveguides on Silicon and their Application to integrated optic components". Opt. Quant. Electron. Vol. 22, pp. 391-416, 1990.

13. M. Hanabusa, Y. Fukuda. "Single-step fabrication of ridge type glass optical waveguides by laser chemical vapor deposition". Appl. Opt., Vol. 28, pp. 11-12, 1989.

14. M. Giudice, F. Bruno et al.. "Structural and optical properties of silicon oxynitride on silicon planar waveguides". Appl. Opt., Vol. 29, pp. 3489-3496, 1990.

15. D. W. Hewak, J. Y. Lit. "Fabrication of tapers and lenslike waveguides by microcontrolled dip coating procedure". Appl. Opt., Vol. 27, pp. 4562-4564, 1988.

16. D. Jestel. "Integrated optical Michelson-interferometer in glass". SPIE Vol. 1014, pp. 31-34, 1988.

17. R. Fuest. "Integrated optical Michelson-interferometer with quadrature phase demodulation in glass for displacement measuring" tm - Techn. Messen, Vol. 58, pp. 152-157, 1991.

18. W. Gleine, J. Müller. "Integrated optical interferometer on silicon

substrate". ECO 1, Hamburg, SPIE Vol. 1014, pp. 9-16, 1989.

19. D. W. Stowe, T.-Y. Hsu. "Demodulation of interferometric sensors using fiber-optic quadrature demodulators". J. Lightw. Techn. Vol. LT-1, p. 519 1983.

20. P. Roth. "Passive integrated optic mixer providing quadrature outputs". SPIE Vol. 1141, pp. 169-173, 1989.

21. R. Stierlin, F. Horst, R. Fuest, N. Fabricius. Birefringence in buried waveguides". OSA Annual Meeting, Boston, 1990.

22. W. Konz, A. Brandenburg, R. Edelhäuser, W. Ott, H. Wolfschneider. "A refractometer with a fully packaged integrated optical sensor head". in H. J. Arditty (editor) Optical fiber sensors, Springer, Berlin, pp. 443-449, 1989.

23. U. Hollenbach, C. Efstathiou, N. Fabricius, H. Oeste, H. Götz. "Integrated optical refractive index sesnor by ion-exchange in glass". ECO Conf. Hamburg, SPIE Vol. 1014, pp. 77-80, 1989.

24. R. G. Walker, C. D. W. Wilkinson. "Integrated optical ring resonator made by silver ion-exchange in glass". Appl. Opt. Vol. 22, pp. 1029-1035, 1983.

25. A. Mahapatra, J. M. Connors. "High finesse ring resonators - fabrication and analysis". SPIE Vol. 651, pp. 272-275, 1986.

26. J. J. Ingenhoff. "Charakterisierung organischer, anorganischer und biologischer Superstratdeckschichten mit Hilfe integriert optischer Interferometerbausteine". Thesis, University of Tübingen, 1994.

27. G. Gauglitz, J. J. Ingenhoff. "Integrated optical sensors for halogenated and non halogenated hydrocarbons". Sensors and Actuators B Vol. 1, p. 207, 1993.

Commercial glass waveguide devices

Martin Mc Court

Corning Opto-Electronic Components
15 rue du Marechal Juin
77000 Melun, France

1 INTRODUCTION

Integrated optics is a generic term used to describe guided wave devices combining several functions on a single substrate. Passive integrated optics, or "planar", refers to a family of devices that requires no external power source to perform their functions. Their most promising short-term application is fiber-based telecommunications systems. Examples are power splitters and wavelength division multiplexers (WDMs). For the purposes of this review, we will consider devices in which light enters and exits via optical fiber pigtails.

A new technology, such as integrated optics, succeeds when it reaches technical maturity in time to meet a market requirement. Improved performance over existing technologies is necessary to attract the investment required to industrialise. Promising technologies can remain at the prototype stage due to the lack of one of these elements.

The objective of this paper is to review the status of single-mode passive integrated optic technology with respect to technical maturity, level of industrialisation and performance with respect to the emerging market requirements of fiber in the loop (FITL).

2 PASSIVE INTEGRATED-OPTIC TECHNOLOGY

There is a perception that there are two passive integrated-optic technologies; glass or silicon devices. This classification is based on the availability of commercial devices. This terminology only addresses the choice of substrate material. The reality is more complex. In both cases waveguides are formed from silica containing oxides, i.e. *glass*. In addition, the waveguide fabrication process, or pigtailing approach, is *not* uniquely linked to the choice of substrate. An alternative frame of reference is to divide the process into two steps, wafer processing and pigtail attachment.

2.1 Wafer Processing : Ion Exchange and CVD

The two most common wafer fabrication processes are chemical vapour deposition (CVD) and ion exchange. CVD is a *Deposition* process, Ion Exchange a *Diffusion* process:

Planar CVD is illustrated in Fig 2.1. A chemical reaction in the gaseous phase produces silica and oxydes particles that are deposited on the substrate. Silicon, widely used in micro-electronics, is the most common substrate, though silica is also used (1). Precision V grooves can be chemically etched in silicon due to its crystalline structure, a useful feature for fiber attachment.

The CVD reaction can be initiated either by high temperatures (Flame Hydrolysis or FHCVD above 1000 C) or by a plasma at lower temperatures (PECVD, < 500).

Due to the relatively large particle size produced by the reaction, FHCVD layers are porous and must be consolidated at high temerature (~1000 C) to produce waveguides.

With PECVD, the deposed layer is already "glassy" and consolidation is not mandatory. By totally avoiding high temperature processing, PECVD enables the fabrication of hybrid opto-electronic circuits. Electronic circuits can be diffused into the silicon before depositing the waveguide as they are not adversely affected by the further processing.

In both cases, channel waveguides are formed by etching ridges in the deposited layer. The structure is symmetrised by depositing a uniform "overcladding" layer of pure silica.

Ion Exchange in glass is illustrated in Fig 2.2. Ion exchange is a diffusion process. The substrate is an optical glass containing alkaline oxydes. The glass substrate is immersed in a dopant salt bath at elevated temperatures, typically > 400 C (2). Common dopants ions are Thallium and Silver. Driven by the concentration gradient, dopant ions diffuse into the substrate. To preserve charge neutrality, glass metal cations diffuse out. The result is a glass composition locally modified to produce waveguides. The waveguides can be buried by applying an electric field (the cations migrate under the action of the electric field).

Low-loss waveguides have been produced both by CVD and ion exchange. Losses as low as 0.1 dB/cm for FHCVD (3) and < 0.05 dB/cm for ion exchange (4) have been reported. The difference in propagation losses may be due to differences in waveguide "wall" roughness. With CVD, the guide walls are directly formed by etching the waveguide layer. There is a direct mapping of the photolithographic mask to the waveguide. Mask roughness is directly transferred to the waveguide wall. With ion exchange, diffusion "smoothes" the walls.

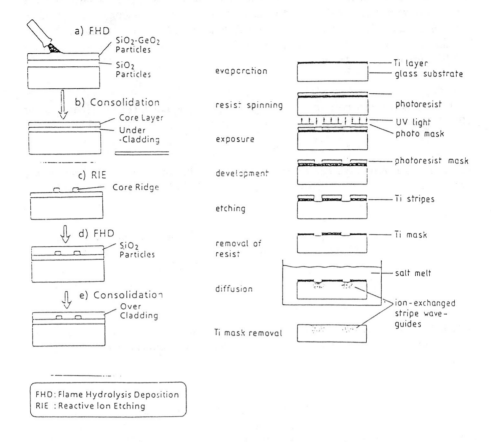

Fig 2.1 CVD process steps   Fig 2.2 Ion exchange process steps

Coupling losses to single-mode fibers of < 0.1 dB have been reported with both processes. Differences in polarisation dependant loss (PDL) have been measured on commercial devices. PDL < 0.1 dB is typical of ion exchange in glass; however, PDL up to 0.5 dB has been measured with CVD devices. This may be due to stress birifringence caused by expansion coefficient differences between the silicon substrate and the CVD layer.

2.2 Fiber Attachment

In mechanical terms, pigtailing is a process in which the fiber optical axis must be aligned with the waveguide optical axis and maintained in this position. In other words, the fiber occupies a specific (x,y,z) coordinate in space. Three approaches are possible:

(a) *Active Alignment*: The substrate contains no precise alignment features (x,y,z coordinates not defined by the substrate). The fiber pigtails are individually aligned by power peaking. A material is used to fix the fiber position relative to the chip, Fig 2.3a (5). During pigtailing this material is fluid, allowing fiber movement. When alignment is completed the material solidifies to fix the fiber position. Two types of materials have been used extensively: a) Organic materials that are polymerised by heat or light and b) low melting point metal alloys (heated to melting point during pigtailing then solidified).

Active Alignment is therefore based on two technologies, automatic fiber alignment of multiple fibers and stable bonding materials.

(b) *Semi-Active Alignment*: The fiber alignment is split into two operations. First, the fibers are placed in blocks containing precise alignment features, Fig. 2.3b (i.e. their (x,y) coordinates are defined *with respect to one another*)(6). The fiber blocks then are actively aligned with respect to the chip, typically monitoring the outer ports only. In this way N fibers are aligned with a single power peaking operation. After power peaking, the blocks are fixed in place with respect to the chip. The materials requirements to immobilise the blocks are similar to those for individual fibers.

Semi-Active Alignment is therefore based on three technologies; precision fiber block fabrication, automatic block alignment and stable bonding materials.

Fig 2.3 Fiber attachement : a) active b) semi active c) static

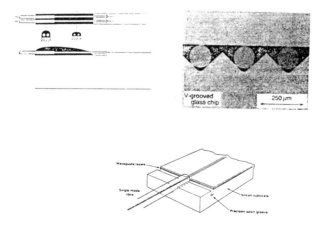

(c) *Static Alignment*: The substrate contains precise mechanical alignment features, fixing the fiber position in space, Fig. 2.3c ((x,y,z) defined). The features are typically V-grooves etched or molded into the substrate (7,8). With CVD on silicon the grooves can be formed by preferential etching and ultimately the intention is to fusion splice the pigtails and waveguides together, as both are composed of doped silica (8).

Static alignment is conceptually attractive. However, all process steps are performed on the same element, giving an overall yield that is the product of the individual step yields. A second drawback is that the fibers must be positioned to enter the alignment feature.

Static Alignment requires two key technologies: an alignment feature fabrication process (.i.e. silicon V grooves) and a fiber bonding material.

All three approaches have been developed and applied to practical devices. No single approach has demonstrated a clear advantage at this time.

3 Industrial Maturity

Three phases have been used to classify industrial maturity:

R+D	Corporate or institutional research and development Laboratory prototypes, low commercial activity
Pilot Production	Products defined and commercialised in low volumes
Manufacturing deployment	Products qualified, industrial supply contracts, real

The following table is an attempt to classify current integrated optic activity. The objective is not to compare companies in competitive terms, rather to indicate the general level of activity. The classification is based on a cross section of commercial literature and scientific publications. It is clear from this table that no one technology predominates.

Fig 3 Industrial Integrated Optic Activity

Company	Wafer Process	Pigtailing	Status
ATT	CVD (FH and PE)	Static	R+D
CSEM	Ion exchange		Pilot
Corning	Ion exchange	Active	Manufacturing
GEC	PECVD	Semi-active	R+D (1)
IOT	Ion exchange	Semi-active	Manufacturing
LETI	PECVD		Pilot
BNR Europe	PECVD	Static	R+D (8)
NSG	Ion exchange	Semi-static	Pilot
NTT	CVD		R+D
PIRI	FHCVD	Semi-static	Pilot (3)

(*) Numbers refer to published references

4 Markets for Passive Integrated Optics in the 1990s

Historically, fiber optics have been used for high bit rate, point to point (PTP), transmission systems. The opportunity to use branching components is limited. Passive components were first introduced to these systems to allow capacity increases, either by upgrading to bidirectional or duplex operation at the same wavelength using directional couplers or by adding a second wavelength with WDMs. Alternative passive technologies include fused biconic taper (FBT) for 1x2's and micro-optics for high isolation WDMs. Planar integrated optics have been deployed with success in this application (9). Cable plant trouble shooting is facilitated by the excellent reproducibility of planar 1x2 insertion loss (differences in insertion loss between different devices is measured in tenths of a dB).

In 1992, significant commitments were made to fiber in the loop (FITL). Ameritech, a Regional Bell Operating Company (RBOC), committed to tripling its fiber base by 1996. British Telecom tests deploy fiber to the business in 1993/94 and plans to install more than 3,000,000 km of new fiber by 1997. In Germany, Deutsche Telekom installed 220,000 lines over fiber in 1993, with a further 500,000 lines/year contracted for 1994.

These point to multipoint (PTM) or passive optical network (PON) applications are an ideal opportunity for passive planar devices. Couplers are used to share expensive head end equipment among many subscribers. Optical splits of 16 and 32 are typical. The cable plant, including the splitter, must operate in both the 1310 and 1550 nm fiber windows. to ensure a future-proof network, optical passbands should be as wide as possible, all the way from the head end to the subscriber. In the United States, Bellcore requires 100 nm passband in both windows (10). Uniformity also is critical, especially when the network carries AM cable TV. AM cable TV receivers require precise power levels to operate correctly. Too little power and the detected signal is noisy, too much signal and the receiver saturates. The uniformity requirements are even more critical when cascaded fiber amplifiers (EDFAs) are used. Values as low as 1 dB for a 16-way splitter are requested.

In FITL, trials both planar and FBT couplers have been used. With FBT technology, up to 15 1x2 devices must be fusion spliced together with 14 splices to make a 16 way splitter array. It is probable that the "*complex and intricate handling required to assemble these products will always tend to be costly*" (11). Reliability is also a concern. In comparison, a single planar monolithic 1x16 device is required, "*giving a major advantage in reduced physical size over the fused fiber devices*" (9). Optical performance of a planar 1x16 is shown in Fig 4. Package size is only 56 x 7.5 x 5 mm.

Fig 4 Spectral attenuation of planar monolithic SM1X16

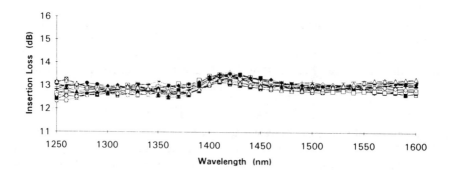

5 Conclusions

Each planar wafer process has specific strengths and limitations. Ion exchange appears ideal for power splitters and WDMs due to the low loss and process robustness. FHCVD has potential in waveguide engineering (designing different waveguides for different applications). PECVD is especially suited to hybrid opto-electronic devices.

Substantial progress has been made in the area of pigtail attachment in the last decade. Integrated optic devices have been in use for almost five years. Long-term reliability of the devices has not been proven, but both field and lifetest data are encouraging (5,12).

Planar technology for optical components is comimg of age in the 1990s. Superior optical performance has been demonstrated and industrial manufacture has begun. With the advent of FITL deployment a market now exists. After years of scientific debate the market place will decide which process gives the best performance at the lowest cost.

6 References

(1) J Allen, N Nourshagh, "Laser Welded Fiber-Waveguide Interfacing", EFOC/LAN 91
(2) L Ross, "Integrated optical components in substrate glasses", Proceedings of 62nd annual meeting of the German Society of Glass Technology, 1988.
(3) Piri commercial literature
(4) M Mc Court, A Cucalon, A Beguin, P Laborde, C Nissim, "Optical and environmental performance of packaged SM 1xN couplers made by ion exchange in glass", OFC'90
(5) J Matthews III, C Giroux, JL Malinge, "Reliability studies of single-mode optical branching devices", SPIE OE/Fibers, Boston, 1991
(6) K Grosskopf, N Fabricius, U Nolte, H Oeste, "Integrated optical multiport splitters in glass for broadband communicatons networks", EFOC/LAN 92, p 148-152
(7) P Barlier, C Nissim, L Dohan, "Passive integrated optic components for fiber optics components for fiber optics communications in moldable glass", IOOC/ECOC 85.
(8) M Grant, R Bellerby, S Day, G Cannell and M Nelson, "Self aligned multiple fiber coupling for silica-on-silicon integrated optics", EFOC/LAN 91, p 269-272
(9) " Mercury doubles trunk capacity with FO couplers", Telephony, April 1, 1991.
(10) Bellcore TR 1209, "Generic `requirements for fiber optic branching components"
(11) S Hornung, P Frost, J Kerry and J Warren, "Flexible architecture and plant for optical access networks", International Wire and Cable Symposium Proceedings 1992.
(12) F Jean, J Matthews III, K Murphy, M McCourt, "Planar Single-mode couplers for fiber in the loop", Europto, Berlin 1993.

Nonlinear Fibers and Waveguides I

Quantum dot glass integrated optical devices

N. Peyghambarian,[1] H. Tajalli,[1,*] E. M. Wright,[1] S. W. Koch,[1,†]
S. I. Najafi,[2] D. Hulin,[3] and J. MacKenzie[4]

[1]Optical Sciences Center, University of Arizona, Tucson, AZ 85721
[2]Ecole Polytechnique, P.O. Box 6079, Montreal, Quebec H3C 3A7, Canada
[3]Ecole Polytechnique, ENSTA, 91120 Palaiseau, France
[4]University of California, 6532 Boelter Hall, Los Angeles, CA 90024

ABSTRACT

Semiconductor quantum dots are being investigated as a new material system for a variety of optoelectronic applications. We will discuss applications of quantum dots for optical switching in waveguide structures, optical data storage, and potential light sources. We emphasize why quantum dots with discrete spectra are expected to be superior in many applications compared with other systems, such as bulk and quantum wells with continuum spectra. We also report on our fabrication of cadmium sulfide quantum dot sol-gel glass channel waveguides using the potassium ion-exchange technique. The waveguides were optically characterized and cross-correlation measurements showed significant pulse shaping, both spectrally and temporally, after femtosecond laser pulses were propagated through the waveguide.

1. INTRODUCTION

The possibility of manipulating the optical properties of semiconductors through various degrees of dimensional or quantum confinement has attracted considerable attention during the last decade. Even though the best-known examples are still quantum-well structures which confine one motional degree of freedom of the excited electron-hole pair, quantum wires and quantum dots are becoming increasingly important since they yield two-dimensional and three-dimensional quantum confinement, respectively.

Three-dimensional quantum-confinement effects in semiconductor microcrystallites occur when the particle size approaches the bulk exciton Bohr radius. This confinement gives rise to interesting new effects, leading to novel optical properties. These properties of semiconductor microcrystallites make them potentially attractive for applications in optoelectronic devices, such as optical data storage and high-speed optical communication.

There has been a growing interest in searching for systems that exhibit such three-dimensional confinement effects and in understanding their behavior (Alivisatos et al. 1988; Arakawa and Sakaki 1982; Bawendi et al.

1990; Banyai et al. 1992; Banyai and Koch 1993; Efros and Efros 1982; Ezaki et al. 1993; Hanamura 1988, 1992; Itoh 1990, Kang et al. 1992; Masumoto et al. 1988, 1992; Krishna et al. 1991; Roussignol 1989; Hiroyuki et al. 1992; Takagahara 1985; Vahala 1990; Weisbuch et al. 1990; Woggon et al. 1993; Yumoto et al. 1990; Nomura and Kobayashi 1994). A number of laboratories have attempted to fabricate quasi-zero-dimensional structures using a variety of techniques, including colloidal suspension of semiconductor particles (Brus 1984, 1986), electron-beam lithography (Kash et al. 1986, Reed et al. 1986, Cibert et al. 1986), and semiconductor microcrystallites in glass matrices (Ekimov et al. 1985, Borrelli et al. 1987, Peyghambarian and Koch 1987, Jain and Lind 1983, Yao et al. 1985, Roussignol et al. 1985, Nuss et al. 1987, Olbright et al. 1987).

It has been shown that special glasses doped with CdS, CdSe, CuCl, or CuBr crystallites (Ekimov et al. 1985) can be fabricated that clearly exhibit quantum confinement. The microcrystallites in these glasses form out of the supersaturated solid solution of the basic constituents originally brought into the glass melt. The crystallites are more or less randomly distributed in the glass matrix. Ekimov et al. (1985) report crystallite growth following the growth law $R \propto t^{1/3}$, where R is the crystallite size and t is the duration of the heat treatment during which the crystallites actually grow. Average crystallite sizes from around 10 Å up to several 100 Å have been obtained.

Semiconductor nonlinear optics has been studied extensively for possible applications in photonic switching and communication (Peyghambarian et al. 1993). Most of the emphasis has focused on quantum well devices including lasers, detectors, modulators, and switches. Quantum dot materials have received less attention for these optoelectronic applications. Quantum confinement in the dots leads to narrowing of the optical spectra (gain and absorption), making them attractive for the development of superior integrated lasers and switches.

We describe the potential application of quantum dots as new switching elements and optical storage devices. We explore the merits of using quantum dots for all-optical waveguide switching applications. Basically, most nonlinear all-optical switching devices require the accumulation of a nonlinear phase difference between two waveguides. However, absorption clearly limits the useful device size, and there is a basic tradeoff between nonlinear refractive index and absorption. Here we explore this tradeoff for quantum dots by introducing a material figure of merit (Stegeman and Wright 1990).

We then describe the potential for using quantum dots for optical data storage applications. Spectral hole burning is a mechanism for this application. We show experimental results demonstrating room-temperature spectral hole burning.

2. OPTICAL CHARACTERIZATION

2.1. Linear optical properties

To understand the optical absorption in quasi-zero-dimensional semiconductors, we consider spherical semiconductor microcrystallites with radii R and background dielectric constant ϵ_2 embedded in another material with background dielectric constant ϵ_1. The radius of the quantum dot is usually on the order of a few tens of Angstroms (Å), comparable to the bulk exciton Bohr radius. This closely models the system of semiconductor crystallites in glass which has previously been studied extensively.

For microcrystals with radius R in the range $a \ll R \simeq a_B$, where a is the lattice constant of the semiconductor and a_B is the exciton Bohr radius, the single-electron properties are determined by the periodic lattice. Hence, the quantum dot has a macroscopic size in comparison to the unit cell, but it is small compared with all other macroscopic scales. In such microcrystallites, which are usually categorized as *mesoscopic* structures, the effective-mass approximation is valid. The electrons and holes are, thus, assumed to have the effective masses m_e and m_h, respectively, of the bulk material. Optically excited electron-hole pairs are influenced by the small size of the microcrystals, leading to quantum-confinement effects.

For the experimentally relevant case of spherical quantum dots, the single particle Schrödinger equations for the electron and the hole in the absence of Coulomb interaction can be written as

$$-\frac{\hbar^2}{2m_i} \nabla^2 \zeta_i(\mathbf{r}) = \mathcal{E}_i \, \zeta_i(\mathbf{r}), \qquad (1)$$

where i = e or h. The boundary condition of ideal quantum confinement dictates that

$$\zeta_i(\mathbf{r}) = 0 \qquad \text{for} \quad r = R. \qquad (2)$$

The solution of the Schrödinger equation (1) with the spherical boundary condition, Eq. (2), is

$$\zeta_i(\mathbf{r}) = \sqrt{\frac{1}{4\pi R^3}} \, \frac{j_\ell\left(\alpha_{n\ell} \frac{r}{R}\right)}{j_{\ell+1}(\alpha_{n\ell})} \, Y_\ell^m(\theta,\phi), \qquad (3)$$

where j_ℓ is the ℓth order spherical Bessel function, $Y_\ell^m(\theta,\phi)$ are the spherical harmonics, $\alpha_{n\ell}$ is the nth root of the ℓth order Bessel function. The quantum numbers for the particle are n, ℓ, and m. The boundary condition (2) is satisfied if

$$j_\ell\left(\alpha_{n\ell}\frac{r}{R}\right)\Big|_{r=R} = 0 \qquad (4)$$

or

$$j_\ell(\alpha_{n\ell}) = 0 . \qquad (5)$$

Equation (5) is satisfied for

$$\alpha_{10} = \pi,\ \alpha_{11} = 4.4934,\ \alpha_{12} = 5.7635,\ \alpha_{20} = 6.2832,$$

$$\alpha_{21} = 7.7253,\ \alpha_{22} = 9.0950,\ \alpha_{30} = 9.4248,\ \text{etc.} \qquad (6)$$

The denominator in Eq. (3) comes simply from the normalization of the wave function. Inserting Eq. (3) into Eq. (1) gives the discrete energy eigenvalues as

$$\mathcal{E}_i = \frac{\hbar^2}{2m_i}\left(\frac{\alpha_{n\ell}}{R}\right)^2 . \qquad (7)$$

It is customary to refer to the nℓ eigenstates as 1s, 1p, 1d, etc., where s, p, d, etc., correspond to $\ell = 0, 1, 2, ...$, respectively ($\alpha_{10} = \alpha_{1s}$, $\alpha_{11} = \alpha_{1p}$, etc.). Note the somewhat unusual notation for atomic spectroscopists, for whom a 1p state would not be possible. The difference here comes from the fact that we are not dealing with a Coulomb potential, but a spherical confinement potential.

The lowest energy states are those with the lowest $\alpha_{n\ell}$ values. Examination of Eq. (6) points to these lowest energies as those with α_{1s}, α_{1p}, α_{1d}, etc. Taking the zero of energy at the top of the valence band, the electron and hole energy levels are then given by [using Eq. (7)]

$$\mathcal{E}^e = E_g + \frac{\hbar^2}{2m_e}\left(\frac{\alpha_{n_e \ell_e}}{R}\right)^2 \qquad (8)$$

and

$$\mathcal{E}^h = -\frac{\hbar^2}{2m_h}\left[\frac{\alpha_{n_h}\ell_h}{R}\right]^2. \quad (9)$$

The lowest two energy levels are plotted schematically in Fig. 1. We see from this figure that the usual three-dimensional band structure is drastically modified and has become a series of quantized single-particle states.

As we mentioned before, the single-particle spectrum does not correspond to the optical absorption spectrum, since the electron-hole Coulomb effects are excluded. The Schrödinger equation for one electron-hole pair is written as

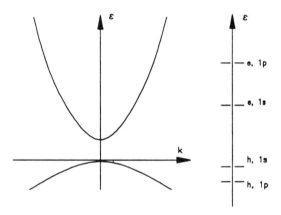

Figure 1. Schematic plot of the single-particle energy spectrum in bulk semiconductors (left) and in small quantum dots (right).

$$\left[-\frac{\hbar^2}{2m_e}\nabla_e^2 - \frac{\hbar^2}{2m_h}\nabla_h^2 + V_c\right]\phi(\mathbf{r}) = \mathcal{E}\,\phi(\mathbf{r}), \quad (10)$$

with the usual spherical boundary condition $\phi(r = R) = 0$. Here V_c is the Coulomb potential. The Schrödinger equation (10), along with the corresponding boundary condition, can be solved analytically in the absence of the Coulomb interaction, resulting in

$$\mathcal{E} = \mathcal{E}^e + \mathcal{E}^h = E_g + \frac{\hbar^2}{2m_e}\left[\frac{\alpha_{n_e}\ell_e}{R}\right]^2 + \frac{\hbar^2}{2m_h}\left[\frac{\alpha_{n_h}\ell_h}{R}\right]^2 \quad (11)$$

and

$$\phi(\mathbf{r}_e,\mathbf{r}_h) = \varsigma(\mathbf{r}_e)\,\varsigma(\mathbf{r}_h)\,, \tag{12}$$

where

$$\varsigma(r) = \sqrt{\frac{1}{4\pi R^3}}\;\frac{j_\ell\left[\alpha_{n\ell}\dfrac{r}{R}\right]}{j_{\ell+1}(\alpha_{n\ell})}\,Y_\ell^m(\theta,\phi)\,. \tag{13}$$

Equation (11) shows that the absorption is blue shifted with respect to the bulk bandgap E_g. The shift varies with crystal size R, like $1/R^2$, being larger for smaller sizes. Furthermore, this equation states that the energy spectrum consists of a series of lines corresponding to the electron-hole transitions. Figure 2 exhibits the schematic representation of the one-electron-hole-pair states. The selection rules for the dipole-allowed interband transitions are $\Delta\ell = 0$ in the absence of Coulomb interaction. For example, the $\mathscr{E}_{1s-1s}$-transition, where electron and hole are both of 1s-type, is allowed.

When the Coulomb interaction is included, the problem can no longer be solved analytically and a numerical approach is needed. The absolute value of the one- and two-pair states is only weakly shifted by Coulomb effects, since the kinetic energy terms dominate for dots with $R \simeq a_B$. The electron-hole-pair wave functions are, however, modified enough to strongly influence the nonlinear effects. The selection rules stated earlier are no longer valid, and transitions with $\Delta\ell \neq 0$ become weakly allowed (see Sec. 3 for more details).

The linear absorption spectra of two samples of CdS quantum dots in glass at room temperature are shown in Fig. 3. The spectrum labeled "bulk" refers to a glass with semiconductor microcrystallites large enough to retain the three-dimensional bulk properties. This spectrum is typical of bulk CdS absorption spectrum at room temperature with a sharp band edge and structureless Coulomb-enhanced continuum absorption at higher photon energies. The average crystallite size becomes smaller for lower heat-treatment temperatures. The spectrum labeled "QD" refers to a sample with small crystal sizes. The quantum-confinement effects are clearly observable in this sample. The absorption has shifted to higher energies as expected for the confinement effect. Furthermore, discrete quantum-confined electron-hole-pair states appear in the "QD" sample.

Figure 4 displays the absorption spectra for similar samples at a temperature of T = 10 K. The lower temperature reduces the phonon broadening and, consequently, the transitions become narrower. However,

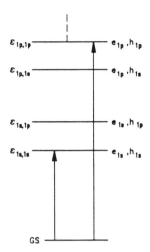

Figure 2. Schematic representation of the one-electron-hole-pair transitions in a semiconductor quantum dot. The notation e_{1s}, h_{1p}, etc. refers to the electron being in 1s state, the hole being in 1p state, etc.

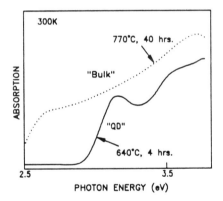

Figure 3. The linear absorption spectra at room temperature of two samples of CdS microstructures in glass. The crystal sizes in the sample labeled "bulk" are large enough to have bulk behavior. In the "QD" sample the crystals have very small sizes (the samples were grown by S. Risbud et al.).

the main features of the spectra have been retained. The quantum-confined transitions are more clearly observed. In this figure, three samples with different quantum dot sizes are displayed.

The transition lines in Fig. 4 are much broader than transitions in bulk materials. The width of the transitions is a result of homogeneous and inhomogeneous broadening mechanisms. The homogeneous component is

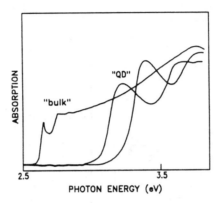

Figure 4. The linear absorption spectra at T = 10 K of CdS crystallites in glass. The two spectra labeled "QD" refer to two samples with small crystal sizes in the quantum-confinement regime. The "bulk" sample has large crystal sizes, exhibiting bulk CdS properties (the samples were grown by S. Risbud et al.).

due to phonon and other scattering mechanisms. The inhomogeneous broadening comes from the fact that the crystallites do not have the same radii. Each radius has its own transition frequency ($\mathcal{E} \sim 1/R^2$), making the effective linewidth broad. This broadening can be taken into account theoretically. Using a density-matrix approach, the optical susceptibility and, consequently, optical absorption can be computed. The absorption coefficient of a single quantum dot as a function of photon frequency in such a calculation is given by

$$\alpha(\omega) = \frac{4\pi\omega}{\hbar c \sqrt{\epsilon_2}} \sum_i |d_{oi}|^2 \frac{\gamma_i}{\gamma_i^2 + (\omega_i - \omega)^2} , \qquad (14)$$

where ϵ_2 is the background dielectric constant of the semiconductor and d_{oi} is the transition dipole matrix element for the transition o → i. The index o refers to the ground state (a state without any electron-hole pairs), while the index i refers to the one-electron-hole-pair state. The energy eigenvalue of the quantum-confined electron-hole transition is $\hbar\omega_i$, as given by Eq. (7), $\mathcal{E}_i = \hbar\omega_i$. Thus, $\hbar\omega_i$ explicitly depends on the size of the crystallites R. The homogeneous linewidth of the transition is γ_i. Equation (14) shows that the absorption spectrum of a single quantum dot consists of a series of Lorentzian peaks centered around the one-electron-hole-pair energies $\hbar\omega_i$. The inhomogeneous broadening may be taken into account by assuming that the particles have a size distribution given by f(R) around a mean value $\bar{R}$. Since $\alpha(\omega)|_R$ in Eq. (14) is the absorption coefficient for a given radius R, the average absorption is then

$$\alpha(\omega)\bigg|_{\text{average}} = \int_0^\infty dR\, f(R)\, \alpha(\omega)\big|_R \ . \tag{15}$$

Using a Gaussian distribution around the mean radius, $\bar{R} = 20$ Å, for $f(R)$ and different Gaussian distribution widths, we calculate the results shown in Fig. 5 for CdS quantum dots. It is clear that the quantum-confined transitions broaden and merge to a continuous structure with increasing width of the size distribution. The spectrum shown by the full line closely resembles the observed spectrum in Fig. 4. This suggests that the samples in Fig. 4 have a size distribution of 15% to 20%.

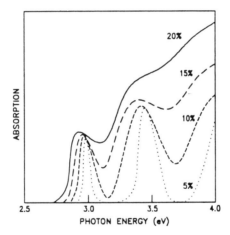

Figure 5. The calculated linear absorption spectra for CdS quantum dots with a Gaussian size distribution around a mean radius of 20 Å. The different curves correspond to different widths of the Gaussian distribution. For example, a 20% width corresponds to a 4 Å width for the Gaussian size distribution around a mean radius of 20 Å.

2.2. All-optical switching using quantum dots

For all-optical switching applications using semiconductor microcrystallites, a large nonlinear refractive-index effect is sought while at the same time minimizing the detrimental effects of absorption. In material terms this calls for a system whose resonance conditions can be varied so that the tradeoff between refractive index and absorption can be adjusted for a given laser frequency. This degree of flexibility is realized in quantum dots (QDs), since the quantum confinement leads to a tunable blue shift in the lowest exciton resonance which increases with decreasing dot radius. In addition, the operating device length can be scaled by varying the QD concentration. Here we theoretically examine the tradeoffs presented by QDs from the perspective of all-optical switching. In particular, we consider the prototypical all-optical switch, namely the nonlinear directional coupler (NLDC) (Jensen 1982, Maier 1982).

The linear directional coupler consists of two identical optical waveguides which are in sufficiently close proximity that their evanescent fields overlap and, hence, there is periodic power transfer between the two waveguides with propagation distance (Marcatili 1969). For initial excitation of only one waveguide and a beat length device $L = L_b$, the output appears only in the input or bar waveguide, whereas for a half-beat length device, $L = L_c = L_b/2$, the output appears in the adjacent or cross waveguide, where L_c is the coupling length. In linear integrated optics the half-beat length directional coupler is operated as an optical switch using the electro-optic effect (Kurazono et al. 1972, Tada et al. 1974): Applying a dc field to the bar waveguide causes a wave vector mismatch, $\Delta\beta = k_0\Delta n$, between the waveguides which inhibits the coupling, and the output exits through the bar state with the dc field applied. The key idea of all-optical switching using an NLDC is that the input field itself generates a nonlinear change in propagation wave vector, $\Delta\beta^{NL} = k_0\Delta n^{NL}$, where Δn^{NL} is the nonlinear change in refractive index so that the input intensity determines the output state (Jensen 1982, Maier 1982). The basic requirement for efficient all-optical switching to occur is that the light-induced nonlinear phase-shift for an isolated waveguide obeys (Stegeman et al. 1987, Wright et al. 1988, Stegeman and Wright 1990)

$$k_0 \Delta n^{NL} L = \Delta\beta^{NL} L > 4\pi , \qquad (16)$$

where L is the device length. Equation (16) will form the basis of our discussion of all-optical switching of an NLDC using QDs.

The basic operation of an NLDC can be modeled theoretically in the framework of coupled-mode theory (Jensen 1982, Maier 1982). Specifically, we expand the field in the coupler as a coherent superposition of the transverse modes of the individual waveguides,

$$\mathbf{E}(\mathbf{r},t) = \mathbf{x}\frac{1}{2}[\mathcal{E}_1(z)u_1(x,y) + \mathcal{E}_2(z)u_2(x,y)]e^{i(\beta z - \omega t)} + c.c , \qquad (17)$$

where z is the propagation direction, $\mathbf{x}$ is the unit polarization vector, $u_j(x,y)$ is the transverse mode of waveguide j, $\mathcal{E}_j(z)$ is the amplitude of the jth mode, ω is the central carrier frequency of the field, and β is the propagation wave vector of the identical waveguides. Here the propagation wave vector β is assumed to account for the optical properties of the host glass, which is treated as a transparent dielectric. The linear and nonlinear optical properties of the QDs are incorporated through the following polarization source term for each individual waveguide (nonlinear coupling between the waveguides is neglected):

$$P_j(z) = P_j^{(1)}(z) + P_j^{(3)}(z)$$
$$= \chi^{(1)}(\omega)\mathcal{E}_j(z) + \chi^{(3)}(\omega)|\mathcal{E}_j(z)|^2\mathcal{E}_j(z) , \quad (18)$$

where $\chi^{(1)}(\omega)$ is the linear susceptibility of the QD system and $\chi^{(3)}(\omega) \equiv \chi^{(3)}(-\omega,\omega,-\omega,\omega)$ is the third-order nonlinear susceptibility. Then, substituting Eqs. (17) and (18) into Maxwell's equations yields the following coupled-mode equations in the usual slowly varying envelope approximation (Stegeman et al. 1987, Caglioti et al. 1988),

$$\frac{d\mathcal{E}_1}{dz} = \frac{i\pi}{2L_c}\mathcal{E}_2 - \frac{1}{2}(\alpha_0 - \alpha_3|\mathcal{E}_1|^2)\mathcal{E}_1 + \Delta\beta_1^{NL}\mathcal{E}_1 \quad (19a)$$

and

$$\frac{d\mathcal{E}_2}{dz} = \frac{i\pi}{2L_c}\mathcal{E}_1 - \frac{1}{2}(\alpha_0 - \alpha_3|\mathcal{E}_2|^2)\mathcal{E}_2 + \Delta\beta_2^{NL}\mathcal{E}_2 . \quad (19b)$$

The first term on the right-hand side of Eqs. (19) describes the linear mode coupling due to the evanescent field overlap. In the absence of the QDs these terms correctly predict the periodic energy exchange between the waveguides. The next two terms involving α_0 and α_3, defined as

$$\alpha_0 = \frac{4\pi\omega^2\eta}{\beta c^2}\text{Im}(\chi^{(1)}(\omega)) \;;\; \alpha_3 = -\frac{4\pi\omega^2\eta}{\beta c^2}\text{Im}(\chi^{(3)}(\omega)) , \quad (20)$$

describe linear absorption and absorption saturation ($\alpha_3 > 0$), respectively. Finally, $\Delta\beta_j^{NL}$ is the nonlinear change in the modal wave vector of the jth waveguide,

$$\Delta\beta_j^{NL} = \frac{2\pi\omega^2\eta}{\beta c^2}\text{Re}(\chi^{(3)}(\omega))|\mathcal{E}_j|^2 . \quad (21)$$

Furthermore, the parameter η in Eqs. (20) and (21) is the volume fill fraction of QDs: $\eta = 0$ corresponds to the absence of QDs.

A material figure of merit for the NLDC can now be obtained using Eq. (16) and Eqs. (19)-(21) as follows: It is clear that linear absorption places a limit on the useful device length, so we require $\alpha_0 L_c \simeq 1$ for a half-beat length coupler. Also, it follows from Eqs. (19) that the nonlinearity will saturate when the field strength approaches $|\mathcal{E}_s| \simeq \alpha_0/\alpha_3$. Substituting this in Eq. (21) and using Eq. (20), we obtain the peak phase shift over the device length as $\Delta\beta^{NL}L_c \simeq \text{Re}(\chi^{(3)}(\omega))/2\text{Im}(\chi^{(3)}(\omega))$, where we used $\alpha_0 L_c = 1$. According to Eq. (16), we require this phase shift to exceed 4π for efficient all-optical switching, so we define the material figure of merit (Stegeman and Wright 1990),

$$W = \frac{1}{8\pi}\frac{\text{Re}(\chi^{(3)}(\omega))}{\text{Im}(\chi^{(3)}(\omega))} > 1 \quad . \tag{22}$$

This figure of merit is identical in form to that previously obtained for the case of two-photon absorption as a limitation to all-optical switching (Mizrahi et al. 1989). Note that W is geometry-independent, and is also independent of the volume fill fraction η. The dependence on the fill fraction η is contained in the device length which is derived using $\alpha_0 L_c = 1$ as

$$L_c = \frac{\beta c^2}{2\pi\omega^2\eta\,\text{Im}(\chi^{(1)}(\omega))} \quad , \tag{23}$$

that is, the device length is inversely proportional to η.

From the perspective of all-optical switching in NLDCs, one wants to make the material figure of merit as large as possible, $W \geq 1$ (switching may still occur for $W < 1$ but at the expense of poor transmission). In other terms, for a material to be suitable for fabricating an NLDC, it is desirable that $\text{Re}(\chi^{(3)})$ is 25 or more times greater than $\text{Im}(\chi^{(3)})$. Whether or not this is possible for a given material depends on its resonance structure. We have evaluated the material figure of merit W for the case of QDs embedded in glass using the theory of Hu et al. (1990) which provides both the linear susceptibility $\chi^{(1)}$ and the third-order susceptibility $\chi^{(3)}$. Figure 6 shows the frequency variation of (a) $\text{Im}(\chi^{(1)})$, and (b) $\text{Re}(\chi^{(3)})$ and $\text{Im}(\chi^{(3)})$ for the semiconductor CdS and a QD radius $R = a_0$, with a_0 being the bulk exciton radius. Here the frequency is represented by the dimensionless detuning $\Delta = (\hbar\omega - E_G)/E_R$, where E_G is the bandgap of the bulk material and E_R is the exciton Rydberg. For CdS, $E_G = 2520\,\text{meV}$, $E_R = 27\,\text{meV}$, and a homogeneous line broadening $\gamma = 2E_R$ was included in the calculations. The $\chi^{(3)}$ values in Fig. 6(b) can be converted to units of $m^2 V^{-2}$ by multiplying by $10^{-16}\eta$. Then, for example, for a detuning of $-20E_R$ we have $\text{Re}(\chi^{(3)}) \simeq 10^{-20} m^2 V^{-2}$ and $\text{Im}(\chi^{(3)}) \simeq 2\times 10^{-21} m^2 V^{-2}$ for $\eta = 10^{-2}$, in reasonable agreement with the measured values reported by Cotter et al. (1992).

Figure 7 shows the material figure of merit W versus detuning Δ for CdS QDs and two values of the dot radius. For $R = a_0$, W remains less than 1 all the way out to $-20E_R$. In contrast, for $R = 0.5a_0$, W is greater than 1 for detunings below $-16E_R$. The variation of W with detuning Δ may be

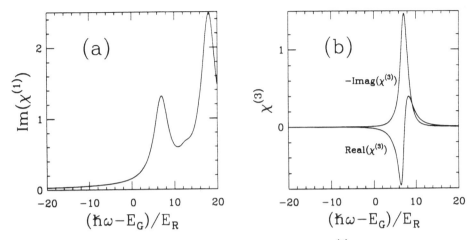

Figure 6. Frequency variation of (a) Im($\chi^{(1)}$), and (b) Re($\chi^{(3)}$) and Im($\chi^{(3)}$), for the semiconductor CdS, R = a_0, E_G = 2520meV, E_R = 27meV, and a homogeneous line broadening γ = $2E_R$.

understood by realizing that the lowest exciton resonance saturates like a two-level system in which the effective detuning is that between the field and the 1s exciton, $\delta = (\hbar\omega - E_{1s})/E_R$ (Hu et al. 1990). In this case, Re($\chi^{(3)}$) varies as $\delta/(1+\delta^2)$, and Im($\chi^{(3)}$) varies as $-1/(1+\delta^2)$, so that according to Eq. (22), W varies as

$$W \simeq \frac{1}{8\pi}\left[-\Delta + \frac{(E_{1s}-E_G)}{E_R}\right] . \tag{24}$$

On the basis of this scaling argument we expect that the slope of the W versus Δ curve should be independent of the dot radius, and this is clearly verified in Fig. 7. Furthermore, since the blue-shift of the 1s exciton energy E_{1s} varies with dot radius as roughly R^{-2} in the quantum confined regime, it follows from Eq. (22) that for a given Δ the figure of merit W is larger for smaller dots. Once again this is verified in Fig. 7.

The results shown in Fig. 7 are for fixed values of the dot radius. These results are intended to show that if the dot radius can be controlled, then it is possible to engineer the material properties so that they are useful for all-optical switching applications. In present QD samples there is invariably a statistical distribution of dot radii, which adds inhomogeneous broadening to the problem. This inhomogeneous broadening is typically of the same order as the homogeneous broadening, and tends to wash out any strong dependence on the dot radius. However, there is no fundamental reason

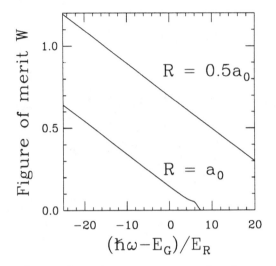

Figure 7. Figure of merit W versus detuning for CdS and R = $0.5a_0$ and R = a_0.

why the inhomogeneous broadening cannot be eliminated, and that the figure of merit of semiconductor QDs can be enhanced by producing small QDs with well-controlled radii.

Figure 8 shows an example of the predicted all-optical switching for CdS QDs and R = $0.5a_0$, and a detuning of $-20E_R$. These results were obtained by the numerical solution of Eqs. (19), incorporating the theoretical

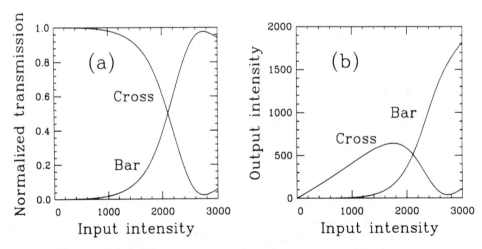

Figure 8. Switching characteristics of a nonlinear directional coupler in CdS quantum dots with R = $0.5a_0$: a) normalized transmission for the bar and cross waveguides versus dimensionless intensity, and b) output intensity for the bar and cross waveguides versus input intensity.

results for the linear and nonlinear susceptibilities in Fig. 6. In this case, W > 1 and all-optical switching is expected. The transmission of both the bar and cross waveguides with respect to the total output intensity is shown versus the dimensionless input intensity $|\mathcal{E}_1(0)|^2$ in Fig. 8(a), and all-optical switching is evident. (The fields are converted to Vm^{-1} by multiplying $\mathcal{E}$ by 10^8.) For an input intensity of 1000, the output exits through the cross waveguide, whereas for an input intensity of 2600 the output exits the bar state. This is also seen in Fig. 8(b) which shows the output intensity versus the input intensity. Here we see that the actual transmission of the device increases at high input intensity due to the saturation of the linear absorption. Assuming a value of $\eta = 10^{-2}$ for the fill fraction, the predicted coupling length is $L_c = 1/\alpha_0 \simeq 5$ mm. In contrast, for a dot size $R = a_0$ and a detuning of $-20E_R$, no all-optical switching is observed in the device simulations, in agreement with the observation that $W < 1$ (see Fig. 7).

2.3. Quantum dot waveguides

In order to realize novel devices, waveguide structures containing quantum dots should be fabricated. Recent developments in the ion-exchanged waveguide fabrication in glass (Najafi et al. 1992) make it possible to build quantum dot slab and channel waveguide devices. We have succeeded in fabricating channel waveguides in cadmium sulfide (CdS) quantum dots in borosilicate glass samples, made by the sol-gel technique (Takada et al. 1994). We have also coupled femtosecond laser pulses through these waveguides and have seen very interesting pulse breakup effects for pulses propagating in the transparency region.

The CdS-doped sodium borosilicate sample was prepared with 15% CdS in the glass. The base glass composition was 5 wt% Na_2O, 15 wt% B_2O_3, and 80 wt% SiO_2. The sample was heat treated at 420 °C for 12 hours. The 8-mm-long waveguides were fabricated in the CdS quantum dot sample by potassium-sodium ion exchange through 5-μm openings in the aluminum mash. Ion exchange was carried out for 10.5 hours in a pure potassium nitrate molten bath. Figure 9 displays the microscope photograph of three waveguides. The inset clearly shows single mode light propagation along the waveguide. The measured transmission spectrum of the CdS quantum dot waveguide is shown in Fig. 10. The transmission drop for short wavelength is due to th elong tail absorption of the CdS quantum dots. The noise comes mainly from the spectral response of the silicon detector used. The sudden drop in transmission near 700 nm corresponds to the cut-off frequency of the quantum dot waveguide. The inset of Fig. 10 shows the absorption spectrum of the quantum dot sample itself. The lowest quantum confined transition appears as a shoulder in the spectrum at 420 nm in contrast to the bulk CdS badgap at 512 nm.

Figure 9. Microscope photograph of potassium ion-exchanged CdS quantum dot glass waveguides. The inset is the near-field mode profile of a waveguide made using a 5-μm-wide mask opening.

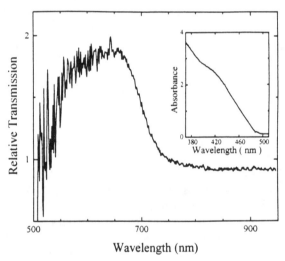

Figure 10. Measured spectral transmission of the CdS quantum dot waveguide. The inset is the absorption spectrum of the quantum dot sample.

Femtosecond pulse propagation through the CdS quantum dot waveguide was studied using a continuously tunable amplified colliding pulse mode-locked (CPM) dye laser system. 60-fs pulses centered at 620 nm from a first amplifier stage are divided into two beams. One beam serves as a reference for the cross-correlation measurement, and the other beam is focused on an ethylene glycol jet to generate the white continuum. Using an interference filter, the test beam spectrum was selected at 687 nm for the pulse to be in the high transmission state, and was reamplified using a second stage amplifier. The full width at half maximum of the test beam is 110 fs. The test beam was end fired onto the CdS quantum dot waveguide. The output surface of the sample was imaged on the iris to block the unguided light. The output of the waveguide and the reference beam was focused on the 300-μm-thick KDP crystal to sum the frequencies. The sum frequency signal was detected with a photomultiplier tube varying the time delay on the reference beam.

The time and spectral profiles of the input (a, b) and output (c, d) pulses are shown in Fig. 11 for a peak input intensity of 12 GW/cm². Pulse breakup is clearly seen in the time domain, as is evidenced by the secondary peak for negative time delays and the shoulder at positive time delay. In the spectral domain this is manifest as spectral broadening and modulation [three peaked spectrum in Fig. 11(d)]. We are currently investigating the physicaal origin of the pulse breakup, potential candidates being coherent effects (Harten et al. 1992) and/or breakup of higher-order solitons arising from the combined effects of the normal group velocity dispersion of the host glass and the negative (Cotter et al. 1992) quantum dot nonlinearity. What is clear from the spectral broadening in Fig. 11 is that the pulse is accumulating a nonlinear phase shift upon propagation. Our results therefore demonstrate that the CdS quantum dot waveguides described here have potential use in all-optical switching applications (Stegeman and Wright 1990); for example, the nonlinear directional coupler. To access this potential, it is important to measure the figure of merit for these waveguides (Stegeman and Wright 1990).

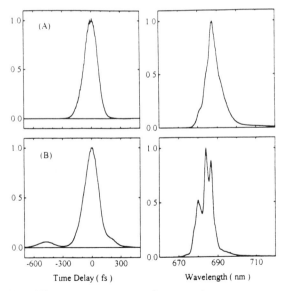

Figure 11. The cross correlation and spectrum of the femtosecond pulse after propagating a 8-mm-long CdS quantum dot waveguide. Input pulse is 110-fs-long with 12 GW/cm² peak intensity.

2.4. Quantum Dots for Optical Data Storage

Spectral hole burning has been employed in the past to investigate frequency-domain optical data storage (G. D. Castro et al., 1978; C. Ortiz et al., 1981). In such an application, a recording material is required that may undergo spectral hole burning with certain well-defined characteristics. Most of the materials that have been used so far have required liquid helium refrigeration. This low-temperature requirement, of course, limits the practical applications of such materials for optical data storage systems.

Figure 12 schematically shows the concept of the spectral hole-burning technique. The absorption spectrum of the recording material has to be inhomogeneously broadened with a total width of γ_{inh}, consisting of homogeneously broadened components with linewidths of γ_{hom}, as shown in Fig. 12(a). When a laser beam with a well-defined wavelength irradiates the sample, only the homogeneous line that is in resonance with the laser can be excited, and consequently gets bleached. Figure 12(b) shows such a bleaching of the homogeneously broadened line, which is referred to as spectral hole burning. For various tuning of the laser frequency inside the inhomogeneously broadened absorption spectrum, holes at different spectral positions can be burned. Each burned hole may be considered as a logic "1," while a no-hole is a logic "0." The number of holes that may be burned (which is related to the storage density) is on the order of the ratio of the inhomogeneous width to the homogeneous width, $R = \gamma_{inh}/\gamma_{hom}$. The larger γ_{inh}, or the smaller γ_{hom}, the larger is the number of burned holes. The laser may be focused to a spot size of $\simeq 1$ μm, and within this small spot a number of spectral holes may be burned.

Quantum dots may be appropriate for data storage. This application stems from the fact that quantum dots in various media have different sizes, leading to an inhomogeneously broadened absorption spectra. The larger the particle size distribution, the broader the total absorption linewidth is. Also, the persistence of quantum-confined absorption peaks at room temperature in quantum-confined structures makes it possible to realize this inhomogeneous broadened lineshape at room temperature.

Spectral hole-burning experiments in CdS and CdSe quantum dots in glass at room temperature were performed. Figure 13 shows a typical result for the CdS dots in glass. The linear transmission spectrum of the sample, the spectral positions of the pump pulses, and differential transmission of the sample (inset of Fig. 13) as a result of the presence of the pump are plotted. The spectral holes burned by the pumps are clearly observable. The spectral hole moves with the pump wavelength, as expected for an inhomogeneously broadened transition. The width of the spectral hole increases as the pump intensity increases. These measurements show that spectral hole burning is possible in quantum dot samples. However, the

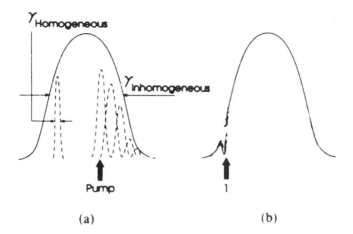

Figure 12. (a) Schematic absorption spectrum of an inhomogeneously broadened absorption line consisting of homogeneously broadened features. (b) The application of a laser beam at the spectral position shown by the arrow results in the creation of a hole, corresponding to binary "1."

width of the burned hole is broad in most of the samples studied so far (Peyghambarian et al. 1989, Spiegelberg et al. 1991), suggesting that not many holes can be burned in the excitation spot by changing the wavelength.

This situation can be improved for samples with better surface quality and for specially prepared samples. For example, spectrally narrow holes have been seen in quantum dots in organic matrices (A. P. Alivisatos et al., 1988). Recently, narrow spectral holes were observed in quantum dots in glass after the sample was strongly illuminated (U. Woggon et al., 1993). An example of such a narrow spectral hole burning measured in our laboratories is shown in Fig. 14. The spectral holes with widths as low as a few meV have been achieved. The data in Fig. 14 was obtained after the sample was irradiated with ~30 MW/cm^2 for two hours. The origin of the narrow holes has been attributed to the growth process, i.e., it is observed only in the samples prepared in the nucleation and normal growth stages, whereas the holes are broad in the sample prepared in the coalescence stage (S. V. Gaponenko et al.).

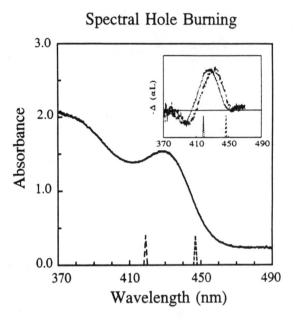

Figure 13. The linear absorption spectrum of a CdS quantum dot sample at room temperature. The sample was pumped at two spectral positions inside the first quantum-confined transition. The positions of the pumps are shown by the small peaks at wavelengths of 419 nm and 447 nm. The inset shows the change in the absorption ($-\Delta\alpha L$) obtained as a result of this pumping. Absorption bleaches and spectral holes are generated, which shift as the pump wavelength is changed. This figure clearly demonstrates transient spectral hole burning at room temperature in a quantum dot sample (sample was grown by S. Risbud et al.).

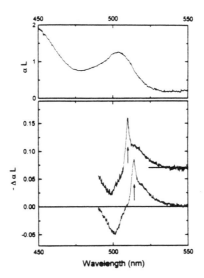

Figure 14. The linear absorption coefficient (upper curve) and the change in the absorption ($-\Delta\alpha L$) for pumping at 515 nm and 520 nm (sample was prepared by S. Gaponenko et al.).

3. ACKNOWLEDGEMENTS

We would like to thank NSF, SDI/ONR/AFOSR, NSERC, and NATO travel for support of some of the work that is reported here. The work has been done in collaboration with B. P. McGinnis, K. Kang, B. Fluegel, A. Kepner, Y. Hu, V. Esch, and S. Gaponenko.

4. REFERENCES

*Permanent address: Physics Department, University of Tabriz, Tabriz, Iran

†Permanent address: Philipps-Universität Marburg, Mainzergasse 33, 35032 Marburg, Germany

Alivisatos A. P., Harris A. C., Levinos N. J., Steigerwald M. L., Brus L. E., 1988, J. Chem. Phys. **89**, 4001.

Arakawa Y, and Sakaki H., 1982, Appl. Phys. Lett. **40**, 939.

Banyai L., and Koch S. W., 1993, *Semiconductor Quantum Dots* (World Scientific, Singapore).

Banyai L., Gilliot P., Hu Y. Z., and Koch S. W., 1992, Phys. Rev. B **45**, 14136.

Bawendi M. G., Wilson W. L., Rothberg P. L., Carroll P. J., Jedjin T. M., Steigerwald M. L., Brus L. E., 1990, Phys. Rev. Lett. **65**, 1623; Brus L. E. 1991, Appl. Phys. A **53**, 465.

Borrelli N. F., Hall D. W., Holland H. J., Smith D. W., 1987, J. Appl. Phys. **61**, 5399.

Brus L. E., 1984, J. Chem. Phys. **80**, 4403.

Brus L. E., 1986, IEEE J. Quantum Electron. **22**, 1909; and references to the author's earlier work.

Caglioti E., Trillo S., Wabnitz S., Stegeman G. I., 1988, J. Opt. Soc. Am. B **5**, 472.

Castro G. D., Haarer D., Macfarlane R. M., Trommsdorff H. P., 1978, U.S. patent no. 4101976.

Cibert J., Petroff P. M., Dolan G. J., Pearton S. J., Gossard A. C., English J. H., 1986, Appl. Phys. Lett. **49**, 1275.

Cotter D., Burt M. G., Manning R. J., 1992, Phys. Rev. Lett. **68**, 1200.

Efros Al L., Efros A. L., 1982 (in Russian), Sov. Phys. Semicond. **16**, 772.

Ekimov A. I., Efros Al. L., Onushchenko A. A., 1985, Solid State Commun. **56**, 921.

Ezaki H., Tokihiro T., Hanamura E., 1993, presented at Phys. Soc. Meeting of Jpn.

Gaponenko S. V. et al., to be published.

Hanamura E., 1988a, Phys. Rev. B **38**, 1228.

Hanamura E., 1992, Phys. Rev. B **46**, 4718.

Harten P. A., Knorr A., Sokoloff J. P., Brown de Colstoun F., Lee S. G., Jin R., Wright E. M., Khitrova G., Gibbs H. M., Koch S. W., Peyghambarian N., 1992, Phys. Rev. Lett. **69**, 852.

Hu Y. Z., Koch S. W., Lindberg M., Peyghambarian N., Pollock E. L., Abraham F. F., 1990a, Phys. Rev. Lett. **64**, 1805.

Hu Y. Z., Lindberg M., Koch S. W., 1990b, Phys. Rev. B **42**, 1713.

Itoh T., Furumiya M., Ikehara T., Gourdon C., 1990, Solid State Commun. **73**, 271.

Jain R. K., Lind R. C., 1983, J. Opt. Soc. Am. **73**.

Jensen, S. M., 1982, IEEE J. Quant. Electron. **QE-18**, 1580.

Kang K. I., McGinnis B. P., Sandalphon, Hu Y. Z., Koch S. W., Peyghambarian N., Mysyrowicz A., Liu L. C., Risbud S. H., 1992, Phys. Rev. B **45**, 3465.

Kash K., Scherer A., Worlock J. M., Craighead H. G., Tamargo M. C., 1986, Appl. Phys. Lett. **49**, 1043.

Kurazono S., Iwasaki K., Kumagai N., 1972, Electron. Commun. Jap. **55**, 103.

Maier, A., 1982, Sov. J. Quant. Electron. **12**, 1490.

Marcatili E. A. J., 1969, Bell. Sys. Tech. J. **48**, 2071.

Masumoto T., Yamazaki M., Sugawara H., 1988, Appl. Phys. Lett. **53**, 1527i.

Masumoto Y., Kawamura T., Ohzeki T., Urabe S., 1992, Phys. Rev. B **46**, 1827.

Mizrahi V., DeLong K. W., Stegeman G. I., Saifi M. A., Andrejco M. J., 1989, Opt. Lett. **14**, 1140.

Najafi S. I., 1992, *Introduction to Glass Integrated Optics* (Artech House, Boston).

Nomura S., and Kobayashi T., 1994, J. Appl. Phys. **75**, 382.

Nuss M. C., Zinth W., Kaiser W., 1987, Appl. Phys. Lett. **49**, 1717.

Olbright G. R., Peyghambarian N., Koch S. W., Banyai L., 1987, Opt. Lett. **12**, 413.

Ortiz C., Macfarlane R. M., Shalby R. M., Lenth W., Bjorklund G. C., 1981, Appl. Phys. Lett. **25**, 87.

Peyghambarian N., Fluegel B., Hulin D., Migus A., Joffre M., Antonetti A., Koch S. W., Lindberg M., 1989, IEEE J. Quantum Electron. **25**, 2516.

Peyghambarian N., Koch S. W., 1987, Rev. Phys. Appl. **22**, 1711.

Peyghambarian N., Koch S. W., Mysyrowicz A., 1993, *Introduction to Semiconductor Optics* (Prentice Hall, Englewood Cliffs, New Jersey).

Rama Krishna M. V., Friesner R. A., 1991, Phys. Rev. Lett. **67**, 629.

Reed M. A., Bate R. T., Bradshaw K., Duncan W. M., Frensley W. R., Lee J. W., Shih H. D., 1986, J. Vac. Sci. Technol. **4**, 358.

Roussignol P., Ricard D., Flytzanis C., Neuroth N., 1989, Phys. Rev. Lett. **62**, 312.

Roussignol P., Ricard D., Rustagi K. C., Flytzanis C., 1985, Opt. Commun. **55**, 1431.

Shinojima Hiroyuki, Yumoto Junji, Uesugi Naoshi, 1992, Appl. Phys. Lett. **60**, 298.

Spiegelberg C., Henneberger F., Puls J., 1991, SPIE Proc. **1362**, 935.

Stegeman G. I., Caglioti E., Trillo S., Wabnitz S., 1987, Opt. Commun. **63**, 281.

Stegeman G. I., Wright E. M., 1990, Opt. Quant. Electron. 22, 95.

Tada K., Hirose K., 1974, Appl. Phys. Lett. **25**, 561.

Takada T., Mackenzie J. D., Yamane M., Kang K., Peyghambarian N., Reeves R. J., Knobbe E. T., Powell R. C., submitted to JOSA B.

Takagahara T., 1989, Phys. Rev. B **39**, 10206.

Vahala K. J., Sercel P. C., 1990, Phys. Rev. Lett. **65**, 239.

Weisbuch C. and Nagle J., 1990, in Science and Engineering of One- and Zero-Dimensional Semiconductors, Plenum, New York, p. 309.

Woggon U., Gaponenko S., Langbin W., Uhrig A., Klingshirn C., 1993, Phys. Rev. B.

Wright E. M., Koch S. W., Ehrlich J. E., Seaton C. T., Stegeman G. I., 1988, Appl. Phys. Lett. **52**, 2127.

Yao S. S., Karaguleff C., Gabel A., Fortenbery R., Seaton C. T., Stegeman G., 1985, Appl. Phys. Lett. **46**, 801.

Yumoto J., Shinojima H., Uesugi N., Tsunetomo K., Nasu H., Osaka Y., 1990, Appl. Phys. Lett. **57**, 2393.

Photosensitive glass integrated optical devices

B J Ainslie, G D Maxwell and D L Williams

B T Laboratories, Martlesham Heath, Ipswich IP5 7RE, UK

ABSTRACT

Photosensitive glasses for applications in integrated optical devices have recently become a popular research topic, due primarily to the success of the germania doped silica fibre counterpart. Whole ranges of devices could in principle be fabricated utilising the photosensitivity phenomenon, from directly written waveguides to more complex grating structures. A key parameter in all applications is the change in material refractive index after light exposure. Considerable effort has been devoted to understanding mechanisms responsible for the phenomena with attempts to maximise the effect. To date index changes achieved are $\sim 10^{-3}$ which is sufficient for the production of highly efficient holographic gratings within a waveguide, but not quite large enough for direct waveguide formation. Various writing schemes have been developed, but most are based on UV laser sources operating in the 240 - 270nm wavelength range to coincide with a germanium related defect absorption. High quality gratings, with reflection efficiencies of 14dB have been written. These gratings have been used as feed back elements in external cavity lasers and when incorporated in a Mach-Zender interferometer, a channel dropping filter has been demonstrated. With continued effort, more compact optical devices utilising gratings and the photosensitivity effect can be expected in parallel with the fibre development work.

1. INTRODUCTION

The concept of using the phenomenon of photosensitivity (change in refractive index on exposure the light) in optics is not new. Substantial (ie ~1%) changes in index were reported in the 1970's in various chalcogenide glasses such as As-S, Ge-Sb and Ge-As-Se[1] upon exposure to argon ion laser irradiation. Chalcogenide systems offer a variety of very interesting and potentially useful photoinduced effects[2] due to structural or physico-chemical changes in the matrix, but in general these systems have not been followed through to practical planar optical devices, most likely for reasons such as toxicity, propagation loss and difficulty with coupling to input and output fibre. In addition, in the mid 1980's Eu^{3+} doped silicate and phosphate glasses were shown to exhibit photosensitivity when irradiated at a wavelength corresponding to the $^7F_0$ - $^5D_2$ absorption band. The mechanism proposed in this case was a change in configuration of network modifier ions in the local environment[3] of the Eu^{3+}. Both permanent and transient gratings were reported. However to the authors knowledge waveguiding structures have not been reported in these glass systems.

Photosensitivity in germanium doped silica optical fibre was first reported in 1978[4] when a longitudinal grating was written at 488nm by forming a standing wave between the transmitted and end reflected power from an argon ion laser. This work was revived in the late 1980's when Meltz et al demonstrated the transverse writing of holographic gratings into the cores of optical fibres using the intereference of two beams of intense UV radiation[5]. Since this time considerable effort has been geared to understanding the photosensitive mechanism to enhance the effect, reduce writing times etc. with a view to fabricating a wide range of practical devices. The Ge doped silica material system is perhaps the most ideal for reasons of stability and compatability with standard fibre and hence worldwide effort and progress with such fibres has been great.

This now begs the question - what can be done with planar waveguides; what is required and what are the critical parameters? This paper reviews the status from the viewpoint of device requirements in general and comments on possible future developments.

2. REQUIREMENTS

Different devices will have differing needs, however we can list some key parameters for practical devices incorporating holographic gratings or direct write waveguides:

- Δn modulation of $> \sim 10^{-5}$ as a reflection element in a device such as a feedback element.
- Δn modulation of $\sim 10^{-3}$ for WDM applications.
- Δn modulation of $\sim 5 \times 10^{-3}$ for direct write waveguides.
- Simple writing technique, ideally with insitu monitoring
- Stable gratings/waveguides with long lifetimes
- Waveguides must be/remain low optical loss
- Efficient, low loss coupling from fibres and devices must be possible
- Writing powers and times must economic

From the list above it is apparant that only modest index modulations are necessary for some important applications. A sinusoidal index pertubation yields a sinc^2 type reflection profile and the relationship between grating bandwidth ($\Delta \lambda$), index modulation (Δn), and grating length (L) can be calculated from:

$$\Delta \lambda = \{\lambda^2/\pi n L\} \cdot \{\pi^2 + (\pi \Delta n L \sigma/\lambda)^2\}^{1/2} \qquad (1)$$

where λ = wavelength of propagating light, n = refractive index of the glass and σ = mode overlap with the index grating. Additionally, the reflectivity (R) of a sinusoidally varying index grating is given by:

$$R = \tanh^2\{\pi \Delta n L \sigma/\lambda\} \qquad (2)$$

From these two equations we can plot reflectivity as a function of bandwidth for different index modulations, as shown in figure 1. It is clear that reflectivities approaching 100% are possible for low index modulations, but bandwidths are narrow and gratings lengths are concomitantly long $\sim$ 10mm. For WDM applications, where a channel might reasonably span ~1nm and spacings between channels might reasonably be few nm, a highly reflecting grating is required. In this case a grating length of $\sim$ 1-2 mm yields the requirement for $\sim 10^{-3}$ modulation.

These index modulations have been shown to be within the reach of the Ge-SiO$_2$ glass system from the extensive work on fibres[6-8]. Direct write guides however require considerably greater index changes and due to the much longer path lengths required for typical devices, loss becomes an increasingly important issue.

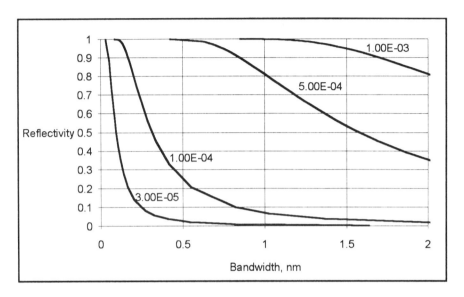

Figure 1 Peak reflectivity for a sinusiodal index grating as a function of bandwidth for different index modulations.

3. MECHANISMS

The materials work which has followed through to successful device fabrication has been restricted so far to doped silica. For this reason this review will limit the mechanism survey to this system.

The most commonly reported phenomenon associated with photosensitivity in Ge doped silica based waveguides is the 5.1eV (240nm) absorption band related to a Ge defect centre, although weak photosensitivity has been reported in Ti-SiO$_2$ planar guides on exposure to intense visible radiation[9]. The proposed structures[10] of the defect are summarised in table 1, clearly suggesting that both Ge concentration and oxygen deficiency will relate to the total defect number. This has been confirmed by showing that the extinction of the 240nm defect band can be enhanced by fabricating Ge-SiO$_2$ fibres in a reduced atmosphere[11].

On exposure to 240nm radiation, the absorption at 240nm decreases. The bleaching suggests that the defect responsible for this absorption band has been destroyed and this is responsible for the index change. The colour centre theory[12] suggests that the index change is caused by an electronic mechanism, with electrons released from the defect site migrating to other sites in the glass structure. An alternative theory[6] suggests that the destruction of the 240nm defect precipitates a more catastrophic structural change, resulting in glass densification. Both theories have as the end state a glass structure with a greater polarisability per unit volume.

Ge doped silica planar material fabricated under conditions to make low loss photonic integrated circuits is remarkably defect free. For this reason untreated waveguides are

Name	Proposed structure	Description and characteristics
An oxygen deficient silica or germanium defect	(Silicon or germanium; Oxygen)	• A neutral oxygen vacancy between two silicon (or germanium) atoms.
	(Silicon or germanium; Oxygen)	• A divalently bonded silicon (or germanium) atom with a lone pair of electrons [Skuja et al., 1984]
	(Germanium; Oxygen)	• A GeO molecule. [Yuen, 1982]

Table 1

barely photosensitive. However, the material can be sensitised by annealing at elevated temperatures in hydrogen. Figure 2 shows the UV spectra of a 6μm thick film of Ge doped silica after various anneal times in H_2 at 750°C, clearly showing the growth of the 240nm defect band with time. Note also the growth of a higher energy band at 195nm.

The bleaching of this band upon exposure at 240nm can be very accurately monitored in the planar geometry. This is shown in figure 3 where the sample depicted in figure 2 was irradiated CW at 244nm for various times. Note that the reduced intensity of the

240nm band with time accompanies an increased absorption at 213nm as predicted[12] by the colour centre model. The first successful application of this hot hydrogenation method in planar waveguides was first reported by Maxwell et al[13] where whole wafers were photosensitised. The local application of a reducing atmosphere by applying a low temperature flame to portion of a device to make limited regions photosensitive has subsequently been reported[14]. Additionally, sol-gel fabricated Ge doped silica films with a high Ge concentration have also been shown to be photosensitised[15] by thermal annealing in H_2. In this case the 240nm defect band was partially bleached by irradiation at 488nm; the bleaching mechanism being two photon absorption into the defect band. This wavelength of irradiation is clearly much less efficient than direct UV excitation. However a modest index modulation was achieved here due to the high Ge concentration of the guide.

Recently very successful work on photosensitising material by "hydrogen loading" has been reported[16]. With this method fibres or planar waveguides are subject to pressures of several tens of atmospheres of hydrogen at ~ room temperature for several days so as to impregnate the glass with molecular hydrogen. On exposure to UV light extremely large increases in index can be achieved; for example with a standard telecoms fibre a change in Δn from 0.005 to 0.017 has been observed[17]. The mechanism for the index change is related to a chemical change in the glass, as irradiated glasses are accompanied by strong hydroxyl absorption at 1380nm. It is interesting to note that the UV induced index change yields much larger changes than thermal treatment following hydrogen loading. However the magnitudes of the index changes being reported in fibres are sufficiently large to make direct writing in planar guides a very exciting prospect. To date the largest index increase in planar guides[18] observed with this technique is 10^{-3}; somewhat lower than that of the fibre results.

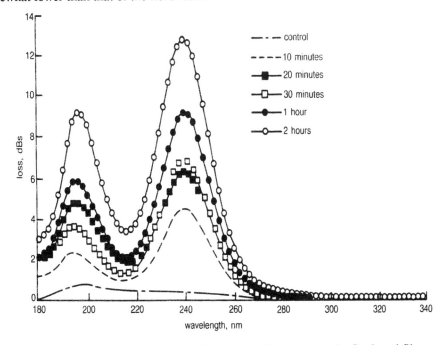

Figure 2 The effect on the UV spectrum of increasing H_2 treatment in Ge doped films.

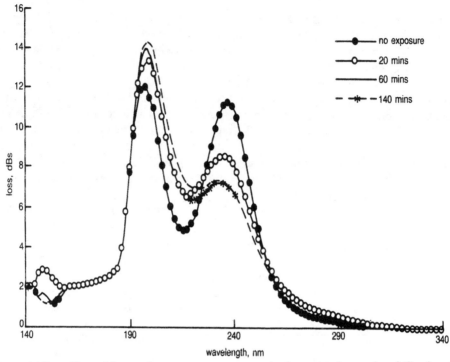

Figure 3 The effect of increasing uv exposure on a hydrogenated sample of Ge doped silica

As with previous work, the largest UV induced index changes are observed in Ge-SiO$_2$ material, however, recently index changes have been observed in P-SiO2 waveguides[19]. In this latter case irradiation at higher energies (193nm) was needed for an observable index change. It is interesting to study the UV spectra of this glass system as clearly a 240nm Ge related defect band cannot be present. UV spectra have been recorded on 30μm thick P-SiO$_2$ waveguides in a number situations: before and after 193nm exposure, before and after H$_2$ loading and 193nm exposure, before and after flame brushing and after flame brushing and 193nm exposure. The results are shown in figure 4 where clearly the most marked changes in absorption spectra (and hence expected refractive index change) after treatment are with the exposed hydrogen loaded samples and the exposed flame brushed samples.

3. UV WRITING SCHEMES

A variety of writing schemes have been reported and these are common for both fibre and planar guides. The essential elements consist of i) a high power UV source, some examples of those used are: a frequency doubled argon ion laser, rare gas halide (RGH) laser, pumped dye lasers, frequency quadrupled Nd:YAG lasers and diode pumped frequency quadrupled Nd:YLF lasers and ii) a means of forming well defined fringes or waveguides.

Clearly the different laser sources provide widely varying output powers and this influences the magnitude of the induced refractive index change. At very high power

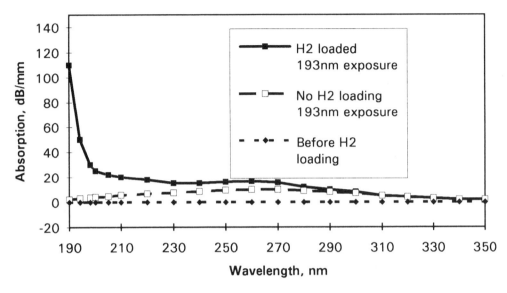

Figure 4a Absorption spectra of 30μm thick P doped silica films at various stages of H_2 loading and exposure at 193nm (after ref 19).

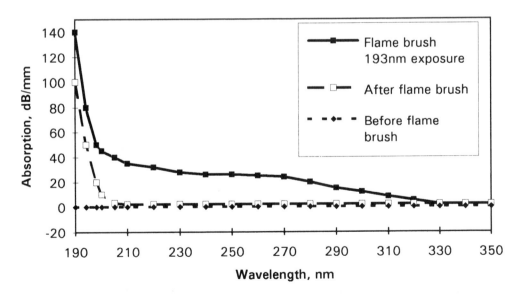

Figure 4b The absorption spectra of 30μm thick P doped silica films at varoius stages of flame brushing and exposure at 193nm (after ref 19).

levels damage gratings can be formed[7]. Here, the index change is confined to the core/cladding interface and this asymetry about the fibre axis leads to out-of-plane reflection.

Two basic writing schemes are commonly used. The first uses two interfering UV beams

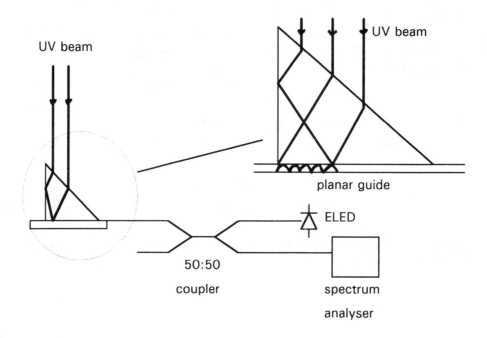

Figure 5 The prism interferometer set-up used for forming gratings in waveguide cores. The inset shows a close-up of the interferometer.

which originated from a laser and were split and then recombined, using mirrors or a prism to form an interference pattern in the core of the waveguide. This is shown in figure 5. The in-situ monitoring of the reflection grating is carried out by coupling broadband light from an E-LED via a 50:50 coupler into the planar waveguide and picking up the back reflection on a spectrum analyser. This writing scheme has been shown to be stable, but requires a high quality coherent light beam. More recently phase masks have been used[20, 21], which allows a reduction of the coherence requirements of the irradiating light. Additionally ease of alignment and a reduced sensitivity to mechanical vibrations are also claimed. Figure 6 details a phase grating based system described by Armitage[20]. The fused silica phase grating was fabricated using E-beam technology. A cylindrical lens was used to form a line focus in the waveguide core to increase uv intensities.

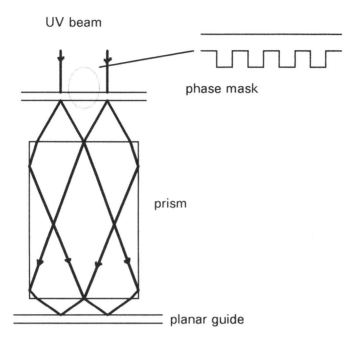

Figure 6 The phase mask arrangement showing the diffraction grating in silica which is used as the transmission hologram.

4. PLANAR DEVICES

4.1 Reflection filters

High quality reflection filters in planar waveguides were first demonstrated using the thermal hydrogen treatment[12,22]. An area of potential concern with this technique is the reduction of the GeO_2 in the glass which results in short wavelength scatter loss and increased UV absorption. It has been demonstrated that hydrogenation times of up to 1 hour at 750°C cause negligable increase in loss at wavelengths longer than 1µm, However, as shown in figure 7 for times longer than this, short wavelength attenuation can result. Using the 1 hour hydrogenation time for sample preparation and writing with ~25mW power at 244nm for 20min yielded the 3mm grating shown in figure 8. Note that a peak reflectivity of 13dB resulted, with a bandwidth of 0.6nm, which translates to an index modulation of 5×10^{-4}. Careful examination of the grating response reveals well resolved secondary peaks either side of the central band, agreeing well with the $sinc^2$ type response predicted for a truncated sinusoidal grating with constant index modulation. This result clearly demonstrated the ability to write high quality gratings into planar guides with negligable harmful effects from reflections from substrate surfaces.

High quality gratings have also recently been demonstrated in silica-on-silicon waveguides using the flame brushing technique[14], with index modulations of 9×10^{-4} reported.

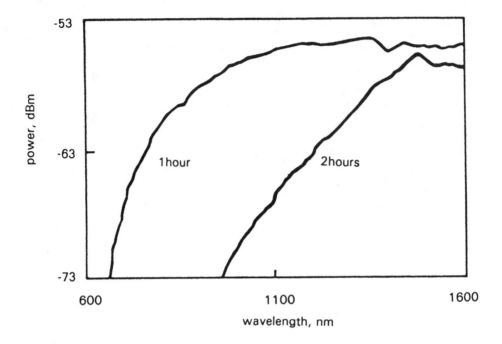

Figure 7 Transmission of 5cm long waveguides after hydrogenation for 1 and 2 hours.

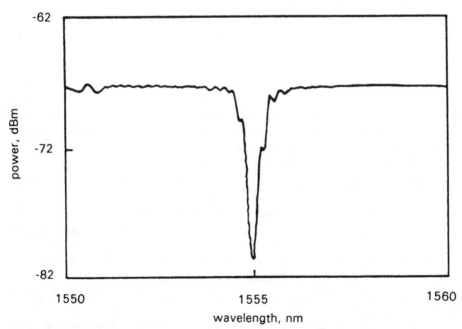

Figure 8 Transmission spectrum of planar waveguide after uv exposure showing holographic grating characteristics.

4.2 Applications of relection filters

4.2.1 External cavity laser

The first demonstation of combining a semiconductor laser with a planar holographic grating to form an external cavity was recently reported by Maxwell et al[23]. The configuration was similar to fibre devices[24] and consisted of front facet AR coated buried heterostructure laser, the output of which was coupled into the planar waveguide via a glass ball lens (see figure 9).

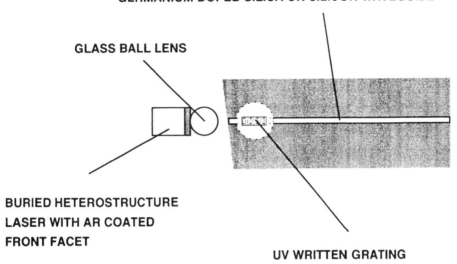

Figure 9 A schematic of the external cavity gating planar laser.

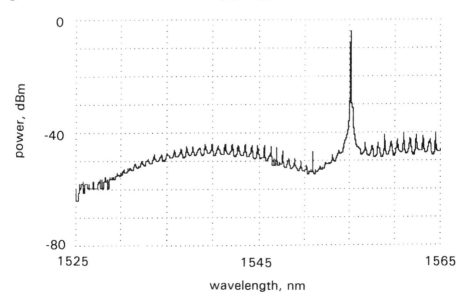

Figure 10a The lasing spectrum for the external cavity grating planar laser

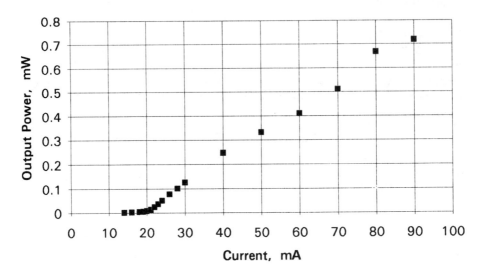

Figure 10b Light output power vs drive current for the external cavity grating planar laser.

The planar guide was polished to an angle of ~2 degrees to prevent reflection problems and ~ 30% coupling efficiency was achieved. The grating reflectivity was 12dB, centred at 1555nm. The lasing characteristics are shown in figure 10a, where a sideband suppression of -40dBm is clear. The threshold of the device was 20mA and an output power approaching 1mW resulted; figure 10b shows the light output power vs drive current characteristics. The low chirp properties of these gratings was demonstrated by directly modulating the device at 2.5Gb/s and transmitting the signal through 100km of nondispersion shifted fibre. The receiver sensitivity for a 10^{-9} bit error rate was -24.8dBm, only 0.3dB lower than the back-to-back power level. This result is considerably better than standard DFB laser structures.

4.2.2 Rare-earth doped planar lasers

The combination of gain from rare-earth dopants and feedback from a holographic grating has recently been reported with the Er^{3+} in P-doped silica[19] and grating writing at 193nm. The waveguide was doped with 0.5wt% Er^{3+}, which yielded a saturated gain of 0.3dB/cm at 1546nm on pumping at 976nm. The single frequency laser was 4cm long and the cavity was constructed by photoimprinting two 5mm long Bragg gratings close the ends of the waveguide. The maximum reflectivity of the gratings was as high as 97% with a FWHM of 0.5nm. The threshold of the laser occurred at an incident power of 60mW and an output power of 340µW resulted with a pump power of 300mW.

4.3 Bandpass filter

Holographic gratings have been written into each arm of a balanced Mach-Zehnder interferometer to demonstrate a planar four port band pass filter[25]. The schematic is shown in figure 11. The interferometer was fabricated in Ge doped silica which was made photosensitive by annealing in H_2 at elevated temperatures. A key feature of the technology was the ability to "UV trim" the device as the propagated light in each arm had to remain in phase for efficient operation. This trimming was achieved by exposing a portion of each arm after grating writing and monitoring in situ the filter response. Figure 12 shows the spectra from the four ports of the device. The complementary nature of the outputs at ports two and three are clear. A 13.5dB rejection within a band stop of 1nm of the output port was achieved. The total fibre to fibre insertion loss of the device was measured to be 1.35dB.

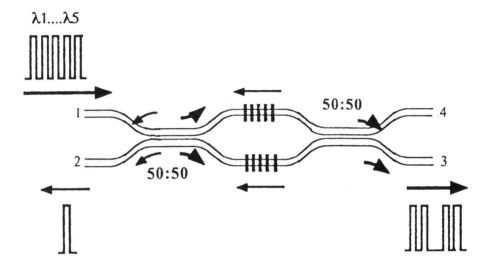

Figure 11 A schematic of the MZ interferometer showing a single wavelength being reflected by the gratings and emerging at port 2.

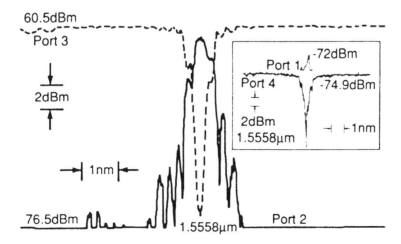

Figure 12 Spectra from the four ports of the MZ interferometer after the trimming process

5. CONCLUSION AND FUTURE DIRECTIONS

Photosensitivity in planar waveguides is a phenomena which has only very recently been shown to be practically useful in the construction of optical devices. The key property of sufficient index modulation in grating formation has now been realised and a range of functions based on reflection elements has been demonstrated. The combination of gain and photosensitivity has also just been shown which will allow more novel optical circuitry to be realised. Key points to be addressed now are durability and lifetime of written gratings and increase of the refractive index modulation for directly written waveguide fabrication.

6. REFERENCES

1. A.M. Andricsh, Y.A. Bykovsky, E.P. Kolomeiko, A.V. Mako-Ukin, V.L. Smirnov, A.V. Shmalko, "Waveguide structures and functional elements of integrated optics systems based on holographic gratings in As_2S_3 films," Sov. J. Quantum. Electron. 7, p347 - 351, 1977.
2. A.E. Owen, A. P. Firth, P.J.S. Ewen, "Photoinduced structural and physicoichemical changes in amorphous chalcogenide semiconductors", Philisophical Magazine, 52, No. 3, p347-362, 1985.
3. F.M. Durville, E.G. Behrens, R.C. Powell, "Laser induced refractive index gratings in Eu-doped glasses", Phys. Rev. B, 34, pp4213 - 4218, 1986.
4. K.O. Hill, Y. Fujii, D.C. Johnson and B.S. Kawasaki, "Photosensitivity in optical fibre waveguides: application to reflection filter fabrication", Applied Phys. Lett., 32, No. 10. p647-649, 1978.
5. G Meltz, W.W. Morey, W.H. Glenn, "Formation of Bragg gratings in optical fibres by transverse holographic method", Optics Lett., 14, No. 15, p823 -825, 1989.
6. D.L. Williams, B.J. Ainslie, R. Kashyap, R. Campbell, J.R. Armitage, "Broad bandwidth highly relecting gratings formed in photosensitive boron co-doped fibres", Proceedings vol3, Post deadline papers 18th European conference on optical communication (ECOC 92), p923-926, 1992.
7. J. Archambault, L. Reekie, P. St. J. Russell, "100% reflectivity Bragg reflectors produced in optical fibres by single excimer laser pulses" Electron. Lett., 24, p453-455, 1993.
8. P.J. Lemaire, R.M. Atkins V Mizrahi, W.A. Reed, "High pressure loading as a technique for achieving ultra high UV photosensitivity and thermal sensitivity in $GeO2$ doped optical fibres", Electron Lett., 29, p1191-1193, 1993.
9 Y. Hibino, M. Abe, T. Kominato, Y. Ohmori, "Photoinduced refractive index changes in TiO2-doped silica optical waveguides on silicon substrate", Electron. Lett. 27, No. 24, p2294-2295, 1991.
10. D.L. Williams, B.J. Ainslie, R. Kashyap, G.D. Maxwell, J.R. Armitage, R.J. Campbell and R Wyatt, "Recent advances in photosensitivity in germanosilicate fibre and planar waveguides", to be published in conf preceedings. SPIE, 2044, Quebec, 1993.
11. D, L. Williams, B.J. Ainslie, J.R. Armitage, R.Kashyap, "Enhanced photosensitivity in germania doped silica fibres for future optical networks, "Proceedings, 1, 18th European conference on optical communication (ECOC 92), p425-428, 1992.

12. D.P. Hand, P. St.J. Russell, "Photoinduced refractive index changes in germanosilicate fibres", Optics Lett., **15**, No. 2, p101-104, 1990.

13. G.D. Maxwell, R Kashyap, B.J. Ainslie, "UV written 1.5µm reflection filters in single mode planar silica guides", Electron. Lett., **28**, No. 22, p2107-2107, 1992.

14. K.O. Hill, F. Bilodeau, B. Malo, J Albert, D.C. Johnson, Y. Hibino, M. Abe, M Kawachi, "Photosensitivity of optical fibre and silica on silica/silicon waveguides", Optics Lett., **18**, No. 12 p 953-955, 1993.

15. K.D. Simmons, G.I. Stegman, B.G. Potter, J.H. Simmons, "Photosensitivity of sol-gel derived germano-silicate planar waveguides", Optics Lett., **18**, No. 1, p25 -27, 1993.

16. R.M. Atkins, P.J. Lemaire, T. Erdogan, V. Mizrahi, "Mechanism of enhanced UV photosensitivity via hydrogen loading in germano-silicate glasses", Electron. Lett., **29**, p1234 - 1235, 1993.

17. P.J. Lemaire, A.M. Vengsarkar, W.A. Reed, V Mizrahi, K.S. Kranz, "Refractive index changes in optical fibres with molecular hydrogen", Technical Digest of Conference on Optical Communication (OFC'94), p47, 1994.

18. V. Mizrahi, P.J. Lemaire, T Erdogan, W A Reed, D J DiGiovanni, R M Atkins, "UV laser fabrication of ultrastrong optical fibre gratings and of germania-doped channel wavguides", Appl. Phys. Lett., **63**, No. 13, p 1727-1729, 1993.

19. T. Kitagawa, K.O. Hill, D.C. Johnson, J. Albert, S. Theriault, F. Bilodeau, K. Hattori, Y. Hibino, "Photosensitivity in P2O5-SiO2 waveguide and its application to Bragg grating reflectors in single frequency Er3+ doped planar waveguide laser", ibid, Post deadline session, p79 -82, 1994.

20. R. Kashyap, J. R. Armitage, R. J. Campbell, D.L. Williams, G.D. Maxwell, B.J. Ainslie , C.A. Millar, "Light sensitive fibres and planar waveguides", BT Technol J, **11**, No.2, p150-160.

21. K.O. Hill, B. Malo, F. Bilodeau, D.C. Johnson, J. Albert, "Bragg gratings fabricated in monomode photosensitive optical fibre by UV exposure through a phase mask", Applied Phys Lett, **62**, No. 10, pp1035-1037, 1993.

22. G.D. Maxwell, B.J. Ainslie, D.L. Williams, R Kashyap, "UV written, 13dB reflection filters in hydrogenated low loss planar silica waveguides", Electron. Lett., **29**, No. 5, p425-426, 1993.

23. G.D. Maxwell, R. Kashyap, G. Sherlock, J.V. Collins, B.J. Ainslie, "Semiconductor external-cavity laser with a UV-written grating in a planar silica waveguide", Technical Digest of Conference on Optical Communication (OFC'94), p151, 1994.

24. D.L. Williams, B.J. Ainslie, R. Kashyap, G. Sherlock, R.P. Smith, J.V. Collins, "Temperature stable 1.3µm laser with Bragg fibre grating external cavity for access networks", Proceedings vol2, 19th European conference on optical communication (ECOC 93), p209-213, 1993.

25. R. Kashyap, G.D. Maxwell, "Laser trimmed four-port bandpass filter fabricated in single-mode photosensitive Ge-doped planar waveguide", IEEE Photonics Technol. Lett., **5**, No2, p191-194, 1993.

Nonlinear Fibers and Waveguides II

Fused bitapered fiber devices for telecommunication and sensing systems

Suzanne Lacroix
Département de génie physique
École Polytechnique de Montréal
P.O. Box 6079, Station "Centre-ville"
Montréal (Québec), H3C 3A7, Canada

ABSTRACT

An overview of all-fiber passive devices and components made by the fusion and tapering technique is given. Single tapered fiber are examined as well as 2×2, 3×3 and other simple star couplers, and interferometric structures made of these components. Some relevant applications are described with their theory of operation and illustrated with typical experimental results.

1. INTRODUCTION

The most common all-fiber component used in single-mode fiber optic communication is the 2×2 coupler made by the fusion and tapering technique: two identical optical fibers are fused laterally together and elongated with a heat source. The geometrical parameters which characterize the resulting structure, and thus the coupler behavior, are the cross section, determined by the degree of fusion, and the longitudinal biconical profile. In particular, the abruptness of the slopes of the conical regions is a crucial parameter because it determines the coupling to higher modes. This process, which is undesirable in the 2×2 coupler because it causes losses, may be exploited, in particular, to customize the responses of other star couplers. In a single tapered fiber, this effect may be used for spectral filtering or sensor applications. The properties and applications of tapered single fibers are described in the following section. Couplers and their standard applications, power splitters and wavelength division multiplexers, are examined in the third section, while the fourth section is devoted to composite structures such as all-fiber interferometers.

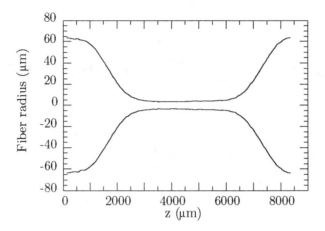

Figure 1: Profile of an abruptly tapered fiber. The untapered fiber parameters are: cladding diameter $\phi_{cl} = 127$ μm and core diameter $\phi_{co} = 4.5$ μm; second mode cutoff wavelength $\lambda_c = 578$ nm.

2. TAPERED FIBERS

2.1. Basic concepts

Locally heating and stretching a single-mode fiber creates a biconical structure such as that of Fig. 1. When the taper angle is everywhere small enough to ensure negligible loss of power from the fundamental mode as it propagates along the structure, the propagation and, by extension, the taper itself is said adiabatic: the fiber transmission is not affected by the tapering process [1]. However when the slopes are abrupt, such as those of the structure shown in Fig. 1, one may observe, as the fiber is elongated, large oscillations in the transmitted power. For a given elongation, similar oscillations are seen in the transmission as a function of wavelength (Fig. 2) or of the refractive index surrounding the tapered part of the fiber [2]. This behavior may be explained in terms of coupling and beating of the structure modes [3]. As a matter of fact, the fundamental core mode is "cut off" for a fiber parameter $V_{co} \approx 0.7$ (corresponding to a cladding radius $r = r_c \approx 21$ μm, at $\lambda = 633$ nm, for the fiber of Fig. 1). For smaller V, i.e., smaller radius, the mode becomes a cladding mode: in the central region of the structure, the guidance is thus ensured by the cladding which plays the role of a core, the external medium, usually air, being then the cladding. Note that, unless its diameter is very small, the central region, when surrounded by air, is multimode. In the downtaper part, when non-

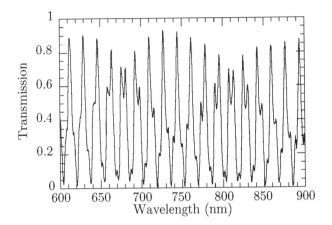

Figure 2: Fundamental mode transmission of the tapered fiber of Fig. 1 as a function of wavelength. After Ref. [2].

adiabatic, the fundamental mode HE_{11} is converted in a superposition of modes of same symmetry (HE_{11}, HE_{12}, HE_{13}, ...) which then propagate in the central region without further coupling. They accumulate relative phase differences before they couple again in the uptaper region. The power is then recovered in the core, totally if the excited modes are in phase, only partially otherwise. As a result from this coupling-beating-coupling process, the taper behaves as a modal interferometer with a transmission approximately given by [4]

$$T = \sum_m T_m + \sum_{m>n} \sum_n \sqrt{T_m T_n} \cos \varphi_{mn} \qquad (1)$$

φ_{mn} being the phase difference accumulated by the modes HE_{1m} and HE_{1n} in the central region and T_m the HE_{1m}-mode transmission. Modes HE_{11} and HE_{12} are responsible for the large oscillations seen in Fig. 2, while the smaller, more rapid superimposed oscillations come from the contribution of HE_{13} and possibly higher order modes. In addition, when the core may be neglected in the cladding guiding regime, one shows that the accumulated phase differences φ_{mn} are proportional to the wavelength λ [5] corresponding to sinusoidal variations as seen in Fig. 2.

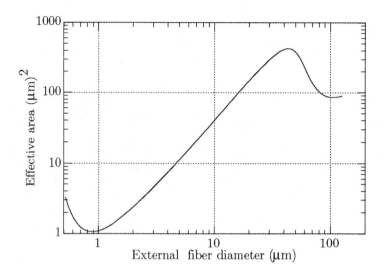

Figure 3: Effective area of the fundamental mode of a three-layer fiber (core/cladding/air) calculated at $\lambda = 1.55\,\mu\text{m}$. The fiber used for calculation is a standard telecommunication matched-cladding fiber: cladding diameter $\phi_{cl} = 125\,\mu\text{m}$ and core diameter $\phi_{co} = 9\,\mu\text{m}$; index step $n_{co} - n_{cl} = 4.5 \cdot 10^{-3}$. After Ref. [7].

2.2. Applications

2.2.1. Adiabatic tapers

Adiabatic tapers, i.e., tapers in which guidance is ensured by the local fundamental mode without coupling to higher modes, is the basic structure of most couplers, as discussed in Section 2.2.2. When the fiber is adiabatically tapered, diffraction makes the fundamental mode escape from the core so that the modal diameter is first enlarged. Then, the field reaches the cladding-air boundary and the modal diameter is determined by the fiber diameter. If the fiber diameter is sufficiently reduced, the field will finally escape from the cladding.

This effect is well described by the effective mode area defined by

$$A_{\text{eff}} = \frac{[\iint_\infty |\psi|^2 dA]^2}{\iint_g |\psi|^4 dA} \qquad (2)$$

where ψ is the field; the integration of the numerator is over the infinite cross section and the integration of the denominator is over of the guide section. Actually, the effective area A_{eff}, shown in Fig. 3, is the relevant parameter for Kerr nonlinear effect perturbative calculations. From this effective area, one may not only follow the field shape evolution of the fundamental mode

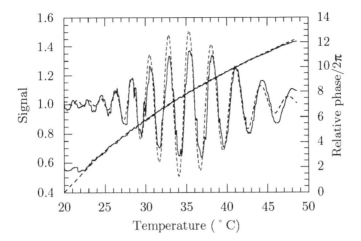

Figure 4: Experimental response of a tapered fiber temperature sensor (solid line) compared to the theoretical response (dashed line). After Ref.[8].

in a tapered fiber, but also calculate its propagation constant as a function of incident power and thus calculate the self phase modulation effect (SPM). Adiabatic tapering, which has been used to adapt the modal fields at the splice between two different fibers and thus greatly reduce the coupling loss [6], is now exploited to condense light and thus locally increase the field intensity [7]. Fig. 3 shows that, compared to an untapered fiber, the effective area of the fundamental mode may be reduced by a factor of the order of 100 at a diameter of the order of 1 μm. This reduction in guide area is equivalent, from the SPM point of view, to an increase of either the power or the fiber length by the same factor. From a practical point of view, tapering the fiber thus potentially decreases the device length or power requirement by two orders of magnitude. It is however important to note that tapering also largely affects the guide dispersion, which may result in unexpected dramatic effects [7]. These combined light condensing and dispersion variation effects may be exploited to design compact all-fiber all-optical switches based on structures such as Mach-Zehnder or Sagnac interferometers, as described in Section 4.

2.2.2. Nonadiabatic tapers

The interferometric response of nonadiabatic tapers, given by Eq. 1, has been exploited for sensing as well as filtering purposes.

- **Sensors**

  The sensitivity of interferometers, which is determined in a classical interferometer by the fringe period (i.e., the wavelength $\lambda \approx 1$ μm) is determined in a modal interferometer by the beat length (i.e., typically $z_b \geq 10$ μm in a tapered fiber). As a result, the classical interferometer is ten times more sensitive to environmental parameters than the modal interferometer. In addition, the length of a tapered fiber is limited to a few centimeters, which also limits its sensitivity. Nevertheless, the tapered fiber is a compact and inexpensive interferometric device with no need for a reference arm nor for any mechanical adjustment which makes it very convenient in sensing applications. Due to its transmission dependence on the external medium, the tapered fiber may also be used as a refractometer or a temperature sensor. Fig. 4 is an example of the signal extracted by the coherence multiplexing technique [8]. The dependence of the propagation constants on the fiber curvature also make it possible to envision tapers as curvature or displacement sensors [2].

- **Wavelength filters**

  Two types of spectral all-fiber filters are based on tapered fibers: selective and large band filters.

  Selective filters are obtained by concatenating several tapers (typically four) with quasi-sinusoidal wavelength responses such as that of Fig. 2 but with different periods. If the absorbing jacket is not removed between the tapers, it traps the cladding modes. The resulting transmission is then obtained by multiplying the transmissions of the four tapers. With careful choice of the periods and the maxima, one can obtain a transmission similar to that of Fabry-Perot interferometer centered on a prescribed wavelength, with typical FWHM of 5 nm [9].

  Wideband filters are designed to increase the isolation of couplers used as demultiplexers (see Section 3.3.1.). For communication applications, the taper transmission must be maximum typically at $\lambda = 1300$ nm (or $\lambda = 1550$ nm) and zero at $\lambda = 1550$ nm (or $\lambda = 1300$ nm) with the widest band possible. This result is not achievable with only one taper on a matched cladding fiber; but it is possible to create such complex profiles with two successive tapers on the same strand of unjacketed fiber. This structure is then equivalent to three interferometers in series, which provides a sufficient number of adjustable parameters to synthesize the desired response [10]. The resulting structure is compact and can be included in the WDM coupler packaging. An example of the transmission of this type of filter is given in Fig. 5.

Mechanisms involved in tapered single-mode fibers, whether adiabatically tapered or not, are well understood and only some applications in the field of

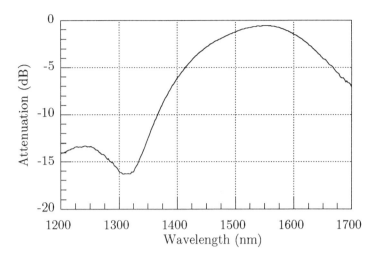

Figure 5: Transmission of a wideband spectral filter for additional demultiplexer isolation. After Ref. [11]

all-fiber components have been described here. The following section is devoted to the most important application of adiabatic tapers: the fiber couplers.

3. COUPLERS

3.1. Basic concepts

The power exchange along a star coupler consisting of N individual identical single-mode guides can be described either in terms of the coupling of the N fundamental modes of the individual guides [12] or in terms of the beating of the first N supermodes [13]. In the case of a fused coupler, the individual guides at both ends are optically separated and the fundamental modes of the N individual guides are orthogonal. Each supermode then coincides with a linear combination of these individual guide modes. The coefficients of this combination are readily obtained using symmetry conditions of the mode with respect to that of the coupler cross section. When launching power in one (or more) fiber at the coupler entrance, one then excites a superposition of N supermodes. In the power exchange region, the concept of individual guides is no longer valid and the supermode beating approach is necessary. Assuming no loss along the coupler, one may keep the fundamental assumption of a N-dimension basis; N supermodes then accumulate phase differences in this central region. At the coupler output, the spatial overlap integral between the N supermodes interference pattern and the fundamental mode of each individual guide dictates the relative power coupled back in each individual fiber. In this light, the coupler, like a non adiabatically tapered fiber, is a modal interferometer.

A key element in the analysis of the coupler is the transfer matrix, i.e., the matrix that relates the individual guide output modal amplitudes to the input amplitudes. As an example, the transfer matrix of a lossless 2 × 2 symmetric fused fiber coupler of length L is given by

$$\mathbf{M}_{2\times 2}(\alpha) = e^{i\overline{\alpha}} \begin{bmatrix} \cos\alpha & i\sin\alpha \\ i\sin\alpha & \cos\alpha \end{bmatrix} \quad (3)$$

where

$$\overline{\alpha} = \int_0^L \overline{B}(z)\,dz \quad \text{with} \quad \overline{B}(z) = \frac{B_{01}(z) + B_{11}(z)}{2} \quad (4)$$

and

$$\alpha = \int_0^L \frac{B_{01}(z) - B_{11}(z)}{2}\,dz \quad (5)$$

Here $B_{01}(z)$ and $B_{11}(z)$ are the propagation constants of the fundamental LP_{01} and first antisymmetric LP_{11} local supermodes. The coupler limits, $z = 0$ and $z = L$, are defined as the points where the individual fibers may be considered to be optically independent. The parameter 2α thus represents the phase difference accumulated by the two modes along the coupling region.

From the transfer matrix one can, for a given excitation corresponding to a given input column vector, calculate the power transmissions in both branches. For example, an excitation in branch 1 corresponds to the particular input vector

$$\begin{bmatrix} 1 \\ 0 \end{bmatrix} \quad (6)$$

and results in a power transmission in the same branch

$$T_1 = \cos^2\alpha = \frac{1}{2}[1 + \cos 2\alpha] \quad (7)$$

When the modes are in phase at $z = L$ ($2\alpha = 0$), the transmission is maximum in the main branch ($T_1 = 1$). When they are out of phase at $z = L$ ($2\alpha = \pi$), the transmission is minimum ($T_1 = 0$), which corresponds to full power transfer from main to secondary branch.

One can also easily calculate the transmissions of concatenated 2 × 2 couplers arranged in structures such as an all-fiber Mach-Zehnder interferometer, as illustrated in section 4.1. It is thus a very useful formalism for the analysis of structures made of 2 × 2 couplers and is readily generalized to other star couplers [14]. As for tapers, the α-parameter defined by Eq. 5 is dependent on elongation, wavelength, and external refractive index, making the coupler transmission oscillatory as functions of these parameters. The response of a given coupler depends not only on the longitudinal profile determined by the elongation process but also on the degree of fusion. These fabrication parameters have been modeled and a program has been developed to calculate the scalar supermodes and their polarization corrections [15]. The α-parameter is then obtained by integration over the longitudinal profile (Eq. 5). Experimental responses vs. elongation and wavelength, including the polarization effects, have been theoretically reproduced for different degrees of fusion [16], which validates each step of the theoretical simulations. The programs developed to simulate the behavior of fused couplers are also powerful tools to design couplers with a prescribed behavior.

3.2. Power splitters and switches

Unlike tapered fibers, 2×2 fiber couplers can exhibit an output response ranging from 0 to 100%, which corresponds to complete power transfers from one branch to another. It makes it possible to fabricate power splitters of any branching ratio at a given wavelength. From a practical point of view, during fabrication the transmission is monitored at the said wavelength and elongation stopped when the desired ratio is obtained. Couplers are often designed for 50/50 coupling ratio (3 dB couplers) but other ratios, e.g. 90/10, are sometimes desirable. Equipartition may also be obtained with higher order star couplers, e.g. 3×3 in a flat or triangular cross section arrangement, and 4×4 in a square arrangement [14].

A switch is basically a coupler elongated until a complete power transfer is observed. It is sometimes referred to as a half-beat length coupler because for a z-invariant structure one would have $L = \pi/(B_{01} - B_{11})$. Power may then be driven back to the main branch, either by some external perturbation of the structure symmetry, e.g. bending, or by using an optical nonlinearity to realize all-optical switching. However, up to now, because of the very small value of the silica Kerr coefficient ($n_2 \approx 3 \cdot 10^{-20}$ m^2/W) [17], such half-beat length structures made of standard fiber are inappropriate. On the other hand, longer couplers, in which the nonlinear phase difference is accumulated over longer distances, are technically impossible to manufacture. New developments in this area may however come soon from the fabrication of highly nonlinear glasses.

3.3. Wavelength Division Multiplexing

Wavelength division multiplexing (WDM) is a widely known technique that takes advantage of the bandwidth of the optical fiber. Mutiplexers are devices which are able to launch on the same line two (or more) signals with different wavelengths which are then separated at the output end of the line by a demultiplexer. The quasi-sinusoidal wavelength dependence of the coupler transmission is the basis of many wavelength division multiplexing couplers. As they are reciprocal devices, the same coupler can be used either as a multiplexer or a demultiplexer, the difference being only the performance required for isolation. The most common one is the 1300/1550 nm WDM, which permits to double the capacity of existing links, but the spectral response of a coupler can also be manipulated to be used in other WDM applications such as the Er^{3+}-doped fiber amplifier/laser. These particular applications are described in the following subsections. The problem of dense multiplexing (wavelength separation of the order of 1 nm) is also examined from the viewpoint of all-fiber device.

3.3.1. WDM for 1300 and 1550 nm wavelengths

To multiplex two given wavelengths λ_a and λ_b (usually around 1300 nm and 1550 nm) one has to adjust the sinusoidal wavelength response of a moderately elongated coupler by carefully controlling the degree of fusion and the elongation parameters [18]. Fig. 6 shows as an example the experimental response of such a coupler. In this particular case, the structure was elongated until a complete power exchange cycle was observed at the first wavelength $\lambda_a \approx 1280$ nm and one and a half at $\lambda_b \approx 1550$ nm. Other recipes may be used with different degrees of fusion and longitudinal profiles [19]. Multiplexers are characterized by their isolation at a given wavelength defined as

$$I(\lambda) = 10 \, |\log T_1(\lambda) - \log T_2(\lambda)| \qquad (8)$$

T_1 and T_2 being the coupler transmission in each branch. The coupler isolation may be different at λ_a and λ_b, and one also defines for each branch ($n = 1, 2$)

$$Y_n(\lambda_a, \lambda_b) = 10 \, |\log T_n(\lambda_a) - \log T_n(\lambda_b)| \qquad (9)$$

These parameters I and Y are of the order of 18 to 20 dB, which is not sufficient for communication needs (35 dB or more). Additional isolation can be obtained with in-line spectral rejection filters (see Section 2.2.2.).

3.3.2. WDM designed for Er^{3+}-doped fiber applications

Slightly elongated couplers which exhibit no power exchange oscillations in the low wavelength range may also be used to fabricate special multiplexers [20]. Recent needs for all-fiber amplifiers have prompted the fabrication of couplers to multiplex the signal wavelength with the pump wavelength. One of the most attractive dopant is erbium which provides amplification around 1550 nm when pumped at 800, 980, or 1480 nm. A WDM coupler for 1550 nm and one of these pump wavelengths, except 1480 nm, may be obtained with a recipe similar to that of 1300/1550 nm WDM couplers. However, a single, very weakly elongated coupler may multiplex any of these wavelengths with 1550 nm. It is based on the following principle. The coupling process, being related to the overlap of the fundamental mode of one coupler branch to the other, is expected to be easier as the wavelength increases. For a given elongation, we can thus define

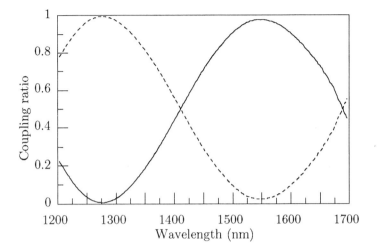

Figure 6: Wavelength division multiplexing in a fused coupler designed for 1280 nm and 1550 nm: coupling ratio (solid line) and its complement (dashed line). The fiber used in this experiment was a standard matched-cladding telecommunication fiber, designed to be single-mode at 1300 and 1550 nm, with an LP_{11} cutoff around 1200 nm.

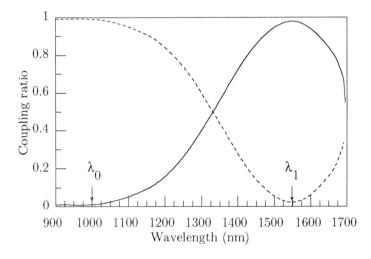

Figure 7: Wavelength division multiplexing coupler response for Er^{3+}-doped fiber amplifiers. The fiber used in this experiment was the same as that used for the 1280/1550 nm WDM.

a minimum coupling wavelength, λ_0, as the wavelength below which there is no power transfer between the individual guide fundamental modes as shown in Fig. 7. In other words, for wavelengths shorter than λ_0, the power remains in the main guide all along the coupler length but for wavelengths longer than λ_0, the two guides are coupled, creating an oscillatory response as the phase between the modes changes with wavelength. Let λ_1 be the first minimum transmission wavelength above λ_0 (see Fig. 7). During the elongation process, as the coupler waist decreases, λ_0 and λ_1 shift towards smaller wavelengths. If the two wavelengths to be multiplexed, λ_p of the pump and λ_s of the signal ($\lambda_p < \lambda_s$), are not too close to each other, e.g. 980 and 1550 nm, the longitudinal profile can be designed so that the wavelength response has a maximum power transfer at λ_s ($\lambda_s = \lambda_1$) keeping λ_0 greater than λ_p. As a result, the wavelength response will be flat around λ_p and the coupler will be able to multiplex λ_s with any wavelength shorter than λ_0. Although the difference ($\lambda_1 - \lambda_0$) cannot be made arbitrarily small, it is possible to produce a coupler for fiber amplifier application, i.e. able to multiplex the wavelength to be amplified (in our case $\lambda_s = 1550$ nm) and any pump wavelength λ_p below 1000 nm. The experimental result is shown in Fig. 7 where the fundamental mode coupling ratio and its complement are plotted as functions of wavelength.

3.3.3. Dense WDM

While 1300/1550 nm all-fiber WDM based on the technology described in Section 3.3.1. are commercially available, dense multiplexers, i.e., with a wavelength spacing $\delta\lambda \approx 1$ nm or less, which could exploit the bands around each of these two wavelengths, are more difficult to manufacture. Most of the dense WDM reported to date are integrated optics devices based on grating dispersion and thus have significant losses (≈ 5 dB per channel) [21]. This performance could be greatly improved by using cascaded 2×2 all-fiber couplers. However, to reduce the wavelength period $2\delta\lambda$ of the coupler transmission given by Eq. 7, one has to increase the accumulated phase difference 2α (defined by Eq. 5) and thus the coupler elongation. This, in turn, increases the polarization sensitivity of the coupler response [22] and, as a result, the isolation performance is degraded. Several solutions have been envisioned to overcome this problem which is related to the strong guidance conditions due to the large core-cladding index difference in the tapered region. One solution would be to etch or polish the fibers prior to the fusion and tapering process [23]. This makes the cores close enough to allow the power exchange between them before the field spreads out in the cladding region. These couplers, however, are difficult to manufacture. Others structures, such as unbalanced Mach-Zehnder, may be used instead (see Section 4.1.).

3.4. Wavelength flattened couplers

Wavelength dependent couplers are fundamental components in multi-wavelength systems. These systems also require wavelength independent splitters. The most common solution to fabricate couplers with small wavelength dependence is to make asymmetric 2×2 couplers [24]. They are made by fusing two fibers of different diameters, a difference which is obtained by pretapering or etching one of the fibers. This asymmetry limits the power exchange between the two guides and the maximum power transferred depends on the diameter difference, the degree of fusion and the longitudinal profile. If this maximum is close or slightly greater than 50%, it is possible to realize couplers which are approximately 50/50 around 1310 and 1550 nm. However, the fabrication is delicate and the flattened response still remains sinusoidally wavelength dependent. The wavelength range is thus limited and, as a result, all coupling ratios cannot be achieved with this technique. An alternative solution to the wavelength flattened response problem is to build a Mach-Zehnder structure, composed of two symmetric 2×2 couplers in series [25]. A detailed description of this device and its performance is given in Section 4.1.

Supermode coupling may also be exploited to flatten the spectral response of tapered couplers. Since the fused coupler geometry is $z-$dependent, modes of same symmetry may couple with each other whenever the adiabaticity conditions are not satisfied [1, 26]. In the case of an adiabatic symmetric 2×2 coupler, as explained in Section 3.1., the power exchange process is described in terms of the beating of the LP_{01} and LP_{11} supermodes which are respectively even and odd with respect to one of the symmetry axis of the coupler cross section. No coupling is then allowed between them whatever the longitudinal geometry of the coupler: if the adiabaticity criterion is not satisfied, coupling occurs with radiation modes and the coupler is lossy. However, in some cases, the supermodes involved in the beating can couple with each other: the 2×2 asymmetric coupler is an example [27], but this supermode coupling may also occur for symmetric star couplers whenever two supermodes involved in the power exchange process have the same symmetry. One of the simplest examples of the structure in which this phenomenon may occur is the 1×3 coupler with guides arranged in a linear array configuration as shown in Fig. 8 for a fiber coupler. In this case, the supermodes involved are named LP_{01}, LP_{11}, and LP_{02} after their analogs in a circular fiber. Assuming the power is launched in the central guide, the supermode beating analysis (without coupling) predicts the following expressions for the transmissions in the main branch

$$T_1 = \cos^2 \alpha \qquad (10)$$

and in the lateral branches

$$T_2 = T_3 = \frac{\sin^2 \alpha}{2} \qquad (11)$$

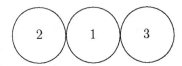

Figure 8: A 1 × 3 coupler made of separate individual guides in a linear array geometry.

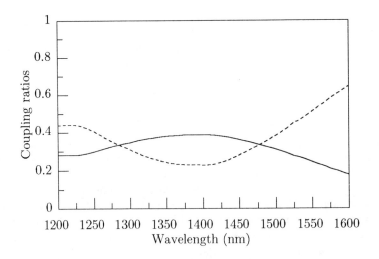

Figure 9: Wavelength response of a 1 × 3 fused fiber coupler in a linear array geometry. The coupling ratios, identical in branches 2 and 3, are in solid line and their complement in dashed line. The incomplete power transfer is the consequence of the LP_{01}–LP_{02} supermode coupling.

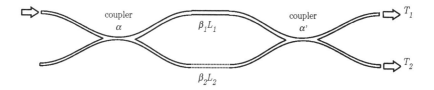

Figure 10: All-fiber Mach-Zehnder interferometer.

where the accumulated phase difference is defined by

$$2\alpha = \int_0^L (B_{01} - B_{02})\, dz \qquad (12)$$

B_{01} and B_{02} being the LP_{01} and LP_{02} supermode propagation constants. These expressions predict total power transfer from the central branch to the side branches as observed experimentally [28]. However, such 1×3 fused fiber couplers may exhibit a variety of responses depending on the degree of fusion and the slopes of their longitudinal profile, as shown in Fig. 9. This may be explained by the LP_{01} and LP_{02} supermode coupling participating to the power exchange process in the slope regions. Other structures such as the 1×4 fiber coupler in triangular configuration and the hexagonal 1×7 coupler (see reference [12, Chapter 29]) may exhibit the same flattened wavelength response since the even LP_{01} and LP_{02} supermodes are likely to couple if the longitudinal geometry of the coupler does not obey the supermode adiabaticity criterion.

4. COUPLER BASED STRUCTURES

4.1. Mach-Zehnder structures

Consider an all-fiber unbalanced Mach-Zehnder structure (M-Z), as shown in Fig. 10. The transfer matrix of such a device is given by the matrix product

$$\mathbf{M}_{MZ} = \mathbf{M}_{2\times 2}(\alpha') \begin{bmatrix} \exp i\varphi & 0 \\ 0 & 1 \end{bmatrix} \mathbf{M}_{2\times 2}(\alpha) \qquad (13)$$

where α and α' characterize respectively the first and the second coupler and $\varphi = \beta_1 L_1 - \beta_2 L_2$ is the phase difference between the two arms of the M-Z interferometer. For power launched in the main branch, the intensity transmissions of the device are calculated to be in the main branch

$$T_1 = \frac{1}{2}\left[1 + \cos 2\alpha \cos 2\alpha' - \sin 2\alpha \sin 2\alpha' \cos\varphi\right] \qquad (14)$$

and, in the secondary branch,

$$T_2 = \frac{1}{2} [1 - \cos 2\alpha \cos 2\alpha' + \sin 2\alpha \sin 2\alpha' \cos \varphi] \qquad (15)$$

which are complementary. These expressions depend on α, α' and φ which, in turn, depend on wavelength; but the particular values $\alpha = \pi/2 + n\pi/2$ and $\alpha' = \pi/4 + n'\pi/2$ (n and n' integers) make T_1 and T_2 equal $1/2$ for any value of φ. This property may be exploited for the fabrication of 50/50 flattened splitters as discussed in Section 4.1.2.

Note now two particular cases for φ.

- $\varphi = 2p\pi$ with $p = 0, 1, 2, 3, ...$, corresponding to the transmission

$$T_1 = \frac{1}{2} [1 + \cos 2(\alpha + \alpha')] = \cos^2(\alpha + \alpha') \qquad (16)$$

- $\varphi = (2p+1)\pi$ with $p = 0, 1, 2, 3, ...$, corresponding to the transmission

$$T_1 = \frac{1}{2} [1 + \cos 2(\alpha - \alpha')] = \cos^2(\alpha - \alpha') \qquad (17)$$

The transmissions are then extremum and, to first order, independent of φ. In these two cases, the M-Z transmissions are identical to those of single couplers characterized by $\alpha + \alpha'$ for Eq. 16 and $\alpha - \alpha'$ for Eq. 17. This is the basis of what is respectively referred to as additive and subtractive structures [29].

The main application of additive structures concerns slightly elongated couplers for which the wavelength dependence of the parameters α and α' is not linear. Concatenating two identical couplers which exhibit no power exchange in the short wavelength range, creates a transmission with the same characteristic feature in the short wavelength range but with a wavelength oscillation more rapid in the longer wavelength range than that of a single coupler. This property is used to devise multiplexers for particular applications, as described in the following section.

Subtractive structures have several applications, the main one being the wavelength insensitive fiber splitter discussed in Section 4.1.2.

4.1.1. Additive structures

Adding the phases of two couplers in series has little practical implication in the domain where $\alpha(\lambda)$ and $\alpha'(\lambda)$ are linear functions of the wavelength because making a longer coupler would be equivalent. However, it can be exploited in the nonlinear domain of their wavelength dependence, i.e., when the couplers are weakly elongated such as those described in Section 3.3.2. As an example structure, we consider three identical couplers ($\alpha = \alpha' = \alpha''$) with $\alpha(\lambda < 1300 \text{ nm}) = 0$ (corresponding to no power transfer $T_1 = 1$) and

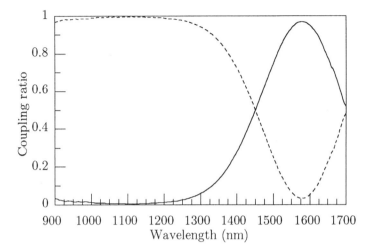

Figure 11: Wavelength division multiplexing coupler response made of three identical couplers in series.

$\alpha(\lambda = 1550$ nm$) = \pi/6$ corresponding to a 25% power transfer for a single coupler and 100% for the whole 3-coupler structure. This makes it possible to realize a multiplexer with a very wide band at 1300 nm which is impossible with a single coupler. The experimental results in Fig. 11 readily shows, though the response is not perfectly centered at 1550 nm, the potential of this technique.

4.1.2. Subtractive structures

Subtractive structures are unbalanced M-Z interferometers with $\varphi = \pi$, which may be used for the design of wavelength flattened splitters.

- **Wavelength Insensitive Couplers (WIC)**

    When the couplers of a M-Z structure such as that of Fig. 10 are not too different, the difference $\alpha - \alpha'$ is by far less dependent on wavelength than α and α'. In particular, it is possible, by playing with the fabrication parameters, to make two couplers having the same wavelength period on a certain range but with $\alpha \neq \alpha'$, i.e involving different number of power cycles. This makes the difference $\alpha - \alpha'$ wavelength independent, even though α and α' vary with wavelength. When compared to α and α', the phase difference φ, when set at its minimum value $\varphi = \pi$ in Eq. 17, is almost wavelength independent. The splitting ratio which then depends only on the difference $\alpha - \alpha'$, may be fixed to any arbitrary value.

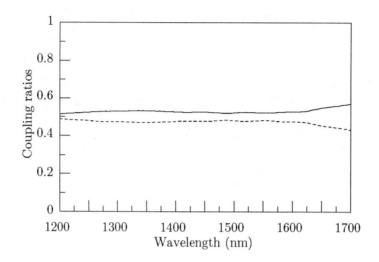

Figure 12: Wavelength insensitive coupler response. The coupling ratio is (52.5± 0.5)% from 1250 to 1600 nm.

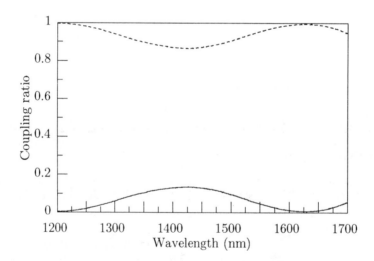

Figure 13: Wavelength insensitive coupler response. The coupling ratio is (5 ± 2)% from 1280 to 1320 nm and from 1520 to 1560 nm.

Experimental results relative to 52.5/48.5 and 5/95 ratios are presented in Figs 12 and 13 respectively.

Note that the theory which has been developed above for 2×2 symmetric couplers is applicable to several $1 \times N$ couplers: flat 1×3, triangular 1×4, rectangular 1×5 and pentagonal 1×6. For those couplers consisting of N fibers located at the vertices of a regular polygon while the main fiber is located at the center, the coupling between two adjacent surrounding fibers is negligible: they may then be considered as a single guide with the same propagation constant and the problem is formally identical to that of a symmetric 2×2 coupler [12, Chapter 29]. However, to obtain the flattened response, the π phase shift must be applied to all the surrounding fibers which may be obtained by twisting the structure.

Note, in addition, that this phase shift between two guides is equivalent to a coupling between supermodes. In the case of the 2×2 coupler, since supermode LP_{01}–LP_{11} coupling cannot take place in a symmetric structure, one has to break the symmetry, e.g. by bending the structure, to create this supermode coupling. In the case of $1 \times N$ couplers, abrupt slopes may induce this coupling as mentioned in Section 3.4.; controlling the supermode coupling in a single coupler without inducing loss may however be more delicate than controlling the phase between the two couplers of a M-Z type arrangement.

- **Multiplexers**

Subtractive structures may also be useful for multiplexers based on slightly elongated couplers. In this case, the Mach-Zehnder widens the long wavelength bandpass (around 1550 nm in the Section 3.3.2. example) without affecting the short wavelength range flat transmission. Here, we propose another application with slightly elongated couplers. A design setting $\alpha - \alpha' = \pi/2$ around 1550 nm (e.g., $\alpha = 2\pi/3$ and $\alpha' = \pi/6$) while $\alpha\,(\lambda < 1000$ nm$) = 0$, permits to widen the passband of WDM couplers around 1550 nm. The experimental result is shown in Fig. 14. The bandwidth of this spectral response is twice as large as that of the single coupler shown in Fig. 7, and may be exploited for Er^{3+}-doped fiber amplifiers.

4.1.3. Polarization insensitive dense multiplexers

As mentioned in Section 3.3.3., the responses of largely elongated single couplers are too polarization dependent to be used for dense wavelength multiplexing purposes. To overcome this problem, one can exploit the spectral response of an unbalanced Mach-Zehnder. The 3 dB couplers may be slightly elongated, corresponding to $\alpha = \alpha' = \pi/4$ in Eqs. 14 and 15, to obtain a negligible polarization dependence. The overall spectral response is then mainly determined by the path difference of the interferometer. If the fibers are not birefringent the corresponding phase difference $\varphi = \beta_1 L_1 - \beta_2 L_2$ is polarization independent.

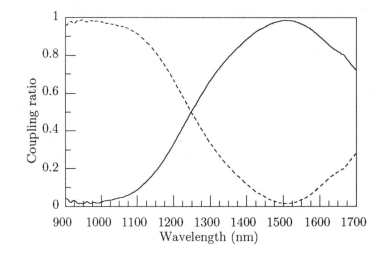

Figure 14: Response of a Wavelength Division Multiplexer made of a M-Z subtractive structure for Er^{3+}-doped fiber amplifiers.

Two approaches are then possible to realize such a M-Z structure: either the geometrical lengths L_1 and L_2 of the interferometer arms are equal and the corresponding propagation constants β_1 and β_2 different; or the lengths are different and the propagation constants identical. In the first case, the asymmetry may be induced by adiabatic tapering of one arm, in which case the wavelength spacing is limited by the taper length. Keeping a reasonable length for the overall device (≈ 5 cm) we have obtained a spacing $\delta\lambda \approx 10$ nm [30]. The experimental result is shown in Fig. 15. Using the second technique, with an arm length difference of the order of 0.7 mm, we obtained a spacing $\delta\lambda \approx 1.25$ nm, which could be further reduced by using a larger path difference (Fig. 16). The latter technique is thus promising for arbitrarily dense WDM but it results in less compact devices (≈ 10 cm), although we hope to reduce these dimensions in the future. Note in addition that, for an ideal performance, M-Z interferometers require two identical 3 dB wavelength independent couplers; otherwise the transmission or/and the isolation may be affected.

4.2. Loop mirrors

Consider a loop structure, as shown in Fig. 17. The transfer matrix of such a device, which is an all-fiber Sagnac interferometer, is given by the matrix product

$$\mathbf{M}_S = \mathbf{M}_{2\times 2}(\alpha) \begin{bmatrix} 0 & 1 \\ 1 & 0 \end{bmatrix} \begin{bmatrix} e^{i\varphi} & 0 \\ 0 & e^{i\varphi} \end{bmatrix} \mathbf{M}_{2\times 2}(\alpha) \qquad (18)$$

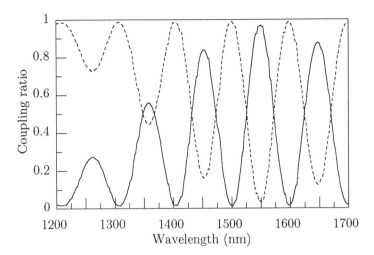

Figure 15: Spectral response of a Mach-Zehnder interferometer for dense wavelength division multiplexing application. The phase difference comes from the adiabatic tapering of one of the arms.

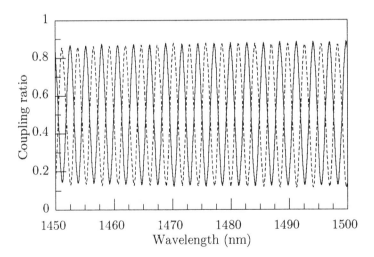

Figure 16: Spectral response of a Mach-Zehnder interferometer for dense wavelength division multiplexing application. The phase difference comes from a difference in the arm lengths.

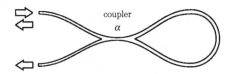

Figure 17: All-fiber Sagnac interferometer.

where α characterize the coupler and $\varphi = \beta L$ is the phase accumulated in the loop.

$$\mathbf{M_S} = e^{i\varphi} \begin{bmatrix} i\sin 2\alpha & \cos 2\alpha \\ \cos 2\alpha & i\sin 2\alpha \end{bmatrix} \qquad (19)$$

For power launched in the main branch, the intensity transmission of the device is calculated to be

$$T = \cos^2 2\alpha \qquad (20)$$

and the reflection

$$R = \sin^2 2\alpha \qquad (21)$$

which is equal to 1 for the particular value $\alpha = \pi/4$, i.e., for a 3 dB coupler. An all-fiber Sagnac interferometer may thus be used as an all-fiber mirror, wavelength selective or flattened depending on the coupler response. It has been used as high and partial reflector in fiber lasers [31]. It is also a popular nonlinear device because it is a more stable interferometer than M-Z's. However, because the fiber is bent and possibly twisted in the loop, the latter is in fact birefringent making the device strongly polarization dependent even though the fiber and the coupler are not.

5. CONCLUSION

Single-mode fiber devices made by the fusion and tapering technique are all-fiber, very low loss, compact, and environmentally stable optical components. The fabrication technique is simple and provides great flexibility in customizing the responses by combining tapered single fibers and couplers in interferometric arrangements. Most of the applications presented here, ranging from sensing to communication networks, involve 2 × 2 couplers but other potential applications involving star couplers can be considered. Finally, the emergence of new materials which may be deposited on the device or which may be drawn in fibers also widens the field of potential applications towards various sensing devices and ultrafast nonlinear switches ultimately used in multigigabits/s communication and/or optical computing.

6. ACKNOWLEDGEMENTS

I am most grateful to people from the Fiber Optics Laboratory of Ecole Polytechnique for their assistance in preparing this article and also to François Gonthier from Alcatel-Canstar who provided me some of the experimental results presented here. Finally, I wish to thank Michel Digonnet for his helpful suggestions on the manuscript.

7. REFERENCES

[1] W. J. Stewart, and J. D. Love, "Design limitation on tapers and couplers in single-mode fibres," in *5th Int. Conf. Integrated Opt. & Opt. Fiber Commun., 11th European Conf. Opt. Commun., IOOC/ECOC'85*, ed. Instituto Internazionale delle Comunicazioni, Venice, Italy, 1985, pp. 559-562.

[2] S. Lacroix, F. Gonthier, and J. Bures, "Fibres unimodales effilées," *Ann. télécomm.*, vol. 43 (1-2), pp. 43-47, 1988.

[3] J. D. Love, W. J. Stewart, W. M. Henry, R. J. Black, S. Lacroix, and F. Gonthier, "Tapered single-mode fibres and devices: Part 1. Adiabaticity criteria," *IEE Proc. Pt.J: Optoelectronics*, vol. 138 (5), pp. 343-354, 1991.

[4] S. Lacroix, R. Bourbonnais, F. Gonthier, and J. Bures, "Tapered monomode optical fibers: understanding large power transfer," *Appl. Opt.*, vol. 25, pp. 4424-4429, 1986.

[5] F. Gonthier, J. Lapierre, C. Veilleux, S. Lacroix, and J. Bures, "Investigation of power oscillations along tapered monomode fibers," *Appl. Opt.*, vol. 26, pp. 444-449, 1987.

[6] K. P. Jedrzejewski, F. Martinez, J. D. Minelly, C. D. Hussey, and F. P. Payne, "Tapered-beam expander for single-mode optical-fiber gap devices," *Electron. Lett.*, vol. 22 (2), pp. 106-107, 1986.

[7] P. Dumais, F. Gonthier, S. Lacroix, J. Bures, A. Villeneuve, P. Wigley, and G. Stegeman, "Enhanced Self Phase Modulation in Tapered Fibers," *Opt. Lett.*, vol. 18 (23), pp. 1996-1988, 1993.

[8] H. R. Giovannini, K. D. Konan, S. J. Huard, F. Gonthier, S. Lacroix, and J. Bures, "Modal interference in an all-fiber sensor measured by the coherence multiplexing technique," *Electron. Lett.*, vol. 29 (1), pp. 29-31, 1993.

[9] S. Lacroix, F. Gonthier, and J. Bures, "All-fiber wavelength filter from successive biconical tapers," *Opt. Lett.*, vol. 11, pp. 671-673, 1986.

[10] F. Gonthier, S. Lacroix, X. Daxhelet, R. J. Black, and J. Bures, "Broadband all-fiber filters for wavelength-division-multiplexing application," *Appl. Phys. Lett.*, vol. 54 (14), pp. 1290-1292, 1989.

[11] X. Daxhelet, "Réalisation et étude de filtres spectraux intégrés aux fibres optiques unimodales," *Masters thesis*, École Polytechnique de Montréal, 1990.

[12] A. W. Snyder, and J. D. Love, *Optical Waveguide Theory*, London: Chapman and Hall, 1983.

[13] J. Bures, S. Lacroix, and J. Lapierre, "Analyse d'un coupleur bidirectionnel à fibres optiques monomodes fusionnées," *Appl. Opt.*, vol. 22 (12), pp. 1918-1921, 1983.

[14] F. Gonthier, "Conception et réalisation de coupleurs multi-fibres intégrés à des fibres optiques unimodales," *Ph.D. thesis*, École Polytechnique de Montréal, 1993.

[15] F. Gonthier, S. Lacroix, and J. Bures, "Numerical Calculations of Modes of Optical Waveguides with Two-Dimensional Refractive Index Profiles by a Field Correction Method," *Opt. Quantum Electron.*, vol. 26 (3), pp. S135-S149, 1994.

[16] S. Lacroix, F. Gonthier, and J. Bures, "Supermode analysis of fused fiber couplers: modeling of symmetric 2×2 couplers," *Applied Optics, submitted for publication*, 1994.

[17] K. S. Kim, R. H. Stolen, W. A. Reed, and K. W. Quoi, "Measurement of the nonlinear index of silica-core and dispersion-shifted fibers," *Opt. Lett.*, vol. 19(4), pp. 257-259, 1994.

[18] C. M. Lawson, P. M. Kopera, T. Y. Hsu, and V. J. Tekippe, "In-line single-mode wavelength division multiplexer/demultiplexer," *Electron. Lett.*, vol. 20 (23), pp. 963-964, 1984.

[19] S. Lacroix, F. Gonthier, C. Houle, R. J. Black, and J. Bures, "Coupleurs 2×2 par fusion et étirage des fibres unimodales : réponses spectrales et biréfringence," in *Onzièmes Journées Nationales d'Optique Guidée*, ed. LEMO, Grenoble, France, 1990, pp. 29-31.

[20] F. Gonthier, D. Ricard, S. Lacroix, and J. Bures, "2×2 multiplexing couplers for all-fiber 1.55 μm amplifiers and lasers," *Electron. Lett.*, vol. 27 (1), pp. 42-43, 1991.

[21] M. Zirngibl, C. H. Joyner, L. W. Stultz, T. Gaiffe, and C. Dragone, "Polarization independent 8×8 waveguide grating multiplexer on InP," *Electron. Lett.*, vol. 29 (2), pp. 201-202, 1993.

[22] M. N. McLandrich, R. J. Orazi, and H. R. Marlin, "Polarization Independent Narrow Channel Wavelength Division Multiplexing Fiber Couplers for 1.55 μm," *J. Lightwave Technol.*, vol. 9 (4), pp. 442-447, 1991.

[23] C. V. Cryan, and C. D. Hussey, "Fused polished singlemode fiber couplers," *Electron. Lett.*, vol. 28 (2), pp. 204-205, 1992.

[24] D. B. Mortimore, "Wavelength-flattened fused couplers," *Electron. Lett.*, vol. 21 (17), pp. 742-743, 1985.

[25] F. Gonthier, D. Ricard, S. Lacroix, and J. Bures, "Wavelength-flattened 2 × 2 splitters made of identical single-mode fibers," *Optics Lett.*, vol. 16 (15), pp. 1201-1203, 1991.

[26] K. S. Chiang, "Design criterion for low-loss optical coupler tapers," *Electron. Lett.*, vol. 23 (3), pp. 112-113, 1987.

[27] K. Okamoto, "Theoretical Investigation of Light Coupling Phenomena in Wavelength-Flattened Couplers," *J. of Lightwave Techn.*, vol. 8 (5), pp. 678-683, 1990.

[28] A. Niu, C. M. Fitzgerald, T. A. Birks, and C. D. Hussey, "1 × 3 linear array singlemode fiber couplers," *Electron. Lett.*, vol. 28 (25), pp. 2330-2332, 1992.

[29] F. Gonthier, D. Ricard, S. Lacroix, and J. Bures, "2×2 fused fiber couplers: wideband wavelength response for multiplexing applications," in *Photonics'92: The second International Workshop on Photonic Networks, Components and Applications*, ed. J. Chrostowski, and J. Terry, OCRI, Ottawa, Ontario, Canada 1992, Montebello (Québec) Canada, 1992, pp. 2.12.1-2.12.5.

[30] S. Lacroix, F. Gonthier, D. Ricard, and J. Bures, "Multiplexeurs tout-fibre serrés en longueur d'onde," in *Treizièmes Journées d'Optique Guidée*, ed. ENSPM, S. Huard, Marseille (France), 1993, pp. 41-1 to 41-3.

[31] I. D. Miller, D. B. Mortimore, P. Urquhart, B. J. Ainslie, S. P. Craig, C. A. Millar, and D. B. Payne, "A Nd^{3+}-doped cw fibre laser using all-fibre reflectors," *Appl. Opt.*, vol. 26 (11), pp. 2197-2201, 1987.

Devices for Communication and Sensors I

Glass Waveguides With Grating

Mahmoud Fallahi

National Research Council of Canada and Solid State Optoelectronic
Consortium, Ottawa, Ontario, K1A 0R6 Canada
Phone : (613) 993 6964 , Fax : (613) 957 8734

ABSTRACT

Fabrication of high-quality glass waveguides with gratings is an important step in the use of glass in integrated optics. In this paper, some of the latest attempts in the fabrication of glass waveguides with gratings are reviewed. The design and fabrication of glass waveguides with ion-exchange and gratings are briefly described. A summary of some of the practical devices is given and the future directions are analyzed.

1. INTRODUCTION

Through the development of low loss fiber optics, research activities on optical communications have been intensified. In order to get the full advantage of optical fibers capabilities, simple, fast, low-cost and reliable optoelectronic devices are required. High performance optical components such as waveguides, lasers, detectors, modulators, multi and demultiplexers are needed. III-V materials among semiconductor materials and glass substrates from amorphous materials are considered to be the suitable candidates for the fabrication of these components. While semiconductor materials have been extensively used for the fabrication of active components such as lasers, detectors and electronic devices, the technology for the use of glass in integrated optics has been developed in the last decade. Some of the main reasons for the development of glass are: the low cost of glass compared to semiconductors, excellent transparency, mechanically hard and availability in large sizes. In addition, since its refractive index is very close to that of optical fibers, its coupling loss to fiber is low. The fabrication of low-loss optical waveguides was the first step toward the use of glass substrate in integrated optics. Excellent progress in this area has been made and since a few years ago, passive integrated optics components have become commercially available.

Diffraction gratings are now considered as key elements in the fabrication of photonic components. Gratings are increasingly used in the fabrication of filters, couplers and single mode semiconductor lasers. They are also very promising for wavelength multi/demultiplexing components. In order to increase the use of glass in integrated optics, new functions have to be incorporated. The use of gratings in glass waveguides seems to provide new opportunities. In the last few years, the number of publications on gratings fabricated on glass waveguides to provide novel functions has been increased.

In this paper, gratings in glass waveguide and resulting devices will be discussed and the future orientation will be analyzed. Although excellent review papers on glass waveguides have been published[1-2] it seems appropriate to give a brief review of the fabrication of glass waveguides in section two. Because of the popularity of the ion-exchange technique, the discussion will be limited to this technique. In the 3rd section the design and fabrication of gratings in glass waveguides are discussed. After a review of the design aspects, the different techniques for generating gratings will be described and their fabrication in glass will be presented. Section four will give a summary of the latest components as well as the new attempts for active components.

2. GLASS WAVEGUIDES

Waveguides are considered to be one of the basic but very important parts of integrated optics. The performance of optoelectronic components depends very much on the quality of waveguides. Fabrication of low-loss, single-mode waveguides with a good control on the dimensions is an initial and a major step toward the fabrication of integrated optics.

There are two approaches to fabricate waveguides on a substrate: deposition of a higher index layer or formation of a higher index region in a lower index material. In semiconductors, optical waveguides are mainly fabricated by epitaxial growth of layers with different refractive index and band-gap. In glass substrate, waveguides have been defined either by sputtering and dielectric deposition or by ion- diffusion process (ion-exchange). Ion-exchange appears to be the most popular technique to produce waveguides in glass because of its simplicity, reproducibility and low-cost.

The first ion-exchanged glass waveguides were demonstrated by Izawa and Nakagome in 1972 by exchanging Ti^+ ions in glass[3]. Since then, this field attracted attention and resulted in a rapid development of glass waveguides. The ion-exchange process offers great flexibility in the selection of fabrication conditions and is suitable for large-volume production. In addition, low-loss waveguides are reproducible at low cost.

The ion-exchange process for the fabrication of glass waveguides is usually carried out at temperatures between 200 °C and 650 °C in a molten bath. Nitrate salts have been used because of their low melting temperatures and good stability. In this process, sodium ions of glass are exchanged by larger ions such as Cs^+, K^+, Ag^+, Ti^+. The refractive index profile of ion-exchanged waveguides depends on many parameters. The most important ones are : composition of the glass substrate, the type and concentration of ion, the applied voltage, the temperature and finally the diffusion time[4]. In order to reproduce waveguides with the same characteristics, it is very important to keep the above parameters unchanged. Table 1 summarizes some typical characteristics of the above ions. As it is shown, Ti^+ results in a large index change and a large diffusion coefficient, therefore it is more suitable for the fabrication of multimode waveguides. On the contrary, K^+ results in a small index change making it suitable for single mode waveguides. For channel waveguides, a cover layer such as a gold or aluminum layer is used. Diffusion depth is directly proportional to the applied voltage.

Table 1 : Typical characteristics of common ions used in the fabrication of glass waveguides by ion-exchange process.

Ions	Substrate	Δn	D (μm^2/min)	Loss (dB/cm)
K^+	Corning 0211	0.005	0.14	0.15
Ag^+	0211	0.06	0.025	0.54
Cs^+	BGG21	0.04	0.19	0.2
Ti^+	Borosilicate	0.1	12	0.1

A two-step double ion-exchange process has also been used to produce dual-core waveguides in a glass substrate[5]. In this case, a first step ion-exchange was carried out at 400 °C using potassium ions, then, silver ion exchange was performed at 300 °C. Silver ion exchange in the presence of potassium ions was much slower than single silver ion-exchange. The diffusion coefficient of silver ions in a double ion-exchanged waveguide in the order of 0.008 μm^2/min. has been obtained which is about three times smaller than the single silver ion-exchange waveguide.

For the fabrication of channel waveguides, an aluminum mask was used. In this case after a potassium ion-exchange, an aluminum mask was deposited on the surface and patterned by photolithography to define the channel waveguide region. The silver ion-exchange was made using this aluminum mask. Silver ions diffused predominantly in depth while their lateral diffusion was decreased. The resulting waveguide had a low loss and a large single-mode region with a more symmetrical profile than single silver ion-exchanged waveguides. Buried waveguide in glass has also been demonstrated with this technique[6]. In addition, gratings fabricated on these double ion-exchange process showed high efficiency.

3. DESIGN AND FABRICATION OF GRATINGS IN GLASS

Corrugated gratings are of a great interest in integrated optics. They are used in distributed-feedback lasers, distributed Bragg-reflector lasers, optical filters, beam deflectors, waveguide couplers, wavelength multiplexers and demultiplexers and many others. Diffraction gratings built directly into a dielectric waveguide can be used as input or output couplers. The theory of grating waveguides have been studied for many years and in detail, as part of passive or active optical elements, in literature[7-9]. In this section, a summary of some of those simplified formulations useful in the estimation of critical parameters and design aspects of grating waveguides, is given.

3.1 : Design of gratings :

The principle of the operation of a grating filter is as follow: A single-mode waveguide with a grating period of Λ will reflect a normally incident guided wave, if it satisfies the Bragg condition :

$$\Lambda = m_g \lambda_b / 2 n_{eff}$$

where m_g is the grating order ($m_g = 1, 2,...$), λ_b is the Bragg wavelength and n_{eff} is the effective index of the waveguide. For a first order Bragg grating, the Bragg wavelength is partially transmitted and partially reflected. The maximum reflection is obtained at $\Delta\beta = 0$ and is given by:

$$R = \tanh^2(\kappa L)$$

where L is the grating length and κ is the coupling coefficient. The coupling coefficient κ is determined by the change of the effective index of the waveguide caused by Bragg gratings and depends strongly on the corrugation depth and the line/period ratio. It can be obtained from[7]

$$\kappa = I_g k_o^2 / 2\beta P$$

$$I_g = \int_{corrugation} \Delta[n^2(x,z)] E^2(x) \, dx$$

where $k_o = 2\pi/\lambda_o$ is the free-space wavenumber, β is the TE-mode propagation constant, $E(x)$ is the y component of the unperturbed electric field, $\Delta[n^2(x,z)]$ is the perturbation in refractive index of the corrugation and P is a normalization constant given by:

$$P = \int E^2(x) \, dx$$

The spectral width of the stop-band is proportional to the coupling coefficient and is given by :

$$\Delta\lambda = \lambda_b^2 \, \kappa/\pi \, n_{eff}$$

As a result, the light of other wavelengths passes through without beeing affected by the grating.

Now, if we consider a more general form for a grating coupler with a period Λ (fig. 1), the diffracted angle can be found by writing the phase-match condition as follow:

$$\beta_d - \beta_i = 2 m \pi/\Lambda \qquad m = 0, \pm 1, \pm 2, \ldots$$

where m is the diffraction order. $\beta_i = 2\pi \, n_{eff}/\lambda$ and $\beta_d = 2\pi \, n_a \, \sin\Theta_d/\lambda$. The diffracted angle Θ_d is then obtained from :

$$\sin \Theta_d = n_{eff}/n_a + m \lambda/n_a \Lambda$$

For a second order Bragg grating ($m_g = 2$) at resonance, it can be easily seen that the second order diffraction provides optical feedback while the first order diffraction results in a radiation normal to the junction plane. The strength of each of these diffracted orders depends on the line/period ratio of Bragg grating. A detailed analysis of second order gratings can be found in the literature[10].

Therefore, using an appropriate grating pitch, depth and length with an appropriate order, one can fabricate optical filters, couplers, wavelength demultiplexers, surface-emitting lasers, and more. In section 4 of the paper, some practical devices fabricated using gratings in glass waveguides will be reviewed.

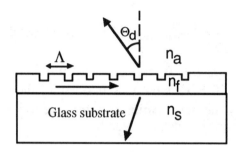

Fig. 1: Schematic view of a waveguide with a grating coupler

3.2 : Fabrication of gratings :

Among the different techniques for producing gratings in glass, etched gratings and ion-exchanged gratings are mainly used. Although ion-exchanged gratings result in a planar structure, their efficiency is low therefore not very useful. Etched gratings have been shown to be very efficient and useful for wavelength selection and filtering. Three techniques have been mainly used for defining grating lines: holographic exposure, electrons/ions beam writing and grating photomask printing.

In a holographic exposure, the waveguide is first coated with a photoresist and baked in an oven or on a hot-plate. A laser source (mainly He-Cd laser) is used for the exposure. The light is first spatially filtered. After expansion, the main beam is split into two equal intensity beams. The two beams are then recombined on the surface of the sample with an angle equal to 2α. This results in the creation of an interference pattern and the formation of a grating with a periodicity of Λ which satisfies the following relation :

$$\Lambda = \lambda_0 / 2 \sin \alpha$$

The resulting period of the grating is then limited to values greater than $\lambda_0/2$. Fig. 2 shows the formation of gratings using a holographic technique. The exposed photoresist is then developed in a developer. The structure is finally transferred into glass waveguides by dry etching in a CF_4.

The second technique for producing grating patterns uses electron-beam or focused ion-beam lithography. These techniques are computer controlled and have the capability for defining very fine features. They are frequently used to produce complex structures such as phase-shifts, curves, dots and circular patterns in sub-half-micron dimensions (below 100 nm).

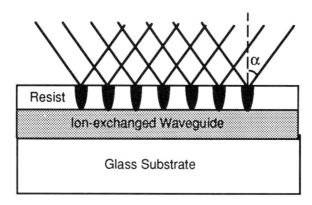

Fig. 2 : Fabrication of a grating in a glass waveguide by the holographic technique.

On the other hand, these systems are costly and the writing time is long. We have used these techniques to produce linear and circular gratings in glass. An electron/ion sensitive resist such as poly(methylmetacrylate) (PMMA) is used.

In order to fabricate gratings in glass, a thin layer of aluminum is first evaporated on the sample. This metal has two main functions, first it prevents the charging up during the electron or ion beam pattern writing, secondly, it is used as an intermediate layer for the transfer of patterns in glass since the PMMA is quickly etched during the plasma etch of glass. PMMA resist with a thickness of 0.25 µm was spin-coated. For ion-beam lithography, a focused-ion-beam (FIB) system was used. Si^{++} ions at an energy of 200 keV and an exposure dose of about 1.3×10^{13} ions/cm^2 was used. After the development, the grating patterns were first transferred into aluminum using BCl_3:He gas mixture, then the patterns were etched into glass waveguides by CHF_3:O_2 using aluminum as a mask. Deep gratings could be obtained since aluminum is not etched in the gas mixture. At the end, the remaining aluminum is etched away by wet or dry etching. Fig. 3 summarizes the fabrication steps of gratings in glass using FIB lithography and dry etching.

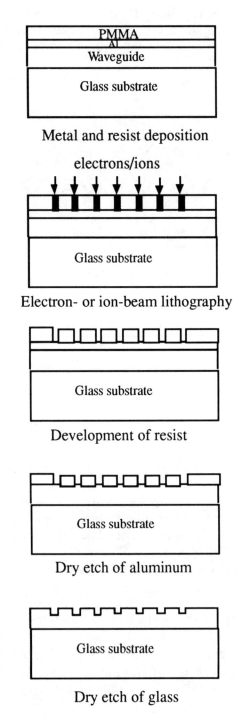

Fig. 3 : Fabrication steps used for the formation of etched gratings in a glass waveguide.

A new technique has recently been developed for reliable fabrication of complex gratings such as phase shifted or multi-period gratings in large volumes and relatively cheap. It is called grating photomask printing technique[11]. In this technique, complex gratings are defined and etched in a fused-silica master mask by one of the techniques described above. A coherent light of a wavelength λ is incident on the grating at an angle θ from the normal. When the periodicity of the gratings satisfies the condition : $\lambda/2 < \Lambda < \lambda$, the beam is then partially transmitted (m=0) and partially reflected (m=-1). The result is the formation of an interference pattern below the mask at the spatial periodicity of the gratings (fig. 4). This is then printed in a photoresist spin-coated on the sample. After the development, the grating pattern can be transferred into a glass waveguide, as described above.

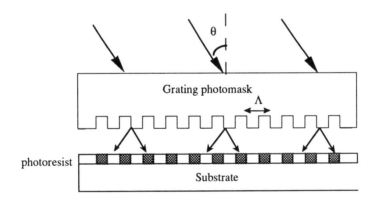

Fig. 4 : Grating photomask printing technique.

Among these three techniques described here, the grating photomask technique seems to be very attractive. This technique has the advantage of rapid printing for mass production as well as the flexibility of fabricating gratings with different periodicities and has been used for the fabrication of multi-wavelengths distributed feedback and distributed Bragg reflector lasers. It can also be applied to glass waveguides without any difficulty.

4. DEVICES AND APPLICATIONS

For most of the applications such as filters or couplers, high-reflectivity of gratings is required. This will then limit the type of the ion-exchange and the gratings. Potassium ion-exchanged waveguides have typically a large depth (5-10 µm deep) which causes a reduction in the efficiency of gratings fabricated on this type of waveguide due to the small interaction between guided mode and grating. In order to obtain high reflectivity, silver ion-exchanged waveguides with a depth of 1μm to 4 μm was used[12]. In this case, a diffracted/transmitted ratio as high as 32 was reported. A higher reflectivity has been achieved by using a double ion-exchanged waveguide. Etched gratings in a dual-core waveguide were fabricated and characterized. A reflectivity as high as 95% at 1.26 μm wavelength with a bandwidth of 7 Å was measured[5].

Circular grating is an unconventional grating structure with a great potential in surface-emission lasers and multi-port couplers. Circular grating lasers fabricated in III-V semiconductors have produced low-divergence high-power surface-emission [13-14]. Circular gratings have recently been fabricated in glass waveguides where it operated as a tap power divider [15]. Fig. 5 shows a schematic view of the device. The device consists of a circular grating patterned on the top of a central circular waveguide. The central waveguide is connected to four tapered waveguides. An input light is directed perpendicular to the structure which couples into the structure through the circular grating and is distributed in the tapered waveguides. For the fabrication, a slab waveguide was first made by potassium ion exchange. Circular gratings were defined by focused ion beam lithography and etched by reactive ion etching (RIE) in the same way as described in the previous section. Finally the channel four-port waveguides were defined by RIE. The device showed to be efficient as a power divider and to be very interesting for three dimensional integration.

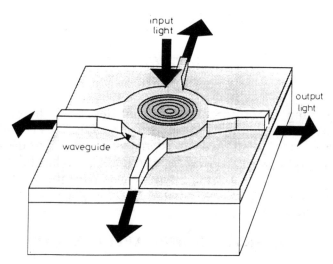

Fig. 5 : Schematic view of a glass waveguide tap power-divider with circular gratings.

Near infra-red laser sources are a major part of optical communication systems. Visible lasers are also very attractive for optical data storage. At the present time, semiconductor lasers are mainly used for these applications. In the last couple of years, some research activities on the fabrication of glass waveguide lasers have been started. They are based on the use of rare-earth doped materials (such as Er and Nd) as an active layer for emission in the 1- 1.5 µm range as well as for up-conversion in the visible range. Green up-conversion laser emission (λ= 551 and 562 nm) in Er-doped $LiYF_4$ and KYF_4 crystals has been obtained[16], but the demonstration on a glass waveguide is yet to be achieved. A success in this area would open new horizons to the use of glass waveguide in integrated optics.

In glass waveguide lasers, the use of gratings for both amplification and surface emission seems to be one possible option. Using this approach, the fabrication of Er-doped sol-gel glass waveguides with circular gratings as a light source was attempted[17]. Fig. 6a shows a schematic view of such a structure. A glass waveguide was produced by an ion-exchange process. A phosphate film doped with Er was then spin coated on the waveguide and baked. Circular gratings were finally defined by FIB lithography and reactive ion etching. One major difficulty with such a structure was how to efficiently couple the pumping laser into the structure. Despite the fact that no lasing was observed, a very good control on the fabrication of the ion-exchanged waveguide combined with the sol-gel technique and the grating definition was demonstrated and future work might result in a major breakthrough.

In the last few years there has been a growing interest in wavelength division multiplexing (WDM) systems in order to increase the transmission capacity of long- and short-haul optical fiber communication systems. One of the major components of such a system is a wavelength demultiplexer with many channels at small channel spacing. Wavelength demultiplexing can be obtained on glass waveguide with gratings.

A demultiplexer with 5 output channels and 30 nm channel spacing in the 1 µm wavelength range has been demonstrated[18]. The waveguide was prepared by silver ion-exchange using a Corning 0211 glass substrate. A chirped grating was fabricated on a Si substrate by electron-beam lithography of PMMA. The grating was 3mm long and 300 µm wide. The grating was then transferred in SiO_2 layer, coated with Al and bonded on the edge face to the glass waveguide. The insertion loss was reported to be less than 5 dB. Although the device showed relatively good performances, it was not a practical device for WDM systems. A monolithic version could probably make the demultiplexers on glass waveguides more attractive. This is possible by defining gratings directly in glass waveguides. The device might use either a multiperiod (chirp gratings) or an echelle grating[19] as it is schematically proposed in (fig. 6b). These variable period gratings can be fabricated by either electron/ion beam lithography or by grating photomask printing technique. By defining gratings with small variations in periodicities, dense wavelength demultiplexing with a channel spacing of a few nanometers can be achieved.

The integration of active semiconductor devices on glass waveguides is also of a great importance. In this case, glass waveguide is used as a substrate as well as waveguide, while the semiconductor device can do the detection or other functions. The integration can be carried out using a lift-off or flip-chip technique. The use of gratings in such systems might greatly enhance the out-coupling from the waveguide to the semiconductor (fig.6c).

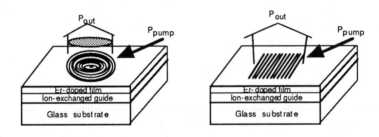

a) Glass waveguide Laser

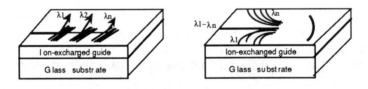

b) Wavelength demultiplexer

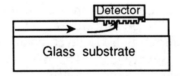

c) Integration with semiconductors

Fig. 6 : Different applications of glass waveguides with gratings.

5. CONCLUSIONS

Progress on the fabrication of low-loss glass waveguides has been significant. Ion exchange process has proven to be a reliable technique to produce large and uniform waveguides with a loss as low as 0.1 dB/cm. Glass waveguides have been used in the production of integrated optic systems. Diffraction gratings on the other hand have been extensively used for the fabrication of active and passive components on semiconductors. Use of gratings in glass waveguides can most probably enhance the use of glass waveguides in integrated optics by providing devices with new functions. Some of the experimental results for such attempts were reviewed. Fabrication of gratings in glass waveguides is simple and reproducible. Gratings in glass waveguides can be used for the fabrication of lasers, couplers, filters, wavelength demultiplexers and more.

6. REFERENCES

1: R.V. Ramaswamy and R. Srivastava, " Ion-exchanged glass waveguides: A review, " J. Lightwave Techn. vol.6, 984, 1988.
2: T. Findakly, " Glass waveguides by ion-exchange: A review," Opt. Eng., vol.24, 244, 1985.
3: T. Izawa and H. Nakagome, " Optical waveguide formed by electrically induced migration of ions in glass plates," Appl. Phys. Lett. vol.21, 584, 1972.
4: S.I. Najafi and M.J. Li, "Ion-exchanged glass integrated optical components" Proceedings of the International workshop on Photonic Networks components and application, vol. 2, 46, 1990.
5: M.J. Li, S. Honkanen, W.J. Wang, R. Leonelli, J. Albert and S.I. Najafi, " Potassium and silver ion-exchanged dual-core glass waveguides with gratings," Appl. Phys. Lett. vol.58, 2607, 1991.
6: R.V. Ramaswamy and S.I. Najafi, IEEE J. Quant. Electon. vol.22, 883,1986.
7: W. Streifer, D.R. Scifres, R.D. Burnham, " Coupled wave analysis of DFB and DBR lasers," IEEE J. Quant. Electron. vol.13, 134, 1977.
8: A. Yariv, M. Nakamura, " Periodic structures for integrated optics," IEEE J. Quant. Electron. vol.13, 233, 1977. couple mode
9: G. Lamouche and S.I. Najafi, " Accurate analysis of ordinary and grating assisted ion-exchanged glass waveguides," Optical Eng. vol.30, 1365, 1991.
10: A. Hardy, D.F. Welch and W. Streifer, " Analysis of second-order gratings," IEEE J. Quant. Electron. vol.25, 2096, 1989.
11: G. Pakulski, R. Moore, C. Maritan, F. Shepherd, M. Fallahi, I. Templeton, G. Champion, " Fused silica masks for printing uniform and phase adjusted gratings for distributed feedback lasers," Applied Physics Lett, vol.62, 222, 1993.

12: S. I Najafi, Ed., Introduction to glass integrated optics, Artech House, Boston.London, 1992.

13: M. Fallahi, M. Dion, F. Chatenoud, I.M. Templeton, and R. Barber, "High power emission from strained DQW circular-grating surface-emitting DBR lasers," Electron. Lett, vol.29, 2117, 1993.

14: S.I. Najafi, and M. Fallahi, " Circular grating surface emitting lasers," SPIE Proceesings vol. 2213 of the Intl. Sysmp. on Integrated optics, Lindau, April 1994.

15: S.I. Najafi, M. Fallahi, P. Lefebvre, C. Wu, and I. Templeton, " Integrated optical circular-grating tap-power-divider," Electron. Lett., vol.29, 1417, 1993.

16: R. Brede, E. Heumann, J. Koetke, T. Danger, G. Huber, B. Chai, "Green up-conversion laser emission in Er-doped crystals at room temperature," Appl. Phys. Lett., vol.63, 2030, 1993.

17: S. I. Najafi, N. Peyghambarian, M. Fallahi, Q. He, " Light emission in optically pumped erbium doped sol-gel waveguides with circular grating," 1993 OSA annual meeting, Technical Digest, Paper FQ4, Toronto, Canada, October 1993.

18: T. Suhara, J. Viljianen and M. Leppihalme, " Integrated-optic wavelength multi-and demultiplexers using a chirped grating and an ion-exchanged waveguide," Appl. Optics vol.21, 2195, 1982.

19: M. Fallahi, K.A. McGreer, A. Delage, I.M. Templeton, F. Chatenoud and R. Barber, " Grating demultiplexer integrated with MSM detector array in InGaAs/AlGaAs/GaAs for WDM," IEEE Photonics Techn. Lett. vol. 5, 794, 1993.

Multimode glass integrated optics

O. Parriaux, P. Roth, G. Voirin

CSEM Swiss Center of Electronics and Microtechnology Inc.,
Maladière 71, CH-2007 Neuchâtel, Switzerland, Fax +41 38 205 580

ABSTRACT

The practical usefulness of multimode guided wave optics has been overshadowed by the powerful trend of optical communications towards single mode technologies. Although multimode integrated optic technology lends itself to a very restricted palette of optical functions, it can perform elementary functions at a cost which falls dramatically as the number of identical samples increases unlike that of all-fibre and micro-optic components. Glass integrated optics brings the additional advantage of stability versus external physical parameters.

1. INTRODUCTION

The last decade has confirmed that optical communications are definitely going single mode. Single mode fibre technology is not only dominating in trunk lines but now also in local networks. Such powerful trend has been a strong incentive for integrated optic (IO) technology to follow since optical communications presently represent by far the main market for IO devices. Going single mode was not a change for active waveguide technologies on $LiNbO_3$ or on III-V semiconductors which were in essence single mode since the beginning. The reorientation was more dramatic in passive technologies which had bet on graded-index fibre networks for a while in the middle of the 80s. This trend to single mode implied renewed efforts for the control of technological parameters and especially for the development of difficult pigtailing techniques. This also implied that the main actors on the scene of passive components left multimode technology aside and did not pursue much further. It is important to remember however that the stage they reached at that time was sufficiently mature to allow mass production[1].

Does this mean that speaking about multimode glass IO technology has become a selected chapter of modern archeology ? This is not the opinion of the authors who will try and show that glass multimode IO technology is bound to an industrial future in application fields, some of which were not foreseen at the early stage of its development. The potential of multimode technology for performing simple sets of elementary optical functions monolithically is likely to enlarge considerably with the trend of a massive use of LEDs and LED arrays as a replacement of light bulbs in a number of domestic applications[6]. The conditions have changed. The waveguide features are not any more fully governed by 50/125 or 62.5/125 graded index fibres. Fibre compatibility means something else today: core diameter of 100 or 200 μm and constant numerical aperture, typical of large index difference step index fibres. Furthermore, fibre compatibility is not a must any more: a number of possible applications in the field of sensors and more generally in microsystems are fibreless. The requirement of fibre compatibility is replaced here by a less demanding and much more elementary list of specifications: a) to trap most of the optical power emitted by a LED and b) to propagate the latter with the least losses along a waveguide circuit which has straight segments, bends and junctions.

The routing of LED light by multimode pipes does not allow for a high level of complexity in the performed functions. However, simple functions such as light confinement along long lengths, beam redirecting by bends, beam splitting and recombination by Y junctions are the most numerous functions met in practice which can just be performed by the sole definition of a planar waveguide layout which then replaces the gear of discrete lenses, mirrors and beam splitters necessitated if such functions were to be performed by free space wave components.

One of the main weaknesses of bulk micro-optic systems among others is their instability versus temperature variations and vibrations. A monolithic approach must bring a plus regarding this issue. This is where glass IO technology has intrinsic advantages as compared with other potentially cheaper passive waveguide technologies such as polymer waveguide technologies which are serious contenders. The stability issue is a critical one indeed since most simple systems will be intensity modulated. The signal delivered by an intensity modulated system must be free of spurious modulation causes such as temperature variations or vibrations. The waveguide functions can be instable if the light guidance is affected: a change of refractive index due to temperature can change the bending loss and the splitting ratio of a Y splitter. A sensor system can of course be designed so as to cancel or to reduce the effects of drifts or intensity fluctuations. This can be achieved by smart bridge configurations[2,3,7,10] or by amplitude to phase conversion[4,5]. However, it is nevertheless a noticeable advantage of glass that the refractive index is little dependent on the temperature: typically $10^{-5}/°C$ against $10^{-4}/°C$ for most organic materials.

Before describing further the technology and the optical functions the latter can perform, a practical example will give a hint of the kind of applications this technology can serve. It is very far away from communications and does not make use of fibres. As illustrated in figure 1, this monolithic smoke alarm sensor makes the sensing of a small air volume between two lenses[7]. The presence of smoke translates into the decrease of the optical power transmitted by the signal arm. All active components, two LEDs and two detectors are contained in a back plane immobilized against the polished side of the multimode waveguide chip. The LEDs are modulated at a different frequency and a smart waveguide bridge configuration allows the compensation of some instabilities. One understands that the sensitivity of such sensor must be high and the probability of false alarm very low. This imposes very demanding stability requirements on the functional elements. The critical ones are the Y splitters and the bends. Glass waveguide technology is stable enough. What mostly matters actually is the modal power distribution at the entrance of each element. As will be discussed in section 4, it is not thinkable and too costly in the power budget to achieve a stable modal power distribution by means of a mode scrambler over such short propagation lengths. Therefore the key stability problem lies in the immobilization of the position of the light sources relative to the guide cross-section.

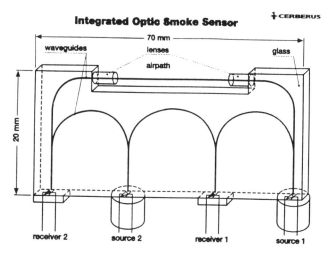

Figure 1

Sketch of a multimode glass integrated optic smoke alarm sensor[7].

2. MULTIMODE GLASS INTEGRATED OPTIC TECHNOLOGY

Single mode glass technology is now a well established industrial technology[9,10]. Both dominant processes, silver[8,9] and thallium[10] ion exchange in a special glass, are in principle capable of providing the type of refractive index distribution needed for a high numerical aperture, large cross-section waveguide. Pilot lines exist for the mass production of IO chips, for their fashioning, pigtailing and packaging; one does not see why these should not play a decisive role in the development of commercial perspectives in heavily multimode applications. However, once such perspectives translate into real market possibilities, it is likely that other glass technologies also have a chance because of the very wide diversity of possible IO structures and applications.

Fibred IO devices will still be small area waveguide chips where the chip amounts to a minor part of the component cost but the promising field of mostly fibreless, monolithic sensor & microsystem distribution platforms will involve larger area IO circuits where the glass and the waveguide fabrication represent a major part in the product cost structure[7]. This means that "window glass waveguide technology" may find its way again. What it can offer is well and widely reviewed by J. Albert in his chapter[11]. There is a new chance, however, only if glass chemistry[12] gives a hand for the selection of the right glass/ion pair.

What follows is an attempt to clarify what are the specific requirements of multimode glass IO technology. As far as the **mask transfer** is concerned, multimode waveguide patterns will mostly be large size circuits requesting moderate resolution. This means that conventional proximity mask aligners[13] are well adapted. A multimode waveguide glass technology must have the following features:

2.1 The numerical aperture and cross-section of multimode waveguide must be large in order to trap a significant part of the light emitted by a LED or transmitted by 100 or 200 μm core silica fibres. The latter are step index fibres of ca 0.04 core-cladding index difference which gives a numerical aperture (NA) of about 0.35. The numerical aperture, NA, is the sinus of the half angle of the beam radiating from the guide opening to the air. $NA \cong n_2\sqrt{2\Delta}$ where n_2 is the guide index and Δ the core-substrate relative index difference. So large refractive index difference and large depth can be obtained in commercially available glasses[14] only by Ag^+ and Tl^+ ion exchange. In contrast with their single mode counterparts, multimode components leave a large freedom as to the refractive index of the substrate. Return loss in a fibred system is not a critical issue since single mode lasers will rarely be used. Return loss will only place a sensitivity limit in reflection type intensity modulated sensors or in bidirectional communication systems.

2.2 The question of **the index profile** is a quite important one regarding fibred devices. Fibreless sensor or microsystem applications can be envisaged regardless of the actual index profile: a graded index waveguide of maximum index equal to that of a step index guide will simply trap less LED power and loose more light in bends but there are many applications where the losses are not the main limitation. As far as 62.5/125 and 50/125 graded index fibres are concerned, the ion exchange waveguide technologies of the eightees have succeeded in fabricating buried waveguides with excellent index matching and propagation loss figures[9,10]. In case of need, these technologies can simply be reactivated. This is no more true concerning large core step index fibres for which some choices and new developments are necessary. The important choice is that of the ion. It is a very interesting and debated question.

2.3 Silver ion exchange from a molten salt (usually a nitrate) at moderate temperature (220 to 400°C) is a very simple technology which can produce deep waveguides after a reasonable number of hours[9]. The penetration depth is proportional to the square root of the process time. The penetration of Ag ions can however be much faster under the application of an external electric field which drives the index distribution into the substrate at a depth which is roughly proportional to the process time. The ion source is generally the very same molten salt but it can also be metallic silver in the form of a film evaporated at the substrate surface[15]. An interesting feature of electric field assisted ion migration in the present discussion is the nearly step index profile this process leads to . The index increase achievable is roughly proportional to the sodium content of the substrate and it depends on the concentration of the silver salt in a mixture of other salts[11]. In soda-lime glasses the maximum index difference is about 0.1. Silver ion exchange technology faces however a stability problem when applied with most commercial glasses. Exchanged silver ions in the glass matrix (0.8 eV redox potential) tend to metallize and to form precipitated colloids by capturing the electron of arsenic As^{+++} and iron Fe^{++} reducing impurities[16]. This provokes a gradual increase of the absorption losses of the ion exchanged waveguide. The latter is enhanced by exposure to light or to higher temperature. The transmission spectrum is also reduced at the blue side of the spectrum: waveguides have exhibited fluorescence when excited by a 514 nm Argon laser beam[17].

It appears that the only way out is the fabrication of a special glass of high purity avoiding especially As, Fe and Sb and offering in addition good diffusion kinetic features for silver ions. This is what has been successfully attempted with the BGG glass series[9]. Figure 2 gives some features of this technology. Thorough glass chemistry investigations, special glass design and fabrication have also taken place in Jena[12] in the direction of highly multimode waveguide and devices compatible with fibres of different core diameter and profile[17,18]. This work has also revealed that Na_2O rich glasses with highly connected network and a low content of Non-Bridging-Oxygen (NBO)[19] lead to a higher Ag^+ mobility due to a viscosity decrease in the exchanged zones and also lead to a considerable decrease of silver reduction. This was evidenced by the suppression of the fluorescence[17] typical for silver colloids.

All this seems to lead to the conclusion that silver ion exchange technology can only be based on a special optical glass. This may however be too hasty a conclusion. It was proved that low NBO content in the glass matrix prevents silver reduction to some extent. However, no systematic study of cheap commercially available glasses regarding this issue has ever been reported. This is an issue that glass chemistry should be able to clear up rapidly. The stakes are again cost questions : a large fibreless monolithic IO platform will not cost the same if made of an optical glass or of "window" glass.

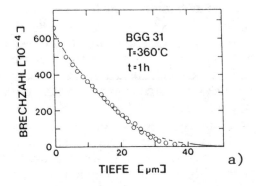 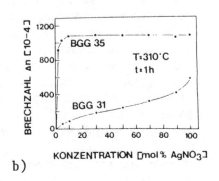

a)   b)

Figure 2

a) *Index profile of a thermally ion exchanged BGG glass with silver*[9]
b) *Dependence of the surface index on the concentration of silver ions in the molten salt for two glasses of the BGG series*[9]

2.4 Thallium ion exchange technology benefits from the 20 year legitimacy of the well known and well sold selfoc lenses[20] of Nippon Sheet Glass. Its alleged toxicity frightens electrical engineers more than chemists. There are much more deterring substances everywhere in the daily life of the "new technologies".

There is no such thing like reduction with thallium ions thanks to their redox potential which is - 0.335 eV. This implies that any type of commercial glass can be considered as the substrate of a low loss waveguide technology provided the kinetics of ion exchange leads to acceptable fabrication times and built in stresses. One expects therefore attractive loss figures and, moreover, a guarantee of stability which is the key issue for product life time.

The polarizability of Tl^+ ions is larger than that of Ag^+ ions. This leads to larger achievable index increase. As an example, a Chemcor glass immersed in a 100 % $AgNO_3$ melt (resp. $TlNO_3$) sees its surface index increased by 0.1 (resp. 0.16)[21]. Although the Tl ion has a rather large ionic radius (1.49 Å against 1.26 Å for Ag^+) its high polarizability allows a comparable diffusion rate. Another remarkable property of Tl^+ ion exchange regarding multimode IO is its capability to give rise to nearly step index profiles by a purely thermal process without the application of an external electric field[22]. This property may have important practical consequences since it allows the fabrication of step index fibre compatible waveguides in a multiwafer batch process. This was demonstrated in a KF3 glass[21]. Under the glass transition temperature, T_g, a high concentration of exchanged Tl^+ ions causes a decrease of the glass viscosity which gives rise to a foreign ion distribution close to a step function. Somewhat earlier, more importantly but insufficiently remarked[23] (figure 3), there was the demonstration that the very same effect exists in the conventional Corning glass B1664 at a significantly higher temperature (thus at a much higher rate) with sulfate salts which are more stable than nitrates. For pure technical reasons and fabrication cost considerations, this should be the way to go.

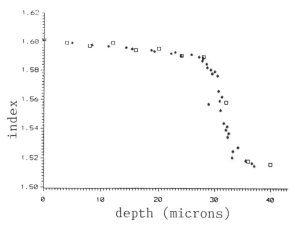

Figure 3

Quasi step index profile obtained by purely thermal ion exchange of thallium ions in a B1664 glass from a sulfate salt mixture[23].

2.5 Waveguide circularization

Low loss fibre-chip pigtailing requires an excellent NA and cross-section match. This has already been achieved industrially in the case of thallium ion exchanged multimode waveguides and graded index fibres[10]. A first thermal exchange is followed by an electric field assisted burying ion migration. Figure 4 illustrates the refractive index mapping of a graded index fibre and that of a buried waveguide. The refractive index profile of ion exchanged waveguides can be accurately and simply measured by using the Refractive Near Field technique for which characterization benches are available[18].

Fibreless systems do not bother about the necessity of circularization of the waveguide cross-section; the captured power is proportional to the product of the square of the numerical aperture and to the cross-section area whatever the cross-section geometry. It seems quite problematic to develop a successful and cost competitive burying technology for very large waveguide cross-sections while maintaining the step index profile feature. This is why the approach consisting of superposing two half-circular cross-sections in a flip-chip configuration may be the preferred solution. Costwise, this has the merit of avoiding the costly individual burying process at the expense of chip on chip manipulation, adjustment and immobilization[47].

Figure 4

Iso-index contours of ion-exchange waveguide and optical fibre[10]

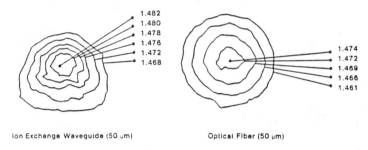

2.6 Protecting the guided light from surface effects

Surface effects remain an essential preoccupation since this will determine the time stability of a whole system. Several possibilities exist. Back-diffusion of the original ion present in the substrate does not allow in general a full recovery of the substrate index at the surface. Back-diffusion under an electric field would do the job but, again, this is bound to be a wafer by wafer technological process. Two practical solutions remain : the first one comes from LCD technology where alcaline ion migration is blocked by an evaporated SiO_2 film; such film, if thick enough (about 3µm thickness), will isolate the propagating light from the outside. Another possibility can be a chemical leaching or de-ionization of the guide surface[24]. Such process decreases the surface index and thus creates within the glass surface the desired buffer layer.

2.7 Alternative technologies

In conclusion, glass technology with both Ag and Tl ions offers all conditions for the achievement of high numerical aperture, step index, large cross-section waveguides compatible with large core fibres and suitable for LED excitation. There are however alternative multimode technologies which should be briefly listed.

Let us start with alternative glass technologies. Waveguide fabrication by surface modification is not the only way. The whole palette of fibre preform Chemical Vapour Deposition technologies can in principle be easily converted to a planar configuration[25] on substrates like silicon or silica. The deposition rates are large enough and the requested high numerical aperture can be achieved by heavy germanium doping for instance. Highly transparent material can be deposited thanks to the tetrachloride transport. The limitation will probably be set by the etching operation: large etching depth is needed and vertical waveguide walls must be very smooth to avoid scattering. Another problem is the inherent coupling loss with circular fibres since the cross-section of such waveguides will essentially be square or rectangular (unlike in the single mode case, where the fibre-guide field overlap can be made close to 100 %, the coupling efficiency here in multimode plumbing is proportional to the geometrical overlap of the cross-sections).

The most contending technologies alternative to glass ion exchange are obviously polymeric technologies. It takes no more time to make a single than a multimode waveguide. Very high accuracy and potentially cheap replication technologies are ready for use as the LIGA technique[26]. The materials are as transparent as conventional glasses except in the 1.3 - 1.55 µm wavelength range where care must be taken to avoid the O-H and C-H harmonics of the vibrational modes by deuteration for instance[27]. Such technologies are bound to have a bright commercial future. The only limitation we see is the functional stability related with the temperature dependence of the refractive index as pointed out in section 1. However, important industrial developments are in progress. The possibility of fabricating polymer passive waveguide structures by moulding and casting techniques has been demonstrated[28]. The laboratory status has now led to pre-series production and fully fibred power splitters are already proposed on the market[26] with excess loss figures lower than 2.8 dB for four plastic fiber channels (figure 5) and temperature characteristics which are very convincing below 100°C as reported in[28]. Important industrial efforts are being made in polymeric technologies for single mode devices[29] which could easily be adapted to multimode devices.

A possible advantage of the technologies that lead to square or rectangular cross-sections and step index profiles is that they can naturally make use of the specific self-imaging and multiple-imaging property of the fabricated waveguides[30]. This should allow the design and the fabrication of multimode power splitters and couplers which are independent of the mode distribution in the waveguide channels as described in section 7.

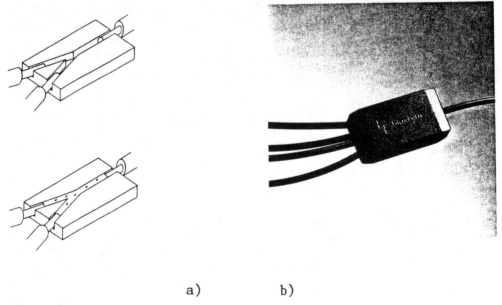

a)　　　　b)

Figure 5

Moulded 1x4 power splitter with plastic fibre pigtailing[26]
a) Sketch of the inside
b) Encapsulated device

3. WAVEGUIDE FUNCTIONAL ELEMENTS

Once light has been coupled into a multimode IO waveguide, it is propagated by the electromagnetic modes of the guide. In high NA and large cross-section guides the number of modes is very high and their propagation constant are very close to each other. Bends, junctions and minute artifacts generate strong coupling between them. Propagation in such structures can be described by ray optics, except in cases where one is able to evidence and use collective coherence properties (section 7).

3.1 In a **waveguide bend**, the very bending of the walls has the effect of decreasing the angle of incidence of the zigzagging rays which at some point along the bend reach the critical angle and get lost. A simplified but very useful rule can be borrowed from the fibre domain: the propagation loss is related to the number ν of propagating modes lost in a bend relative to the total number of modes propagating in a straight waveguide section[31]. In a step index profile, ν writes $\nu = a/(R\Delta)$. In a parabolic profile, ν is just twice this value. a is the guide radius, R the radius of curvature and Δ is the relative core-cladding index difference. A more convenient expression is that of the term $a/(R\Delta)$ leading to an attenuation α expressed in dB: $\nu = 1 - 10^{-10\alpha}$. Figure 6 illustrates this general scaling rule for a multimode bend. This says: the larger the cross-section, the smaller the curvature radius, the smaller the waveguide-substrate index difference, the larger the losses.

Figure 6

Graph showing the value of the normalized opto-geometric parameter $a/(R\Delta)$ leading to a loss α in dB versus α

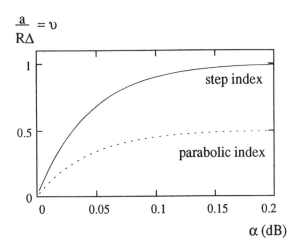

3.2 In a **Y - junction**, the light is expected to be splitted into two equal signals since this is a very symmetrical element. However, the splitting ratio is very much dependent on the modal power distribution of the incoming wave. In case of underfilling of the numerical aperture of a wide and short waveguide, a minute change of the excitation conditions can have a large effect on the splitting ratio. In the reverse regime of excitation by one of the two branches, the same would lead to a change of the excess losses around the 3dB figure which is expectable from radiance preservation considerations[32]. The same type of dependence is expected from N x M star couplers. In the case of IO devices to be used with 50 or 62.5/125 graded index fibres, much has been made to reduce the mode distribution dependence of multimode waveguide couplers. For instance, 4-branch couplers showing a 0.16 dB uniformity of the insertion losses can be fabricated[10].

3.3 Wavelength (de)multiplexing is a very important optical function which allows the use of the very large optical bandwidth between the blue part and the near IR part of the spectrum. The analysis and conclusions of the classical review papers on wavelength (de)multiplexing[33,34] are still valid and should be referred to. There is no property of light propagations in a large NA and large cross-section waveguide which is sufficiently wavelength dependent to give rise to a useable wavelength separation scheme. Mode selection due to geometrical effects in a non-symmetrical Y-junction can lead to the separation of two wavelength channels as analysed in the case of a one-dimension simulation[61]. In practice however the wavelength selective action must be extrinsic to the waveguide and one has to resort to micro-optic assemblies whereby the waveguide is interrupted at some point from where the guided beam propagates as a free space spherical wave, the latter is transformed into a collimated beam which can then be handled by existing or adapted wavelength selective elements such as gratings and interference filters[36]. Such manipulation of the phase space of the beam can be made very efficiently by using the proper optical components. However, the related adjustment costs make a more monolithic approach more attractive for elementary wavelength separation functions such as in two-channel systems: glued selfoc lenses with sandwiched wavelength selective thin plates[36]. It is however always a waste of efforts to operate the double phase space transformation from waveguidance to free space propagation and conversely for just performing an optical function. A more integrated compromis was found where a very thin (about 50 μm) and high quality sawed slit[37] is made across the waveguide in which a very thin interference filter is slidden and glued[38] (figure 7). The filter must be especially designed to have the same wavelength behaviour for all rays within the numerical aperture.

Figure 7

Multimode integrated optic wavelength (de)multiplexer using very thin interference filter in slit[38].

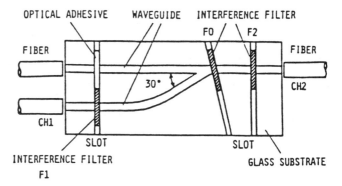

3.4 Active functions: There is some ambiguity as to what the term "active" covers. If it refers to light modulation, it is clear that there is not much that can not be made by simply intensity modulating the light source if the latter is close by. The main interest would rather be in switching capabilities. However, the magnitude of the possible effects is so weak and the spatial coherence in a multimode guide so poor that no waveguided solution can reasonably be expected. One exception could be the use of liquid crystals in the evanescent field of a multimode waveguide[39]. This is not pure speculation since LCs are seriously considered for spatial light modulators in optical computing.

If the term "active" refers to light generation, the perspectives are real when considering rare earth doped glass substrates or waveguides. High dopant concentration can lead to high radiance, broad optical band sources in the spontaneous regime. It can also be imagined that high power waveguide lasers pumped by an array of semiconductor lasers can use the very same waveguide configuration composed of a single mode active waveguide embedded into a large multimode waveguide able to trap most of the multi-source pump[40]. The situation would even be more advantageous here since the multimode pump waveguide can easily be taylored to the geometry of the laser array thus creating more efficient conditions for the exhaustion of the pump over shorter lengths.

4. MULTIMODE WAVEGUIDE EXCITATION

The question of the excitation of a multimode waveguide is an important one regarding two main issues: the efficiency and the stability. As to efficiency, the best that can be made is the complete filling of the numerical aperture and of the cross-section of the excited waveguide. In the case of semiconductor laser excitation, the beam aperture is much larger than the fibre NA and the emitter cross-section much smaller. The excitation efficiency will be enhanced by using a lens which transforms the phase space in reducing the beam divergence and increasing its cross-section. The same can be said in the case of a small area LED: more of the angular spectrum of the source can be captured by using a lens. However, if the emissive surface of the LED is larger than the cross-section of the waveguide, there is no increase of excitation efficiency in using a lens: the best that can be expected is in placing and fixing the LED surface against the guide opening in which case the excitation efficiency I_e is given by[41] $I_e = A_c/A_s \cdot NA^2/n_c^2$ where A_c and A_s are the contacting areas of the waveguide core and of the LED source respectively. The source is assumed to be a Lambertian radiator and the waveguide has the constant index n_c. It is unfortunately not often possible to achieve this optimum butt-coupling configuration in practice because of the presence of the contact wire. It is likely however that LED makers will soon supply less awkward and still cheap sources[42,6].

Such butt-joint configuration is also the best one for a mechanically stable excitation. This is where the thermal expansion differentials cause the least relative displacements within the excitation head under temperature variation. Assuming that the mechanical rigidity of the excitation head is the main optical stability factor, the magnitude of the optical instability still depends substantially on the optical injection conditions. In general, overfilling of the NA and of the cross-section is preferable to underfilling regarding the stability of the optical functions performed by the waveguide elements downstream (branching elements, bends) since no such effect like the zigzagging propagation of a bundle of rays will occur.

The question of the mode dependence of passive components is of prime importance in multimode optical communication systems for the designer who has to plan his power budget. The main subject of concern is the mode dependence of the excess loss and of the insertion loss of branching devices[43]. In sensor applications this question is not less important since the fluctuations of the splitting ratio of a Y splitter in the bridge of an intensity modulated sensor can be taken as a variation of the measurand. These problems have been thoroughly studied and partially solved at the beginning of the eighties in connection with step and graded index fibres in view of rendering the optical function of passive components independent of their position in the fibre network[44]. Specific methods have been designed to characterize multimode passive devices regarding their insertion into multimode LANs.

Basically there are two methods for measuring multimode components[45], one is to use an Over Filled Launch condition (OFL) at the excitation so that the mode Power Distribution is unity for all the modes; the other excitation uses a Limited Phase Space (LPS) condition: 70 % of the numerical aperture of the fibre is used and a surface with a radius of 70% of the core is illuminated. This last launch condition predicts best the insertion loss of the optical components for any system complexity.

Microbending mode scrambling designs have been considered to decrease the mode dependence[46]. This had been envisaged in fibres for characterization purpose with the aim of generating the steady state modal power distribution without resorting to the 1-2 km fibre length necessary to reach the steady state naturally[45]. Such approach was soon recognized as too lossy for a too modest improvement of the mode dependence. Convincing commercial devices were proposed by Corning[10] (Photocor ™) as early as 1988 which exhibited a total excess loss lower than 1 dB (1 to 4 ports), a uniformity between output ports better than 0.5 dB (1 to 6 ports) with a mode dependence of less than 0.16 dB between the OFL and LPS launching conditions.

5. MULTIMODE WAVEGUIDE HYBRIDIZATION TECHNIQUE

As compared with the single mode pigtailing problem[47], that of multimode circuit hybridization is considerably easier, the more so as the fibres which are considered now tend to be large 100 and 200 µm core step index fibres. Furthermore, the sources used in multimodes systems are likely to be mostly Light Emitting Diodes (LEDs) for reasons of cost, stability and simplicity. Such simple systems will use large area detectors. All this means that input/output light coupling is relatively very tolerant. Therefore, micron accuracy batch hybridization techniques can be used such as standard silicon mother board technology[48] as illustrated in figure 8[63]. If components must be placed individually, there are pick and place machines and robots which match the required accuracy.

Accuracy is not a critical issue in the adjustment of the elements composing a multimode optohybrid. More critical are the questions of cost and stability. It is recognized that the pigtailing and packaging of single mode IO devices represents the major part in the cost structure of a completed device. Important developments have been made to reduce this cost substantially[47]. A well accepted alignement strategy is that of the collective relative positioning of individual elements in form of an array followed if necessary by the positioning and immobilization of the array relative to the IO circuit(s). Such kind of approach was adopted for multimode pigtailing in a fully monolithic form whereby the waveguide pattern is defined and ion-exchanged at the surface of a moulded special glass plate already possessing a mesh of accurate V-grooves[62]. Another approach, less monolithic, involving sandwiched IO chip and silicon platform has been developed in a recent Eureka project inclusive of the necessary handling and transfer tools and of the immobilization and packaging technologies[47,63].

In glass integrated optics the instabilities mostly originate from the hybridization technology. Temperature differentials cause multimorphous effects if the expansion coefficients of the parts of the hybrid differ or if the expansion effects are incorrectly uncoupled in the design of the hybrid. A thermally homogeneous multimode hybrid can be achieved by switching to another moterboard material, by using Liquid Crystal Polymers or metal multilayers and by a suitable distribution of the residual stress[63].

Figure 8

Sketch of the common silicon mother board hybridization of an IO chip with an array of fibres[63].

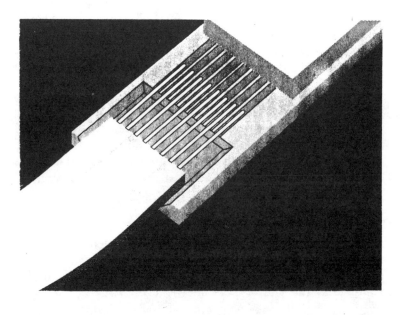

6. FORESEEABLE APPLICATIONS

It is certain that all very local communication systems using plastic optical fibres will make use of injection moulded[28] or casted passive components as in automotive application for instance. This technology, which defines the waveguide network and the fibre registration features in a single moulding operation will also be used in connection with large core silica fibres[26]. Multimode glass integrated optics has a chance in large core silica fibre systems in cases where high stability is desired and also in specific branching and routing applications where the cost of a mould can not be amortized on a very large number of components or modules. The specific merits of glass IO seem to be best exploitable in fibreless monolithic systems as suggested by the example in the introduction.

The following types of applications can be thought of:

- Replacing fibre tip sensors in cases where the source does not have to be remote as for instance in two-fibre proximity sensors[49], fibre tip pressure sensors[50] or others[51].

- Replacing micro-optic systems where the light routing functions are complex or where the number of sources or of output ports is large. There is a wide variety of possible applications such as the monitoring of water cleanliness, of the gaseous and liquid exhausts of chemical plants, of the vapours in bioreactors, by a simple absorption mechanism or by means of suitable dyes incorporated in selective membranes[52] or else in exciting and efficiently trapping fluorescence as in immuno-assays[53].

- Evanescent wave sensing of refractive index, of absorption and fluorescence can also be used with non-buried multimode waveguides[54] (figure 9) and large spectrum light sources although the sensitivity is very much dependent on the mode power distribution. However, a stable excitation will enable reproducible sensing conditions after a first calibration.

These are only a few examples of application domains where the advantages of monolithic multichannel light routing should soon become perceptible: the technologies are available, LEDs will become increasingly present in everyday life[6], some of the collective hybridization techniques that have been developed for highly demanding single mode applications can easily be made useable and cheaper for multimode system assembling and packaging[47].

Figure 9

Cross sectional view of surface waveguide and evanescent wave interaction with organic species enriched in the sensing membrane[54].

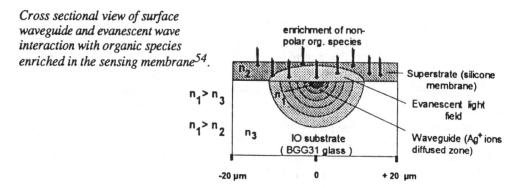

7. MULTIMODE INTEGRATED OPTICS CONDEMNED TO POOR FUNCTIONALITY ?

The brief history of the search for large bandwidth in multimode fibre technology already showed that some degree of spatial coherence preservation can be achieved by properly shaping the core index profile. Going from the confined free space of a step index fibre to a quasi-parabollic self-focusing profile led to a dramatic bandwidth increase; what concerns us here in such index distribution is that the field at some point on the longitudinal abcissa is imaged periodically along the direction of propagation. This period Λ is given by[55]: $\Lambda = 2\pi a/\sqrt{2\Delta}$.

This shows that a suitable choice of refractive index distribution may lead to collective propagation features which can be used on purpose. This infers that in such condition one subtle action on the guide may affect all modes. The most vivid illustration of how a defined perturbation can affect all propagation modes is given by the microbending sensor in which all modes are coupled to their nearest neighbours and eventually to the cladding radiation modes when a microbending is exerted at a spatial period equal to the collective beat length of the modes.

This image formation process is perfect as far as meridional rays are concerned. It experiences aberrations as soon as skew rays are involved which indeed happens when the object point is not much smaller than the core cross-section or when the excitation is off-center and non-paraxial. It was recently shown by S. Krivoshlykov[56] that even in the case of strong aberrations due to a deviation from the parabolic profile, an image of the excitation point is always formed periodically down the fibre lead when paraxial rays are concerned. This period can be much larger than the self-focusing period. The same author in his interesting book[56] studies further multimode waveguides which are non-uniform in the direction of propagation. Such guides propagate "quasi-mode" which will eventually radiate into the substrate more or less rapidly. A given quasi-mode can however propagate without losses within waveguide walls especially taylored whereas all other modes will leak out. This approach gives a further degree of freedom for the selection and the collective handling of the modes of a multimode waveguide. For instance one can then think of a large multimode waveguide to propagate a single very large spot size mode.

The simplest and most beautiful spatial coherence maintaining waveguide is the step index multimode waveguide of rectangular or square cross-section with its property of multiple self-imaging first revealed and demonstrated by Ulrich[30]. The self-imaging effect in step-index optical waveguides is due to phase coincidences of the guided eigenmodes during their propagation in the waveguide. After a propagation length $2L_1$ in a slab waveguide, the eigenmodes have the same phases (modulo 2π) as at the waveguide input. Hence, they interfere in the same way as in the input plane, forming an erect image of the input light source distribution. Because the odd guided modes of an optic guide with symmetric claddings show an odd symmetry, the waveguide forms an inverted image at the propagation length L_1, approximately given by $L_1 = 4n_f w_a^2/\lambda$, where n_f is the refractive index of the waveguide core, w_a is the effective width of the waveguide and λ is the free-space wavelength.

The image forming length is represented in a normalized form in Figure 10 (curve N = 1) versus a normalized waveguide parameter $2w_a\sqrt{n_f}/\lambda$.

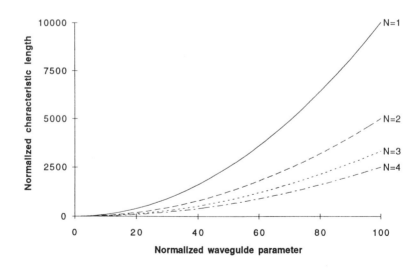

Figure 10 *Normalized characteristic length L/λ as a function of the waveguide parameter, for the formation of an inverted image (N=1), and of a two-fold, three-fold and four-fold image (N=2, 3 and 4 respectively)*

The effective width w_a of the above equation is the waveguide width w augmented by the penetration depth associated with the Goos-Hähnchen shift. Since this effective width depends on the eigenmode order, the characteristic length L_1 is not exactly the same for all guided modes. As a result, the resolution of the self-image is degraded, in particular when the higher order modes of the waveguide contribute to the image formation. The image resolution is further worsened if the profile deviates from a step and if the cross-section differs from a rectangle. The most useful property of the self-imaging effect which is of concern here is that it also produces manifold images at integer fractions of the characteristic length L_1: after a propagation length $L_N = L_1/N$, N subimages of the input light distribution appear. Figure 10 illustrates the waveguide scaling rule for the formation of N images. If we now consider a two-dimension cross-section, the same imaging effect will independently occur in the orthogonal direction. A square cross-section waveguide will therefore act as a fan-out element delivering a matrix of N x N dots from a single excitation point. If the filling of the aperture is complete, a high N figure can be obtained and the uniformity between output dots can in principle be very good and little dependent on the excitation conditions since modes of low and high order all participate in the formation of all images in contrast with what occurs in multimode proximity or asymmetric branching elements. In order to illustrate this, let us take a square waveguide section of 25mm length, of index 1.5, excited at 800 nm wavelength. The single image will be formed at the section output for a guide width of 58 µm. (2x2), (3x3), (4x4), (8x8) dots matrices will be formed for a square guide width of 82, 100, 115 and 163 µm respectively.

So far, such imaging property has been extensively used in one-dimension waveguide technology only where the phase relationship between sub-images[57] has been used for a number of functional devices called MMI (Multimode Interference sections). For instance, choosing a self-imaging waveguide with length $L_2=L_1/2$ provides a 1x2 coupler if a single-mode waveguide is used at the waveguide input[58]. The manifold imaging effect was also used in a flat core fibre, in order to build a 2x4 coupler, mixing the reference and signal beams of an interferometer and providing four output power signals in quadrature[59]. This device has also been demonstrated in the form of an ion-exchanged, single-mode in depth waveguide[60]. In truely multimode applications there was the demonstration of a power splitter using multiple imaging[64].

These last examples are intended to infer that one can find or one can generate large waveguide pipes which have some spatial coherence preserving properties. This field has not been investigated exhaustively yet but a renewed interest in multimode waveguiding may stimulate new attempts and orient them towards propagation properties which would help achieve more complex functions such as wavelength (de)multiplexing, mode filtering, modulation and switching.

8. WHO IS IN THE FOREFRONT ?

The development of multimode glass integrated optic industrial applications is likely to be quite diverse and to rely on small and middle size companies as well as upon the known big ones. In an attempt to shed light on those who can already undertake the development and the fabrication of multimode glass products and to incite those who have identified marketable devices, modules or microsystems to have them realized, we have made a survey of the existing fabrication potential amongst most of the companies and institutes who have published papers related with this technology during the last decade. The list of actors hereunder is probably not exhaustive and only those who have answered the poll are listed. We apologize to those who were involuntarily left aside and encourage them to make themselves known.

Canadian Marconi Company, Montreal, Quebec H3P 1Y9. Fax: + 1 514 340 3120. Contact person: Franz Blaha.

Devices and custom fabrication services based on ion-exchanged soda-lime glass. Waveguides structures can be supplied in the form of processed wafers or fully packaged and connectorized units. Multifibre ribbon pigtailing by means of silicon V-groove arrays available. Available fibred glass IO products: pressure transducer, accelerometers, passive network components.

Corning Optoelectronic Components, F-77015 Melun, France. Fax: + 3316471 3090. Contact person: Martin McCourt.

Most achieved multimode waveguide technology for packaged devices with 50/125, 62.5/125, 85/125 and 100/140 fibres. Devices are buried 1 x N branching elements, N = 2, 3, 4, 6, 8, 12, star N x N devices, dichroic WDMs and bridge circuits[10]. Potential for large scale production available and usable for communication and non-communication applications.

CSEM Swiss Center for Electronics and Microtechnology, CH-2007 Neuchâtel, Switzerland, Fax: + 41 38 205 580. Contact person: Olivier Parriaux.

IO mask writing, prototyping of silver ion exchanged multimode circuits. Complete silicon mother board fabrication with submicron accuracy wafer scale and hybridization technologies. Development and fabrication of optical microsystems in small series with possible integrated optoelectronic signal processing.

Fraunhofer-Einrichtung für Angewandte Optik und Feinmechanik, D-07745 Jena, Germany, Fax: + 49 36 41 52963. Contact person: Rolf Göring.

This institute gathers all about former GDR's research in glass science and fabrication processes for multimode integrated optics: fundamentally low loss glasses for high Δn with silver ion exchange, large cross-section waveguide fabrication modelling, dedicated instrumentation for index profile measurement[18]. Technology available for IO circuit development and prototype and test series production.

GeeO, F-38031 Grenoble, France, Fax: +33 7657 4584. Contact person: Antoine Kevorkian.

Design, industrial development, small-scale production of buried silver ion-exchanged integrated optic multimode waveguide compatible with 50/125 and 62.5/125 fibres. Fibre coupling and 2 cm propagation losses of 0.2 dB currently achieved. Technology extendable to deeper guides.

IOT Integrierte Optik, D-68744 Waghäusel-Kirrlach, Germany, Fax: + 49 7254 925 210. Contact person: Roland Fuest.

Mainly active in the design and production of packaged and pigtailed single mode IO components. Special glasses and multimode waveguide fabrication technologies were developed between 1986 and 88[9]. These can be used for various multimode applications[54].

Optonex, SF-02101 Espoo, Finland, Fax: + 358 0455 1803. Contact person: Seppo Honkanen.

Consultance, design, services and middle-scale production of multimode glass waveguide circuits inclusive of fibre pigtailing. Waveguide Δn up to 0.05, depth up to 100μm[15].

Vavilov State Optical Institute, NITIOM, Optical Glass Laboratory, St Petersburg, Fax: + 7 812 560 10 22. Contact person: Leonid Glebov.

Science, development and small to large scale production of special glasses for integrated optics on a custom basis.

9. REFERENCES

1. M. Mc Court, J.L. Malinge, J.P. Lormeau, T. Dumas, "Applications of photolithographic and ion-exchange techniques for the fabrication of Photocor™ integrated optic couplers and wavelength division multiplexers", Proc. SPIE vol 993, pp. 277-287.

2. Optical Fiber Sensors : Systems and Applications, Ed. B. Culshaw, J. Dakin, Artech House, 1989, Section 12.3

3. B. Culshaw, J. Foley, I.P. Giles, "A balancing technique for optical fibre intensity modulated transducers", 2nd OFS, Stuttgart, 1984, pp. 117-120

4. P. Sixt, L. Falco, J.P. Jeanneret, O. Parriaux, "General self-referencing single-fibre head for intensity modulated sensors", IOOC-ECOC'85, Venise, Oct. 1985, pp. 797-800

5. D.E.N. Davies, J. Chaimowicz. G. Economou, J. Foley, J. Lightwave Technology, LT-2, 1984, p. 387

6. HP press information on TS LEDs, 11 Feb. 1994

7. G. Pfister, P. Ryser, "Sensorik für den Schutz von Menschen und Gütern", Neue Zürcher Zeitung, Forschung und Technik, Nr. 93, April 22, 1992.
J. Werner, P. Ryser, Optischer Rauchmelder, Patent Nr 02 157/92-9, 08.07.1992

8. N. Fabricius, H. Oeste, H.-J. Guttmann, H. Quast, L. Ross, "BGG31: A new glass for multimode waveguide fabrication", EFOC/LAN 88, Amsterdam, 29 June-1 July 1988, pp 59-62

9. L. Ross, "Integrated Optik in Glas", MIOP'88, Wiesbaden, 2-4 March 1988, Paper 6B-4

10. M. McCourt, "Passive integrated optics for optical communications", EFOC/LAN'89, Amsterdam, June 12-16, 1989, pp. 292-296

11. Introduction to Glass Integrated Optics, Ed. S.I. Najafi, Artech House, 1992

12. Ch. Kaps, T. Possner, "Material science aspects for multimode waveguide fabrication in oxide glasses", EFOC/LAN'91, London, June 19-21, 1991

13. E. Cullmann, W. Vach, "Comparison of resolution for standard, mid-UV and deep-UV", Proc. Semicon East, Boston, 1982

14. T. Findakly, "Glass waveguides by ion exchange. A review." Opt. Eng., Vol. 24, 1985, pp. 244-250

15. S. Tammela, H. Pohjonen, S. Honkanen, A. Tervonen, "Fabrication of large multimode glass waveguides by dry silver ion exchange in vacuum", SPIE Vol. 1583, Integrated Optical Circuits, 1991. pp. 37-42

16. R.V. Ramaswamy, R. Srivastava, "Ion-exchanged glass waveguides: A review", J. Lightwave Technology, Vol. 6, June 1988, pp. 984-1001

17. T. Possner, G. Schreiter, R. Müller, Ch. Kaps, H. Kahnt, "Special glass for integrated and microoptics", Glastech. Ber. 64 (1991), Nr 7, pp. 185-190

18. T. Possner, G. Schreiter, R. Müller, "Stripe waveguides with matched refractive index profiles fabricated by ion exchange in glass", J. Appl. Phys. 70, Aug. 1991, p. 1966.
R. Göring, T. Possner, "Optimization of integrated optic components by refractive index profile measurements", EFOC/LAN'92, Paris, June 24-26, 1992, p. 396

19. Ch. Kaps, W. Fliegel, "Sodium/silver ion exchange between a non-bridging oxygen-free boroaluminosilicate glass and nitrate melts", Glastech. Ber. 64 (1991), Nr 8, pp. 199-204

20. K. Nishizawa, H. Nishi, "Coupling characteristics of gradient-index lenses", Applied Optics, Vol. 23, 1984, pp. 1711-1714

21. O. Parriaux, P. Roth, Ch. Kaps, "Generating refractive index step profiles in glass by purely thermal ion exchange", ECIO'93, Neuchâtel, April 19-22, 1993, pp. 9-22

22. Ch. Kaps, J. Non-crystalline Solids, Vol 123, 1990, pp. 315-320

23. J.-F. Bourhis, "Contribution à l'étude des mécanismes de l'échange d'ions thallium/potassium dans un verre aluminoborosilicate. Application à la réalisation de composants en optique intégrée unimode". PhD thesis, May 1992, Institut National Polytechnique de Grenoble

24. Ch. Kaps, Fachhochschule Jena, private communication

25. T. Kominato, Y. Ohmori, H. Okazaki, M. Yasu, "Very low-loss GeO_2-doped silica waveguides fabricated by flame hydrolysis deposition method", Electronics Letters, 1990, Vol. 26, pp. 327-328

26. M. Serizawa, A. Rogner, H. Pannhoff, "Fabrication of passive polymer waveguide devices by injection-moulding", 8th. Int. Microelectronics Conference, April 20-22, 1994, Sony City, Omiya, Japan

27. A. Neyer, T. Knoche, L. Müller, P.C. Lee, J.H. Kim, M.A. Andrade, J. Carvalho, J.L. Figuheirodo, A.P. Leite, "Design and fabrication of low loss passive polymer waveguides based on mass replication techniques", ECIO'93, Neuchâtel, April 19-22, 1994, pp. 9/10-9/11

28. Y. Takezawa, S.-I. Akasaka, S. Ohara, T. Ishibashi, H. Asano, N. Taketani, "Low excess losses in a Y-branching plastic optical waveguide formed through injection molding", Applied Optics, Vol. 33, 1994, pp. 2307-2312

29. W.H.G. Horsthuis, M.M. Klein Koerkamp, J.-L. P. Heideman, H.W. Mertens, B.H. Hams, "Polymer based Integrated Optic Devices", SPIE Vol. 2025, 1993, pp.516-523.

30. R. Ulrich, "Image formation by phase coincidences in optical waveguides", Opt. Commun., Vol, 13, 1975, pp. 259-264

31. D. Glodge, "Bending loss in multimode fibers with graded and ungraded core index", Applied Optics, Vol. 11, Nov. 1972, pp. 2506-2513

32. Y. Kokubun, S. Suzuki, K. Iga, "Evaluation of distributed-index branching waveguides by phase space", Applied Optics, Vol 25, Oct. 1986. pp. 3401-3404

33. G. Winzer, "Wavelength division multiplexing - Status and trends", ECOC'82, Cannes, September 1982

34. H. Ishio, J. Minowa, K. Nosu, "Review and Status of wavelength -division-multiplexing technology and its applications", J. Lighwave Technology, Vol. LT-2, Aug 1984, pp. 448-463

35. J.P. Laude, J. Lerner, "Wavelength division multiplexing using diffraction gratings", SPIE 503-04, San Diego, 1984

36. J.P. Laude, "Wavelength division multiplexer, technological trends", EFOC/LAN 87, Basel, June 3-5, 1987. pp. 85-90

37. J. Watanabe, T. Saito, "Precision machining for microoptical devices with powder-particle collision", CLEO, 1987, paper W04, p. 192

38. Y. Ikeda, E. Okuda, M. Oikawa, "Graded-index optical waveguides and planar microlens arrays and their applications", EFOC/LAN'87, Basel, June 3-5, 1987, pp. 103-107

39. F. Saint-André, "Réalisation d'un polariseur et d'un interrupteur optique", PhD thesis, 1992, Institut National Polytechnique de Grenoble

40. P.L.Bocko, "Rare-earth-doped optical fibers by the outside vapor deposition process", OFC'89/paper TUG2

41. O. Parriaux, "Guided wave electromagnetism and opto-chemical sensors" in Fiber Optic Chemical Sensors and Biosensors, Ed. O.S. Wolfbeis, CRC Press, 1991

42. K. Hayamizu, H. Watanabe, S. Ogata, H. Imamoto, K. Imanaka, "High-power pinpoint AlGaAs LED fabricated by a simple wafer process", CLEO'92 Technical Digest, Papers CTh133 and CTh134

43. Y. Kokubun, T. Fuse, K. Iga, "Optimum length of multimode optical branching waveguide for reducing its mode dependence", Applied optics, Vol. 24, Dec. 1985, pp. 4408-4413

44. A. Béguin, T. Dumas, M.J. Hackert, R. Jansen, C. Nissim, "Fabrication and performance of low loss optical components made by ion exchange in glass", J. of Lightwave Technology, Vol. 6, 1988, pp 1483-1487

45. P.R. Reitz, M.J. Hackert, J.E. Matthews, K.W. Murphy, D. Hanson, M. Hartmann, T. Huegerich, K. Kevern, T. Odderstol, "A comparison of loss measurement methods with observed system loss in optical links" Proc. SPIE 559, p. 24

46. Y. Kokubun, S. Suzuki, T. Fuse, H. Uehara, K. Iga, M. Oikawa, S. Misawa, "Novel mode scrambler for reducing mode dependence in multimode optical waveguides", Electronics Letters, Vol. 19, 1983, pp. 1009-1010

47. J.-F. Bourhis, V. Mignot, R. Hakoun, "Automatic pigtailing of passive integrated optics components with large number of ports", Int. Symposium on Integrated Optics, April 11-15, 1994, Lindau, paper N° 2213-30

48. E.J. Murphy, T.C. Rice, "Self-alignment technique for fiber attachment to guided wave devices", IEEE QE-22, June 1986, pp. 928-932

49. D. Walsh, B. Culshaw, "Novel passive compensation technique applied to a white light interfometric system", 8th OFS, Monterey, January 29-31, 1992, pp. 221-224.

G.J. Philips, "F-O Displacement sensors for dynamic measurements", Sensors, Sept. 1992, Helmers Publishing.

50. P. Sixt, L. Falco, P. Dierauer, H.W. Lehmann, "Microstructure fiber-tip sensor with spectral encoding", SPIE Europtica conference, Hamburg, September 19-23, 1988

51. A. Suhadolnik, A. Babnik, J. Mozina, "Optical fibre reflection refractometer", EUROPT (R)ODE II, Florence, April 16-18, 1994, p. 197

52. D. Freiner, R.E. Kunz, D. Citterio, U.E. Spichiger, M.T. Gale, "Integrated optical sensors based on refractometry of ion-selective membranes", EUROPT(R)ODE II, Florence, April 19-21, 1994, p. 167

53. G.A. Robinson, J.W. Attridge, J.K. Deacon, S.C. Whiteley, "The fluorescent capillary fill device", Sensors and Actuators B, vol. 11, 1993, pp. 235-238

54. J. Bürck, B. Zimmermann, H.-J. Ache, "Integrated-optical NIR - evanescent wave sensor for monitoring organic compounds in water", EUROPT (R)ODE II, Florence, April 19-21, 1994, paper Tu1B.14
B. Drapp, J. Ingenhoff, G. Gauglitz, V. Hollenbach, "Characterisation of integrated optical interferometers for evanescent field sensing", EUROPT(R)ODE II, Florence, April 19-21, 1994, paper P1.5

55. J.E. Midwinter, Optical Fibers for Transmissions, John Willey & Sons, 1979, Chapter 6.

56. S.G. Krivoshlikov, "Quantum-Theoretical Formalism for Inhomogeneous Graded-Index Waveguides", Akademie Verlag, 1994

57. M.K. Smit, Integrated optics in silicon-based aluminium oxide, Ph.D. thesis, ISBN 90-9004261-X, Delft University of Technology, Delft, the Netherlands, 1991

58. L.B. Soldano, M. Bachmann, P.-A. Besse, M.K. Smit, and H. Melchior, "Large optical bandwidth of InGaAsP/InP multi-mode interference 3dB couplers", ECIO'93, April 18-22, 1993, Neuchâtel, Switzerland, pp. 14-10 - 14-11

59. Th. Niemeyer and R. Ulrich, "Quadrature outputs from fiber interferometer with 4x4 coupler", Opt. Lett., Vol. 11, 1986, pp. 677-679

60. P. Roth and O. Parriaux, "Integrated interferometer with phase diversity", IOOC'89, July 18-21, Kobe, Japan, 1989, pp. 120-121

61. S. Safavi-Naeini, S.K. Chaudhuri, A. Goss, "Design and analysis of novel multimode optical filters in dielectric waveguide", J. of Lightwave Technology, vol. 11, 1993, pp. 1970-1977.

62. E. Paillard, "Recent developments in integrated optics on moldable glass", Fiber Optics '87, London, April 29th, 1987.

63. EUREKA project EU-465, ASSYSTO.

64. R. Ishikawa, H. Nishimoto, K. Minemura, S. Matsushita, "Kaleidoscope micro-optic star coupler", Electron. Lett., Vol. 16, 1980, pp 249-251.

GLASS WAVEGUIDES IN AVIONICS

F.A. Blaha
Canadian Marconi Company
Montreal, Quebec, Canada H4M 2S9

ABSTRACT

A variety of glass waveguide devices, including passive components (splitters and couplers), active network components (waveguide switches, lasers and amplifiers), and sensors for the measurement of pressure acceleration, position, rotation, etc, are in various stages of development, or have reached the production stage, at various companies. These components are expected to have a major impact on the implementation of fiber-optic (FO) systems for aircraft. Many new functions and applications can be realized with their introduction. This paper examines factors which have so far delayed the introduction of these products. Applications of FO and integrated-optic (IO) technologies, and the potential of glass waveguide devices in aircraft systems, are discussed.

1. INTRODUCTION

New-generation aircraft will require control, communications, and sensor systems of increasing complexity and sophistication. FO and IO technologies[1] offer substantial performance improvements over conventional electronic technologies, and appear to be prime candidates for these applications.[2]

In spite of their apparent advantages, only a few of the many potential applications of FO and IO technologies in aircraft systems have so far reached the production stage. The reason for the delay is apparent when evaluating the key factors influencing their acceptance:

Key Factors	FO Systems vs Electronic Systems
Features	Implementation of several novel concepts become feasible
Performance	Matches or exceeds electronic system performance
Technical Risk	Still higher, with many unproven components
Reliability	Potentially higher; many components immature
Maintainability	Unknown (lack of experience)
Flexibility	Higher (provides for expansion)
Life-cycle cost	Potentially low but unproven (high acquisition cost at present; MTBF and maintenance costs uncertain)
Survivability (military)	Uncertain
Certification	Unknown; new technology
Weight	Lower
Power	Higher (temperature stabilization; laser/LED power supplies)

2. FIBER AND IO TECHNOLOGIES FOR AVIONIC SYSTEMS

This section discusses some of the more demanding requirements of future aircraft, and the application of fiber and IO technologies, with emphasis on glass IO devices. Some of the simpler applications for which fiber-optic technologies are already in use include passenger entertainment systems, and military data transfer systems (stores management, night vision systems.

2.1 High-Speed and Reliable Data Processing and Distribution

Present-day commercial aircraft carry approximately 200 avionics "black boxes", or line-replaceable units (LRUs), each of which performs a specific function. In order to reduce the high maintenance costs of avionic systems, new-generation military aircraft (eg, the F-22, RAH-66, A-12) as well as commercial aircraft like the B777, are designed around an Integrated Modular Avionics (IMA) concept.

IMA departs from the black-box approach of current integrated systems. The basic functional building blocks are standardized, becoming line-replaceable modules (LRMs). Aircraft-specific IMA cabinets house these LRMs, which replace many LRUs. The cabinets, actuators and sensor subsystems are interconnected via a standardized digital data bus (Figure 1).[3]

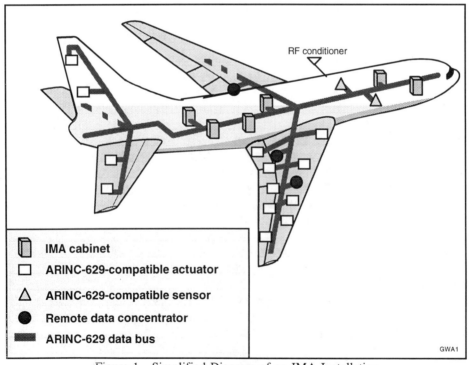

Figure 1 - Simplified Diagram of an IMA Installation

Currently accepted data bus standards are ARINC-429, and its successor ARINC-629, for commercial aircraft, and MIL-STD-1553 for military aircraft. Twisted shielded-pair (TSP) copper wires are the transmission media for these standards. ARINC-429 is unidirectional, and operates at a data rate of either 12 to 14.5 kHz or 100 kHz. ARINC-629 and MIL-STD-1553 bus structures are bi-directional, operating at data rates of 2 and 1 megabits/second (Mb/s), respectively.

An attempt was made to provide a FO substitute for MIL-STD-1553, using the new MIL-STD-1773. Although MIL-STD-1773-based systems offer advantages such as EMI immunity and weight savings, they do not exhibit higher performance, and are more expensive. For these reasons, they have not found acceptance in aircraft (although they have been successfully applied in spacecraft).

A high-speed FO digital data network is preferred for IMA systems. For military aircraft such as the F-22 Advanced Tactical Fighter and the RAH-66 "Comanche" helicopter, hardware for a 50-Mb/s FO high-speed data bus (HSDB) is being developed and qualified. The complete HSDB interface is contained in a 2.5" square module, 0.14" high, weighing 70 g[4,5,6], with optical fibers as the transmission media. Glass IO splitters, and eventually waveguide amplifiers, are candidate network components for future use in HSDBs.

It was originally intended to install an FO high-speed digital data network on the B777[7,8], which would have made it the first commercial aircraft to be so equipped. Unfortunately, the network eventually selected was ARINC-629, due to a number of factors including the lack of proven FO and IO components. The trend, however is clear: data distribution networks in future aircraft and for system upgrades of existing aircraft will employ FO and IO technologies.

2.2 Extensive Use of Sensors

Military or commercial aircraft currently employ several hundred sensors of various kinds. More than half of them are used for the measurement of position, pressure and temperature. This number of sensors will drastically increase with the introduction of FO technologies in integrated propulsion and flight control, and "smart structures and skins".

A large number of FO sensors will form an integral part of smart structures, in which they will be used for the measurement of strain, vibration, temperature, acoustic emission and corrosion, and actuators for modifying the structure's mechanical characteristics. In the near term, a more modest approach, involving tens to hundreds of sensors, will be incorporated in load monitoring systems (LMSs) for tracking structural usage. Loss-free glass IO components (couplers, splitters) would be ideally suited for interrogating this number of sensors in real time. The Canadian Department of National Defence plans to equip its Challenger aircraft fleet with LMSs incorporating FO strain sensors. Economic considerations may result in limiting the number of sensors to 32.

2.3 Improved RF Distribution

Aircraft navigation and communication systems employ coaxial cables or waveguides for connecting antenna systems with the transceiver. Coaxial cables attenuate the signal and are susceptible to RFI, sometimes requiring double-shielding. They are also quite heavy; antenna pre-amplifiers are therefore frequently used. Cable length and routing must be carefully controlled. A case in point is the GPS receiver installation in the B777, where the complete GPS receiver is mounted at the antenna location in order to avoid RF transmission problems altogether.

FO transmission lines, with virtually unlimited bandwidth and low loss (>0.2 dB/km), offer an alternative approach in which the light signal is modulated by the RF. In a radically different concept (still being researched), the EM field directly modulates the light carrier (photonic-field sensor), thus eliminating the conventional antenna, which converts the EM field into an electric signal.[9,10] The use of glass IO couplers and splitters incorporating waveguide amplifiers (when available) can be expected in FO RF transmission networks and phased-array antennas.

2.4 Improved EMI Protection

Fly-by-wire (FBW) is today's flight-control technology, and it is recognized that EMI, electromagnetic pulse (EMP), HIRF, and other new threats such as directed energy weapons, can pose a serious reliability problem for FBW systems. A high degree of shielding is required to sufficiently reduce their vulnerability to microwave radiation, leaving only very-high-power radiation as a threat. Heavy shielding introduces a weight penalty, which can be avoided with fly-by-light (FBL) systems. An FO data bus requires shielding only at the receiver end, where the conversion to an electronic signal takes place. Weight savings of hundreds or even thousands of pounds can be realized with the implementation of FBL.

2.5 Low Maintenance Requirements

The IMA has built-in redundancy[11] and includes a health monitoring system (HMS) for reliable in-flight fault detection and isolation to a specific LRM. The HMS will take corrective steps and re-configure the fault-tolerant IMA system. All details of this corrective action are recorded by an on-board maintenance system. IMA maintenance can be delayed and carried out at the most convenient time and place. Many faults will require repair action only during regular maintenance checks.

IMA requires a very reliable high speed data bus[12]. Although an electronic data bus is currently used, FO networks with built-in redundancy will have to be applied in the future in order to accommodate growing HMS and on-board maintenance demands.

2.6 Cost Savings

IMA acquisition costs are currently high, but the cost of ownership (life-cycle cost) is substantially lower than that of black boxes. Cost savings result from higher reliability, modular design, and reduced maintenance and maintenance-base costs.

Reliability is one of the key factors affecting cost of ownership, and high reliability is therefore an important requirement for the FO network and its glass IO components. An MTBF of greater than 10,000 hours is typically required for equipment housed in IMA cabinets. For equipment or components located in inaccessible areas, MTBF must be higher than 40,000 hours.

3. INTEGRATED OPTICS COMPONENTS IN AVIONIC SYSTEMS

The introduction of IO components is linked to the application of the FO networks in aircraft. An overview of potential fiber-optic technology applications in aircraft has been provided in the previous section. This section will deal with integrated optic components which are either developed[13], or are being developed, for aircraft FO systems. The requirements for this application, as they apply to IO components, are discussed below.

3.1 Environmental Conditions

Table 1 summarizes environmental test conditions which must be met by IO components for aircraft applications. MIL-STD-810 and DO-160 are the relevant documents from which the data have been extracted.

Table 1. Typical Environmental Conditions for Avionics Equipment

		Commercial		Military
		Exterior	Equipment Bay	
Temp Range	Operating	-55 to +70°C	-15 to +70°C	-57 to +71°C
	Non-Oper.	-55 to +85°C		-62 to +85°C
Temp Variation		10°C per min.	2°C per min.	25°C per min.
Humidity		144 h at 95% RH cycled between -55 and +38°C	48 h at 95% RH cycled between -50 and +38°C	240 h at 95% RH cycled between -65 and +28°C
Shock	Oper.	6 g, 11-ms pulse; 3 shocks per orientation		15 g, 11-ms pulse; 3 shocks per orientation
	Non-Oper.	15 g, 11-ms pulse; 2 shocks/orientation		30 g, 11-ms pulse; 3 shocks/orientation
Vibration		Sinusoidal ≤2 g 5 to 2,000 Hz	Sinusoidal ≤1.5g 5 to 2,000 Hz	Sinusoidal ≤2 g 5 to 2,000 Hz

3.2 Integrated Optic Components for Data Distribution Networks

FO data distribution networks currently under development use standard 100/140 micrometer multi-mode graded index fibers. Correspondingly, glass IO circuits for this application are multi-mode couplers and splitters which match this fiber standard. Glass IO devices which meet these requirements are commercially available, and in use in the telecommunications industry.

3.3 Integrated Optic Components for Sensor Systems

Figures 2 and 3 show the systems unit and a block diagram, respectively, of an FO strain sensor system for aircraft structural health monitoring. The system is intended to replace conventional strain gauge systems in military aircraft, and employs FO air-gap sensors. A conceptual diagram and photograph of this sensor are provided in Figures 4 and 5, respectively. A key feature of this sensor is that its thermal expansion coefficient can be matched to that of the host material. Single-mode fibers are required for the sensor network. Sensors and system have passed reliability and environmental tests.

Fused couplers and a 1:4 glass IO splitter are employed in the system. System requirements dictate a flat response over a broad wavelength range (1.3 to 1.5 µm), which can be readily achieved with IO splitters. Future production will incorporate a glass IO device which has the complete splitter network integrated on a single substrate (Figure 6). Detectors could be eventually directly butt-coupled on the substrate.

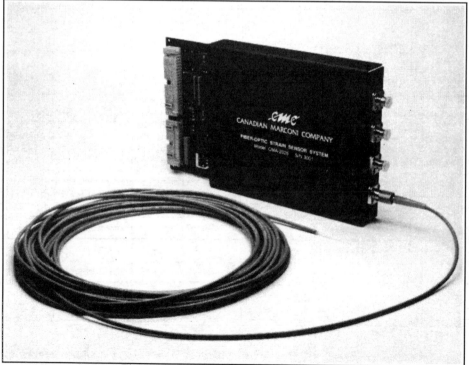

Figure 2. FO Strain Sensor Systems Unit

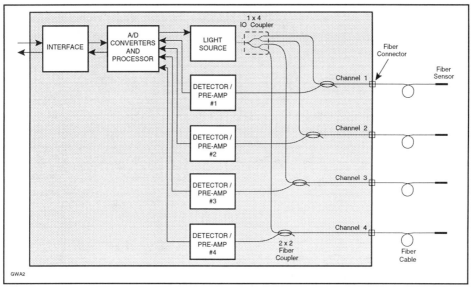

Figure 3 - FO Strain Sensor System Block Diagram

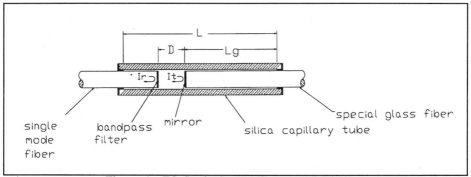

Figure 4. FO Strain Sensor Conceptual Diagram

Figure 5. FO Strain Sensor

A system which is being developed for aircraft loads monitoring employs 32 FO strain sensors and three IO accelerometers. The strain sensor system and network will employ numerous single-mode IO splitters in the systems unit and at network branching points. These devices are commercially available. A typical example of such a device, with a splitting ratio of 1:16, is shown in Figure 7.

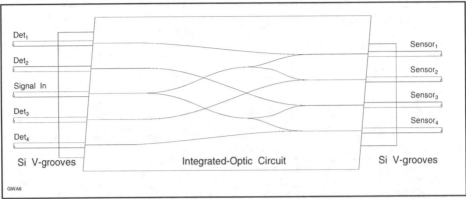

Figure 6 - Glass IO Branching Network for an FO Sensor System

Figure 7. 1:16 Glass IO Splitter (courtesy IOT, Germany)

The introduction of an IO accelerometer is planned for this system; a conceptual diagram and an advanced Development Model (ADM) of the accelerometer are shown in Figures 8 and 9, respectively. Waveguides around the circumference of a glass membrane form a Mach-Zehnder interferometer that senses membrane deformation. The centre of the membrane is loaded with a seismic mass, the size of which determines the measurement range. An IO pressure transducer, based on the same concept, was developed for aircraft propulsion systems (Figures 10 and 11).

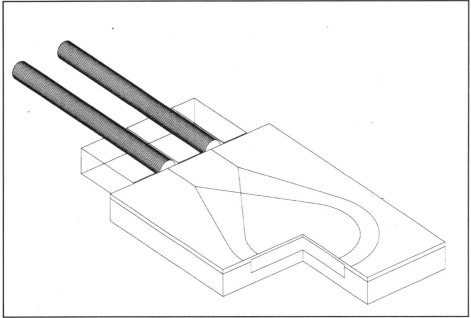

Figure 8. Diagram of IO Accelerometer

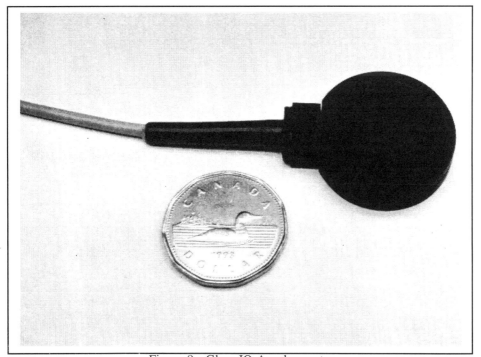

Figure 9. Glass IO Accelerometer

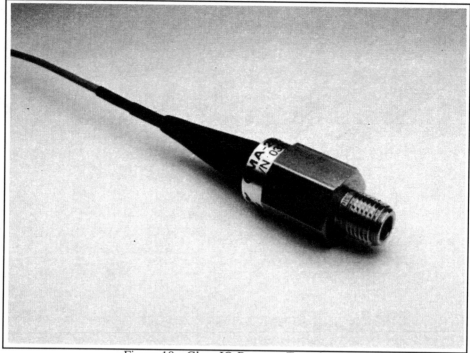
Figure 10. Glass IO Pressure Transducer

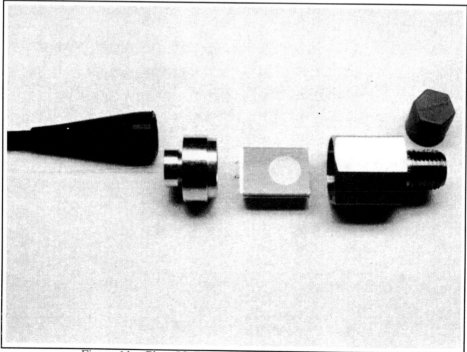
Figure 11. Glass IO Pressure Transducer (disassembled)

3.4 Integrated Optic Components for RF Distribution Networks

The introduction of RF signal distribution in optical fibers removes the limitations of coaxial cable or RF waveguide interconnects. New system architectures, which are free of these physical constraints, can therefore be considered. Loss-free splitters will eventually be required for RF distribution networks, to minimize signal degradation. Of considerable interest is the application of optical fibers for signal distribution in phased-array antennas. This technology makes it possible to realize stable and programmable delays implemented as a compact module. A design under implementation is shown in Figure 12.

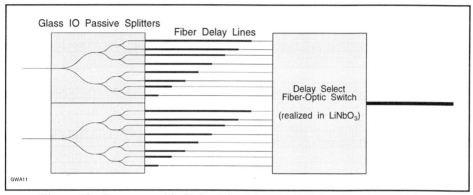

Figure 12. Programmable Delay Line Module Utilizing Glass IO Devices

4. CONCLUSION

FO technology is well-established in long-distance communication systems, but in aircraft systems it has only found limited applications, mainly in non-critical areas. The first critical and demanding applications will likely be for high-speed data transmission networks. The implementation of FO sensor systems for in-flight load monitoring can also be expected in the near future. Fly-by-light, which includes flight and engine control systems, should be introduced by the year 2000.

These new FO sensor systems will require IO components, the acceptance of which will, to a large extent, depend on their performance and reliability as demonstrated during qualification tests. The establishment of industry-wide standards, for the components as well as the systems, would be most helpful at this time.

5. ACKNOWLEDGEMENTS

The author would like to thank his co-workers for useful discussions, and John Russell for compiling the environmental test data. The editorial assistance of Andrew Solkin is also gratefully acknowledged.

6. REFERENCES

1. S.I. Najafi, *Introduction to Glass Integrated Optics*, Artech House, Boston, 1992.
2. G.L. Abbas et al, "Photonics Technology for Avionic Systems", *IEEE Proceedings*, pp 349-356, July 1993.
3. D.C. Hart, "A Primer on IMA", *Avionics*, April 1994.
4. C.S. Figueroa et al, "Fiber-Optics for Military Aircraft Flight Systems", *IEEE Lightwave Communication Systems*, vol 2, no 1, pp 52-65, February 1991.
5. S. Reich and C. Ritter, "The impact of fiber optics (photonics) on future aircraft", *SPIE Proceedings*, vol 1170, pp 77-89, 1989.
6. R. Uhlhorn and J. Snow, "Civil and Military Applications of the High Speed Databus", *Avionics*, pp 24-31, March 1994.
7. J.A. White, "Fiber-Optic Technology for Transport Aircraft", *Aerospace Engineering*, July 1993.
8. J.R. Todd, "Toward Fly-By-Light Aircraft", *Proceedings of the Society of Photo-Optical Instrumentation Engineers*, Boston, September 1988.
9. J.C. Wyss and S.T. Sheeran, "A Practical Optical Modulator and Link for Antennas", *Journal of Lightwave Technology*, vol LT3, no 2, April 1985.
10. S.S. Sriram, S.A. Kingsley, J.T. Boyd, "Electro-Optical Sensor for Detecting Electric Fields" US patent number 5,267,336, November 1993.
11. A.S. Glista, "Fault Tolerant Topologies for Fiber Optic Networks and Computer Interconnects Operating in the Severe Avionics Environment", *IEEE Lightwave Communication Systems*, pp 66-78, February 1991.
12. J.R. Todd et al, "Fly-By-Light Installation and Maintenance on Transport Aircraft", *IEEE Proceedings*, pp 357-362, July 1993.
13. N. Fabricius, "Reliability of Passive Fiber-Optic Branching Devices", *IEE Colloquium on Reliability of Fiber-Optic Cable Systems*, London, January 1994.

Devices for Communication
and Sensors II

Fibre to waveguide connection

Jean-François Bourhis
Alcatel Cable Interface 65 rue J.Jaurès 95870 Bezons FRANCE

ABSTRACT

The technique of the fibre to waveguide connection are described and compared in terms of insertion loss, time requirement, return loss and environmental performance. The most promising techniques are outlined together with the latest technical improvements.

INTRODUCTION

As early as in 1969, S. Miller [1] introduced Integrated Optics (IO) and presented its main potential asset : the reliable mass production through photolithographic transfer, of chips wih a dense and complex combination, in a circuit of several optical functions.

After 25 years of countless promising demonstrations in labs, IO components are coming up only today as commercial products. Most of them are complex in the only sense that several identical functions are duplicated on one chip.

For example, passive 1XN splitters made of either ion exchange [2][16] or silica deposit on silicon [3] are made of N - 1 Y junction. Commercially available splitters exhibit excellent performance in the broad wavelength range of 1.27-1.6 µm for up to 32 output ports. They suit very well to the Passive Optical Networks needs and they are currently installed in large scale fibre distribution projects [17].

External modulators [4][5] made of titanium diffusion in lithium niobate are used in cable TV networks . The same technique allows fabrication of switch matrices up to 32 ports [48][101] and the fabrication of components for fibre optics gyroscopes (fog) [18] which can be found in (air)crafts. DFB lasers [10] are undoubtly among the most popular and well-known IO devices and can be found in arrays.

While those monofunctions components are widespread and are currently mass-produced, monolithically integrated multifunction complex chips are developped in labs. Smart combinations of splitting and demultiplexing optical signals have been demonstrated [6][3] on passive components. A large variety of Opto-Electronic Integrated Circuits (OEIC) can also be found in literature[82][93]. IO duplexers [7][11] could decrease the overall cost of fibre distribution and are

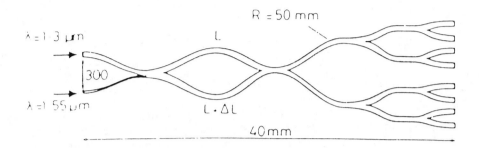

figure 1 : integrated wavelength multiplexer and 1x8 splitter made by ion exchange in glass [6]

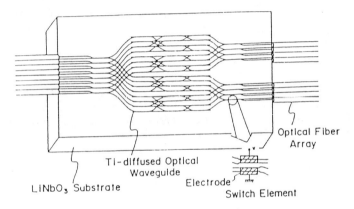

figure 2 : 8x8 matrix in LiNbO3 [85]

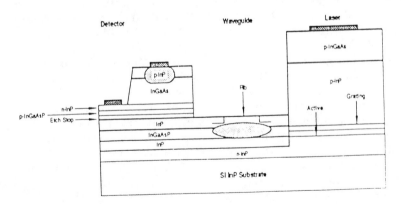

figure 3 : OEIC including detector, laser and waveguide (drawn orthogonal to show its basic structure)

valuable outputs of the RACE project OMAN [7]. Diplexers are also proposed [8]. The partners in the RACE project MUNDI presented transceivers for Dense Wavelength Multiplexing [9]. Laser, amplifier and multiplexer combination [12] is another example of monolithic OEIC integration. Hybrid assemblies [19][20] and such as developped in RACE project MOSAICS would also extend the limits of integration.

Researchers and Engineers are taking full profit of the obvious assets of the several IO techniques. The large choice of materials and structures to realize each optical function, combined with accurate reproducibility of the best configuration, lead to high optical performance components. The demand on ever more complex structures could only increase the needs for IO solutions.

One can expect the compact and in chip embedded structure to have low sensivity to environmental changes. Furthermore the production of IO chips enables mass production at low cost.

In front of those assets, one only wonder why IO components are not spreading faster. One reason is probably the relatively low complexity of the components required until now. But there is also the technical problem of the connection of the planar waveguide(s) on the chip to the circular optical fibre(s) on line. This interface alters optical performances and environmental withstand. It increases the overall cost of the component.

The connection adds coupling and misalignment loss to the waveguide propagation loss (see part 1). The former is divided in fibre and waveguide modes mismatch and Fresnel loss, and can not be reduced during pigtailing but the later has to be minimised by choosing an adapted alignment process (see parts 2 to 5). The Fresnel loss comes from refractive index changes at the interface and may create unbearable backreflections (see part 6).

The fibre to waveguide connection is a complex hybrid structure : there are the IO substrate, the fibre, possibly some positioning element and fixing agent(s). They have different sizes and thermal expansion coefficients. Therefore the interface is sensitive to any environmental change. The induced fibre moves would impair the optical performances of the component. Therefore any pigtailing technique requires a lot of care. Each chip connection to fibres has to be handled separately. In comparison with the batch processing of the IO chip it is a bottleneck and a highly time consuming operation. It is accounting for 40 or even up to 80 % of the total device cost.

In this paper we review the several existing connection techniques which have been developped to overcome these problems. Most of the basic principles were already studied about 5 years ago (14). But recent improvements have been achieved due to combined effects of the larger spread of IO devices and new techniques refinements. They shall lead to viable and low-cost pigtailing solutions.

We first examine in part 1 what are the requirements on alignment accuracy in the only case of single-mode waveguides. For multimode applications, they are relaxed in a factor of 5 to 10. Then we describe in part 2 to 5 the different fibre to waveguide alignment methods and we compare them in terms of eficiency, required equipment complexity, time consumption and cost. We focus on techniques for waveguides made of glass, polymers and lithium niobate but they are also valid for III/V materials. In part 6 we present the requirements on return loss and the means to fulfil them. The part 7 is devoted to the fixing techniques.

1- PIGTAILING-INDUCED EXCESS LOSS

1.1. Excess loss

The excess loss for any fibre(s) pigtailed waveguide is the sum of 4 terms : the propagation loss in the structure, the Fresnel reflection, the mode-size mismatch between the fibre(s) and the waveguide and their misalignment. The propagation loss is beyond the scope of this paper. It depends on the chip fabrication parameters and the structure design. But one should note that guides with large mode size cannot accept strong bendings. The overall waveguide length has to be increased, leading to higher propagation loss (see part 1.2 and 1.3).

The Fresnel reflection depends on the index change at both waveguide and fibre endface. The reflected light (return loss) can be written as

$$\eta_R = \frac{\left(\frac{n_1-n_2}{n_1+n_2}\right)^2 + \left(\frac{n_3-n_2}{n_3+n_2}\right)^2 + 2\left(\frac{n_1-n_2}{n_1+n_2}\right)\left(\frac{n_3-n_2}{n_3+n_2}\right)\cos\frac{4\pi n_2 g}{\lambda}}{1+\left(\frac{n_1-n_2}{n_1+n_2}\right)^2\left(\frac{n_3-n_2}{n_3+n_2}\right)^2 + 2\left(\frac{n_1-n_2}{n_1+n_2}\right)\left(\frac{n_3-n_2}{n_3+n_2}\right)\cos\frac{4\pi n_2 g}{\lambda}}$$ (1)

n_1, n_3 : effective index of the fibre and the guide mode,
n_2: refractive index in the gap g between the fibre and the guide.
λ : wavelength

It is usually extremely low if some index matching material ($n_2 \approx n_1$) is inserted or if endfaces are anti-reflection coated. But even small (0.01 - 0.1%) backreflected light might be a large problem for devices which are being installed in communication systems. Therefore we will detail the Fresnel reflection impacts and the methods to reduce them in part 6.

The mode size mismatch induces loss because the transverse mode coupling can not be complete. The efficiency is computed using the well known overlap integral

$$\eta_0 = \frac{\iint |E_1 E_2|^2 \, dx \, dy}{\iint |E_1|^2 \, dx \, dy \iint |E_2|^2 \, dx \, dy} \qquad (2)$$

It expresses the coupling between the electric fields E1 and E2 of the modes which are propagated respectively in the fibre and in the waveguide. The fibre mode is well described by a circular gaussian with a waist w_0 (typical value of 4.5µm for single-mode fibres at 1.3/1.55µm). The mode in the waveguide strongly depends on the fabrication process. Almost perfectly circular waveguides are obtained by both silica growth on silicon and double step ion exchange in glass [21-23]. This is not the case for lithium niobate. Some diffusion-related asymmetries and eccentricities remain [14] because there is no possibility for deep burying of the waveguide in the substrate. Waveguides in III/V semi-conductors also exhibit strong eccentricity [7].

In the most general case the mode profile can be approximated by a combination of half gaussians (see figure 4) and the overlap integral gives [14]

$$\eta_0 = \frac{\left(\sqrt{\omega_1}\left(\frac{\omega_1}{\omega_0}+\frac{\omega_0}{\omega_1}\right)^{-1/2} + \sqrt{\omega_2}\left(\frac{\omega_2}{\omega_0}+\frac{\omega_0}{\omega_2}\right)^{-1/2}\right)^2}{\frac{\omega_1+\omega_2}{2} \left(\frac{\omega_3}{\omega_0}-\frac{\omega_0}{\omega_3}\right)} \qquad (2\text{-}1)$$

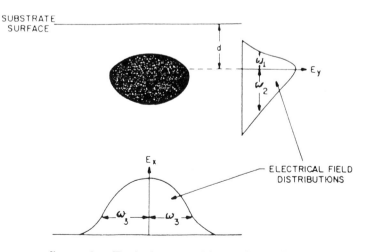

figure 4 : Typical waveguide mode profiles [14]

If the mode profile is eccentric but symmetric ($w_1 = w_2$) one gets [13]:

$$\eta_0 = \frac{4}{\left(\dfrac{\omega_3}{\omega_0} + \dfrac{\omega_0}{\omega_3}\right)\left(\dfrac{\omega_1}{\omega_0} + \dfrac{\omega_0}{\omega_1}\right)} \quad (2-2)$$

It is usually possible to fabricate waveguides with a smaller mode than the fibre mode. It allows to stronger bend the waveguides and therefore to get more compact devices with lower insertion loss but larger coupling loss [24]. We will comment this point in part 1.3.

Let us now detail the pigtailing excess loss which is induced by a non-perfect control of the fibre(s) and waveguide(s) positions during the connection process.

1 2 Sensitivity to misalignment

Misalignment may occur on each degree of freedom. We will first discuss on side-misalignment before to study angular and longitudinal (i.e. along the propagation direction) contributions to excess loss.

The impact of the side-misalignment on the coupling efficiency can be derived from (2). Assuming the fibre core is misaligned with the waveguide by d, one gets the following coupling efficiency [13]:

$$\eta = \eta_0 \, e^{-\dfrac{2d_x^2}{\omega_0^2 + \omega_1^2}} \, e^{-\dfrac{2d_y^2}{\omega_0^2 + \omega_3^2}} \quad (2-3)$$

dx and dy are respectively the vertical and lateral misalignment.

Therefore the sensitivity to side misalignment is decreasing when the mode size is increasing. In order to get a more detailed view on the real meaning of this phenomenon, we will now examine the three distinct cases of "fibre mode-matched" waveguides (like waveguides obtained by double step ion exchange in glass or silica growth on silicon), asymmetric mode waveguides with a mode side width roughly equal to the fibre mode width (i.e. the case of lithium niobate titanium diffused waveguides) and narrow eccentric mode waveguides (III/V devices).

In the case of "fibre mode-matched" waveguides, the equation 2-3 becomes

$$\eta = e^{-\dfrac{d^2}{\omega_0^2}} \quad (2-4)$$

For typical single-mode fibres at 1.3μm, the waist equals 4.5μm. Misalignments below 0.2μm will account for less than 0.01 dB/connection. A 1 μm misalignment adds 0.2 dB excess loss and the coupling efficiency is reduced by 0.8dB (-17%) if the lateral misalignment reaches 2 μm.

In case of lithium niobate devices (and also single step ion exchanged glass waveguides), the diffusion process leads to strongly eccentric and asymmetric waveguide mode with a width over twice the height ($w_1 + w_2 < w_3$). In order to minimise the coupling loss, the most common trade off is to set w_3 larger but close to w_0. Therefore lithium niobate waveguides exhibit stronger misalignment induced excess loss if the offset is vertical (perpendicular to the diffusion plane). While 1 μm lateral misalignment gives 0.2 dB excess loss, the same offset in the vertical direction leads to a twice higher excess loss of 0.45 dB[14]. Birefringence is also an important factor because it may lead to different optimum coupling positions for various polarisation states.

For III/V waveguides, the index change between the core and the cladding is high. Single-mode waveguides have narrow modes with a typical waist of 0.5 to 1 μm and strong eccentricity [15][32]. High excess loss of 6 - 10 dB occurs in fibre to waveguide coupling[15][32][97]. The sensitivity to the side misalignment is only slightly higher than for lithium niobate waveguides although there is a factor of 4 in the mode sizes. This can be seen from equation (2-3). Let us assume that we are coupling a fibre to very small waveguides, i.e. $w_1, w_2, w_3 \ll w_0$:

$$\eta = \eta_0 \, e^{-\frac{2d^2}{\omega_0^2}} \quad (2\text{-}5)$$

A maximum sensitivity is obtained which only depends on the fibre mode size and is about twice (in dB) the sensitivity in the most favourable case of mode matched waveguides (equation 2-4).

The exact sensitivity is clearly depending on the true mode sizes but those examples illustrate the difference in pigtailing various waveguides types. One can draw the conclusion that 0.1μm accurate alignment is satisfactory for any type of waveguides if connected to standard 1.3/1.55 μm single-mode fibre. This requirement can be relaxed in a factor of 2 for waveguides with modes close to the fibre mode.

The longitudinal misalignment broadens the mode. But it can be neglected as long as the gap between optical endfaces is below 10 μm. The main problem is the wavelength dependency of the so formed Fabry-Perot cavity but elimination of the backreflection (see part 6) strongly reduces the phenomenon.

With the alignment methods we will describe below it is relatively easy to maintain the fibres and the waveguides in the same axis. Therefore the angular misalignment brings very little excess loss [13]. One should note that a combination of longitudinal and angular misalignment results in a side offset which can be a real problem.

The side misalignment actually brings the higher excess loss. We will now look over a few techniques to improve coupling efficiency by mode matching. We will discuss how they loose the coupling efficiency sensitivity to side misalignment.

1.3 Fibre and waveguide mode matching

It is not always possible (III/V and lithium niobate devices) or desired (cheap single step ion exchange in glass or compact passive devices) to match the mode propagated along the whole waveguide structure to the fibre mode.

In case of ion exchange in glass, the fabrication of fibre mode matched structures requires a double step process, usually with electric field assistance. It might appear very convenient for cheaper and simpler process to realise waveguides in one single step. The waveguide is similar to lithium niobate waveguides in both terms of eccentricity and asymmetry. As pointed out in part 1.1, it is also of some interest to use narrower modes for more compact devices. Therefore Zhenguang and al.[25] proposed to use tapered waveguides at the endfaces. They are made by a smooth increase of waveguide width. It results in adiabatic expansion of the propagated mode. I.Shanen and al. [26] demonstrated the expansion up to 200 µm of a single mode waveguide width. This shows the potential of the technique. The main problem lies in increasing the mode height by ion exchange process because no material is grown. A double ion exchanged is required and the asset of a simpler process is greatly reduced. Furthermore the taper induces some excess loss and therefore a mode matching on the whole chip remains the best trade-off.

The case of lithium niobate is different because the mismatch coupling losses are higher. Some loss in a matching structure can be accepted. Tapers have also been proposed but the best results are obtained with diffused microlenses [27-28].The beam expansion in height offers correct matching of the guide mode to the fibre mode. The coupling structure loss is around 1 to 2 dB. Mode matched chips exhibit about the same excess loss than standard ones once they are pigtailed. But the requirements on alignment accuracy are lower and similar to mode-matched passive devices (see part 1-2).

In the case of III/V chips, mode matching is highly interesting because one can theorically gain up to 10 dB. The most widespread technique is the use of lensed fibres. They are prepared by either etching, fusion or laser machining [29][30]. Excellent coupling (excess loss reduced to 0.5 dB) has been demonstrated [29] but this method requires individual treatment of all fibres and does not relax the alignment tolerances. Therefore it is attractive to integrate the mode matching structure on the chip in order to take benefit of the photolithography accuracy. Verdiell and al.[31] etched a lens in front of the waveguide. T.L.Koch [35] introduced tapers and demonstrated enhanced tolerance on misalignment (-1dB for 3 µm offset instead of for 1.5µm). The coupling loss remains quite high at - 4.2 dB because the only modes widths are matched. This taper structure is different of the one which is described above. The guide width turns smaller which looses the guiding efficiency and broadens the mode. P.Doussiere and al [30] reported similar results. R.Zengerle and al.[32] demonstrated mode diameter transform from 1 up to 8 µm with excess los of 2 dB. G.Muller [33] obtained a lower loss of 1.4 dB for vertical tapers. T.Brenner and al.[34] combined lensed fibre and tapered waveguides for a total loss of 0.9 dB. Yamaguchi [46] obtained similar results of 0.7/1 dB. In those two cases, the required accuracy in alignment is very similar to lithium niobate because the mode waist is about 2.5 µm.

Glass waveguides may also help to match the III/V mode to the fibre mode. Y.Shani and al. [36] reported such hybrid structure with a silicon optical bench. The coupling loss decreased from 11 dB to 3.1 dB. A very nice asset of the technique is that the required alignment accuracy is exactly the same than for mode matched waveguides described in part 1-2. This hybrid assembly allows to combine the mode matching function with other passive functions as splitting or wavelength multiplexing. Lately J.Buus [37] proposed a multimode guiding structure and demonstrated theorical broadening of a 1µm mode up to 6 µm with excess loss below 1 dB.

In conclusion, the proposed mode matching structures are very promising for coupling loss reduction but also for loosening of alignment requirements. The fibre to III/V waveguide connection turns very similar to the cases of lithium niobate or even the "best" case of passive silica on silicon or ion exchanged glass fibre mode-matched devices.

We have seen that the required alignment accuracy is 0.1-0.5 µm for low connection loss (<0.1 dB). We will now describe the achievements in alignment methods for the later IO techniques.

2 INDIVIDUAL ACTIVE ALIGNMENT

Each fibre is held and positionned in front of the waveguide by the actuator. The intensity of the light coupled through the device is measured and provides a feedback when the actuator moves the fibre. Once the maximum intensity is obtained, the fibre is fixed in position (see part 7). This feedback method is aimed to warrant optimum coupling on each port. It succeeds if moves are controlled within 0.1-0.5µm (see part 1) and if the actuator has an harmless but stiff grip on the fibre.

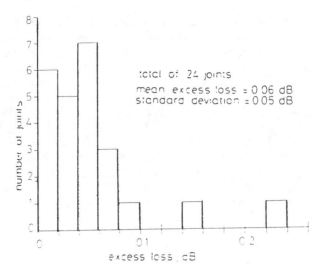

figure 5 Excess pigtailing loss achieved by active alignment of fibre to LiNbO3 waveguides [41]

Several positioners and associated softwares are commercially available [38-40]. Alignment can be optimised in less than 1 s once significant coupling is obtained, i.e. the fibre core is roughly in front of the waveguide.

The obvious asset of the technique is the low excess loss. The multiport chips will keep unchanged their performances in both terms of uniformity (balance) and average insertion loss. The typical excess loss is below 0.1dB (see figure 5) in the quite demanding case of lithium niobate [41].

The method does suffer important drawbacks. It is extremly time-consuming because each fibre is individually positioned and secured. One also needs to search for significant coupling, i.e. a rough alignment of the fibre in front of the related waveguide so that a measured increase in coupling efficiency actually means a better position of the fibre. K.H.Cameron [41] proposed an imaging system to look at the near-field pattern of the output face while moving a fibre in front of the input face. The input guide is found once the near field pattern is the one of an excited output waveguide(s). This step is manpower intensive and troublesome to automatise.

Another problem comes from the proximity of neighbouring waveguides. In order to minimize the bending elements in the structure, the pitch between the ports are kept in a typical range of 250-500 µm if standard 250µm coated fibres are used and 80-150µm for dedicated fibres [48]. Therefore the actuator has to be small enough to align a fibre without breaking adjacent achieved connection(s). A miniature but stiff gripping tool is required. T.Yoshizawa and al.[42] proposed to use glass microstrips. They are permanently attached to the fibres and relax the need for a compact tool. However they can not be removed. The principle is limited to a maximum of 2 ports per face. Other possibilities include mechanical tweezer or vaccuum sucker as proposed by D.J.Albares and al.[43]. The small size pipe can grip the fibre and is easy to remove by switching off the vaccuum pump. Leakage is easy to sense and offers correct feedback on gripping efficiency. This type of gripping does not prevent sliding moves and its force drops as soon as the fibre is slightly moved aside of the pipe. H.P.Zappe reported poor control of the moves using a similar chuck for aligning laser and waveguide [100]. We have demonstrated a more rigid tool [44]. The gripper is a heating head with a base of 0.1x1mm covered by a layer of thermoplastic adhesive. The head is moved so that it touches the fibre. It is briefly heaten up around 150°C. The adhesive melts and wets the fibre before it cools down. A firm gripping of the fibre is obtained which withstands the glue shrinkage during curing. The gripper is removed by heating up. We are currently using this tool with very satisfactory results for 250µm pitch fibre arrays. The method also allows a perfect gripping of the fibre, even with a position offset over 10 µm which simplifies the automation process.

In spite of important drawbacks, the active alignment method is in use for pigtailing active waveguides (DFB lasers) but also some passive 1x16 splitters [45] due to the high level of the obtained optical performances.

3 MONOLITHIC PASSIVE POSITIONING

3.1 description

The fibres are inserted into positioning structures (grooves or ferrules) which have been directly fabricated on the IO substrate. Once the fibres are transferred into these structures, they will stop at a position where their core are self-aligned with the corresponding waveguides.

The assets of passive alignment are simplicity and the short time required for fibre(s) to waveguide(s) connection. Neither complex actuator nor complete optical line are required. It is also possible to connect in one single step all the fibres to the waveguides of one complete face of the IO chip. But the chip processing time and the fabrication complexity do increase. The associated cost should also be taken in account when one compares alignment techniques. It is not only coming from the longer processing time in chip fabrication but also from the increased complexity which lowers the yield.

The main drawbacks are the higher insertion loss and the increase of non-uniformities when compared to insertion loss of the chip. This is due to the lack of feedback. In order to overcome such problem, one should reach an absolute accuracy of about 0,1-0,5 µm (see part 1) in the control of waveguides position, fibres geometry and positioning structures. For example, the fibre geometry control has progressed a lot in past years but the standard fibre may still induce 1,25 dB/connection excess loss.

The key technology for passive positioning is clearly the technique of positioning structures manufacturing which should be adequate for the IO substrate.

3 2 Positioning structures

Several groove and hole-type structures have been demonstrated in various substrates. We are now going to list the most common techniques and the ones we find the most attractive in order to evaluate their potential and also to which IO substrate they may apply.

One can distinguish between V-grooves on one side and U-grooves or holes on the other side. In the later structure the fibre may rest along any part of the whole surface, and its exact position strongly depends on the roughness of the structure. In V-grooves the fibre is touching the grooves edges along two well defined lines and its position only depends on the quality of the surface in those limited areas.

V-grooves can be moulded or coined in metals, glass or polymers. P.Barlier [49] claimed a very good accuracy of the fibre position (0.2µm) when moulding low melting-temperature glass. The technique was only used for multimode waveguides and 50/125µm fibres pigtailing loss was estimated at 0.4dB/face. This indicates a lower accuracy or possible problems at the interface between the grooves and the waveguides. One should underline that the positioning structure is not easy to turn compatible with the waveguides fabrication, in particular with the photolithography which requires flat surfaces. Moulding process is still studied for polymer waveguides. The RACE project POPCORN[50] is aiming to emboss at the same time the tiny groove for the TFPMA waveguiding zone and the large V-groove to hold the fibre. The mould is made by either metal structuring by the LIGA process [60] or by silicon wet etching .

The later technique is also used to produce positioning structures and it offers the asset to include a photolithographic step. The combination of mask patterns for the groove and the waveguide leads to their self-alignment. Wet KOH etching of silicon is anisotropic .

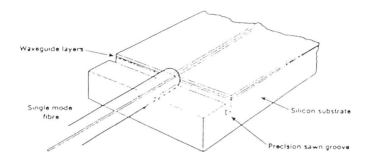

figure 6 : monolithic passive positioning   [45] (see part 3)

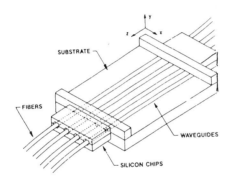

figure 7 semi-active alignment   [14] (see part 4)

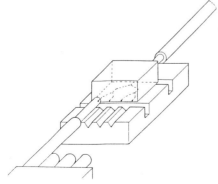

figure 8 hybrid passive positioning   [75] (see part 5)

The etching rate is much higher in one crystal direction than in the other directions. If all wafers have strictly the same crystal orientation, the same V structure will be reproduced whatever the exact fabrication parameters are [55]. The angle of 71° brings a relatively high dependence of the fibre core position to the fibre outer diameter (factor of 0.8). But the structure is extremly stable. By pressing down the fibre in the V-groove by means of a block, it finds its right position with a high repeatability. Low pigtailing excess loss of 0.4 dB/face has been reported [56].

Mechanical machining has alos been proposed. It applies to most of the substrates providing some adaptation of speed and grinding material. +/- 1 µm accuracy on the fibre position was demonstrated in glass [53].

The other techniques lead to U-groove or hole structuring. Ion milling [51] allows to fabricate shallow grooves. One has to etch the fibres down to a diameter of 15µm. The roughness of the groove walls does not allow a very good reproducibility of the positioning. The rough endface of the chip scatters the light which contributes in 0.5 dB excess loss. Mechanical milling of holes in lithium niobate [52] has been demonstrated as well as laser machining [54]. Once again, those techniques don't benefit of a direct self alignment between the positioning structures and the waveguides. Etching is more appropriate. Wet etching of glass was proposed [89] using a HF solution. The U groove had a 0.93 circularity and low roughness.

Reactive Ion Etching (RIE) of silicon also leads to U grooves. G.Grand and al.[57] first demonstrated its interest for chip connection. They pointed out that the technique is more flexible than silicon wet etching : the grooves can be designed in any direction because the etching is crystal-orientation independant. They reported maximum pigtailing excess loss of 0.6 dB/face. In the RACE project 1008 [58], both RIE and wet etching have been evaluated in terms of performance. The later had still the lead because 0.4 dB is a maximum with this technique and 0.2 dB can be obtained in 66% of the cases.

Nowadays the silicon wet etching seems the most mature technique for fabrication of positioning structures. This has been seen as a major asset of silicon over other substrates like silicate glass for low cost pigtailing. But one has to solve the problem of waveguide endface preparation.

3.3 Waveguide endface preparation

In the active alignment technique (part 2), the fibres are buttcoupled to the chip which therefore could be polished. It is not anymore possible in the case of passive self-alignment because the waveguides do not anymore stop at the endface of the chip.

One convenient method is to saw a slot in front of the waveguides. The sawn surfaces have an optical quality [59] if some care is applied. N.Mekada reported excess loss of 0.2 dB or less [61] which demonstrates very low scattering. No endface preparation is required when the silica on silicon waveguides are etched by RIE[57]. Wet etching requires some treatment. Otherwise the fibre does not get to the waveguide angled endface. Therefore mechanical sawing has to be used. Y.Yamada and al. proposed to combine RIE and wet etching in the so-called terraced structure [62]. V-grooves are etched first. Silica glass is grown and the waveguides are fabricated. Finally, the grooves are opened by Reactive Ion etching of the silica.

3.4 Conclusion

The monolithic self-alignment technique is very attractive in principle. The fibres can be treated in arrays and the pigtailing operation is fast and simple. Dense connections can be made. One has to find a suitable structuring technique, and to overcome the overlapping effects with the waveguide processing, in particular in terms of waveguide endface quality. Although many Research and Development works have been carried on, there is at our knowledge no commercial products using the monolithic self-aligned fibre to waveguide connection.

4 SEMI-ACTIVE PIGTAILING
(active fibre-arrays buttcoupling)

It is a combination of both active and passive techniques. The fibres are inserted into a block with accurate positioning structures described in part 3. The fibre array is polished and actively buttcoupled to the chip using the method described in part 2.

It leads to a trade-off between the assets and drawbacks of the former described techniques. It is not as much time consuming as the active technique because there is only one optimisation per face. It allows dense connection just like passive alignment . If the lack of individual fibre tuning increases the non uniformity of the output ports when compared to active alignment, the required positioning accuracy is only relative and not absolute like for passive positioning. This partly relaxes the requirements on positioning structures and leads to excess loss which are theorically in between the performances obtained by active and passive pigtailing.

Both positioning structures fabrication and actuators are key technologies to get satisfactory results. The choice of the positioning structures fabrication does not depend on the IO substrate and there is no increased complexity in waveguide realisation. The actuator is even more complex than for active alignment because one has to control angular misalignment by optimising the position of two fibres on each multiport face. E.J.Murphy and T.C.Rice [69] proposed to get rid of this angle alignment problem by preparing the fibres on longer blocks which rest on the top of the chip. The average pigtailing excess loss they obtained was rather high (0.9 dB/face).

Compared to passive positioning, there is no problem of endface treatment. The fibres blocks and the chip are polished. This preparation is time consuming if compared with fibre cleaving but it also allows angled polishing which is of high interest to get rid of the return loss (see part 6). It is the main reason why it is extensively used for lithium niobate devices.

In this case the reported pigtailing excess loss per face is in the range of 0.3dB to 1 dB [4][63][65] and usually below 0.5dB. The positioning structures are V-grooves made by either machining in glass or wet etching of silicon. The number of connected ports reaches 16 or even 32 [48]. For glass components made either by silica growth on silicon or ion exchange, one can expect even better results because the waveguides mode matches the fibre mode (see part 1). Reported pigtailing excess loss per face values are ranging from 0.2 [66] to 0.6 dB [64] for up to 19 ports. H.M.Presby [64] predicts that 0.1dB could be obtained if the fibres are selected. It is confirmed by recent results [96] for passive glass devices. As we have seen in part 3 1, there is a clear need for fibres with accurate geometry. Otherwise the non-uniformity per face may reach 1.25 dB.

Very dense connections have been demonstrated by the semi-active technique. Dedicated pitches from 80 µm to 200 µm have been proposed [19][48][64]. They allow to keep the devices smaller resulting in more compact and lower loss devices.

The semi-active technique is a valuable trade-off. The excess loss per face is typically 0.1-0.2 dB higher than for active pigtailing for fibre mode matched waveguides and 0.2-0.3 dB higher for lithium niobate or similar waveguides. But one gets major assets in terms of time consumption although one should take in account the exact fibre preparation time. The method is used for standard commercial products in lithium niobate and glass.

5 HYBRID FLIP-CHIP PASSIVE POSITIONING

The use of flip-chip assembly for fibre to waveguide connection was first proposed by H.P.Hsu and A.F.Milton in 1976 [67]. The basic idea remains unchanged. One takes benefit of the best positioning structure fabrication process to fabricate an optical bench , or "platform". This platform is aligned and assembled to the optical chip so that inserted fibres would be in front of the waveguides. Whatever the IO technique is, the waveguides are close to the surface and therefore the chip has to rest on the platform in the so-called flip-chip configuration (figure 8). Once the chip and the platform are assembled a passive positioning takes place as described in part 3.

The pigtailing is fast and simple and the waveguide endface can be polished prior to mounting. Like for semi-active pigtailing, the hybridation allows to discouple the choice of positioning structures and of the IO substrate. The complexity associated with monolithic integration (see in part 3) is eliminated. Hsu and Milton already used a silicon platform with V-grooves and most of proposed devices since then dealt with the same technique. Neither actuator nor optical line is required and the fibre preparation is simpler than for the semi-active technique. However the control of fibre, waveguide and positioning structure positions and geometries has to be better than for either monolithic passive or semi-active pigtailing methods because one has to add the offset between the platform and the chip to the position mismatch . Therefore the flip-chip technique is very demanding on assembling control. It was not succesfull until recently but it takes benefit of recent progress in hybrid integration of OEIC.

The two key technologies are the positioning elements fabrication and the flip-chip assembly. The first one has been described in part 3. We will list the several flip-chip assembling process. We distinguish between the methods which require some feedback (visual alignment) and the passive alignment methods.

5 1 Feedback flip-chip alignment techniques

Hsu and Milton proposed a feedback method based on alignment marks. While the waveguides were fabricated in lithium niobate, some extra stripes were in-diffused. During the structuring of the optical bench, narrow and shallow grooves were etched in the silicon. The diffused stripes and etched grooves were aligned by the operator in a similar manner to masks alignment in microelectronics. Although the authors claimed +/-1 µm accuracy in alignment, the final pigtailing excess loss reached -6 dB. The causes are likely to come from a poor control of waveguides, fibres and grooves geometry and position but also of the true gap between the chip and the bench surfaces.

C.H.Bulmer and al. [68] proposed to add a final active tuning of the fibre position. A slot was machined in the silicon bench along the lithium niobate chip endface. A tapered fibre was inserted in this slot. It was sliding in order to tune the exact position of the fibre tip in front of the waveguide. The pigtailing excess loss was reduced to 0.7-1 dB/face. One should however note that the technique then requires insertion on optical line although there is no need for search of the waveguide. The use of a tapered fibre does not allow to optimise many fibres.

In the Eureka project ASSYSTO, we proposed [75] to take full advantage of the flip-chip principle by positioning the glass chip and the silicon platform using accurate etched marks. A good plane to plane contact is obtained (gap < 0.5µm) thanks to the use of a dedicated alignment machine and the application of vacuum in between the chip and the platform. A pigtailing excess loss of 0.5 dB/face is measured for up to 8 ports /face. This can be compared to monolithic passive pigtailing (see part 3). It demonstrates that very accurate flip-chip assembly can be obtained by marks alignment, providing the gap between the chip and the platform is minimised. We also have the possibility to tune the coupling efficiency thanks to a novel and compact gripper (see part 2). The pigtailing excess loss is then below 0.2 dB for glass and lithium niobate waveguides. The excellent prepositioning allows a fully automated active pigtailing with no search for guide, automated glue dispensing and fibre gripping.

5 2 Passive flip-chip alignement techniques

K.P.Jackson [76] and H.Kaufmann and al.[70][71] proposed to use etched ribs atop the GaAs waveguides and to insert them in large V-grooves in the silicon optical bench. They measured low pigtailing excess loss of 0.25 dB/face for one single fibre per face. From measurements of loss sensitivity to misalignment they conclude they reach a maximum misalignment of +/- 1 µm. Such low excess loss comes from plane to plane contact on a short surface and excellent alignment between the ribs and the waveguides which are simultaneously produced.

The solder bump technique [76] allows to combine alignment and fixing. The microbumps of solder metal are grown atop the platform and the chips. The exact geometry can vary. Destefanis [72] proposes the use of Sn/Pb microballs. They enables a position control better than 3µm which could be improved to 1 µm by reducing the ball diameter from 25 to 15µm. By the use of simple solder bumps, M.J.Wale and al.[73] demonstrated optical fibre connection to lithium niobate waveguides of 1 dB/face. The bumps are also used as electrical connections. J.W.Parker reports 1 µm accurate alignment when reviewing the activities within european project ESPRIT II OLIVES [74].

The solder bump technique is very interesting because it fulfills alignment, fixing and electrical connections. It could be very usefull for any electroactive device. The drawback is the need of a thick metallisation and a good control of oxydation during the soldering.

5 3 Conclusion

The flip-chip passive alignment is a hybrid solution which gathers many assets of the passive alignment without most of the drawbacks. Recent flip-chip assembling demonstrations show its large potential of applications. We also demonstrated a flip-chip assembly on a platform with grooves for MT connector pins [75]. The connectorised 1:8 device exhibited 1 dB/face pigtailing excess loss.

6 RETURN LOSS

6 1 Requirements

In single-mode optical communications systems, limited optical return loss is required. In case of duplexers, it directly induces false informations on receivers. In all cases, it affects the characteristics of laser emitters [78][79]. A shift in emitted frequency and power may occur which strongly reduces the system performances. Analog systems like Cable TV distribution are particularly sensitive. Information rate, total line length and distance between the IO device and the emitter are also of importance. The device induced return loss decreases along the line due to propagation loss while the fibre backscattering increases. Therefore the operator requirements strongly depend on the type of application.

A return loss of -30dB can be low enough in some short range telephony applications or for sensor systems. It is the standard requirement for single-mode optical connectors. In case of branching devices, Bellcore advised [91] -40 dB for intermediate (15 km) range FITL and SONET but -50 dB for long range communications. In Cable TV distribution the requirements may reach -55dB. They are more usually ranging from -45dB[80] up to -50dB[81].

If one considers what the fibre to waveguide connection brings by computing from equation 1, one comes to the conclusion that air gap will lead to about -10/-14 dB. Some process is therefore required to eliminate a large part of this return loss.

6 2 Index matching gel /glue dispensing

In the case of silica grown or ion exchanged waveguides, its effective index matches the index of the fibre (1.46). By dispensing some optical organic material (glue or gel) with the same index in between the cleaved fibre tip and the polished endface of the chip, the air gap is eliminated. One achieves a return loss in the order of -40dB depending on the exact refractive index of the chip and of the gel or glue.

This method obviously applies for any of the alignment techniques described above. The lithium niobate and the III/V materials have respectively refractive index of 2.2 and 3-3.3. therefore one can not adapt the index of the substrate to the fibre index. Gel or glue dispensing does not help to reduce the backreflected light intensity at a level below -10 to -15 dB.

6.3 fibre/guide endface treatments

The first idea is to offer a better match of the surface index by coating them with AntiReflection layers. The technique allows to reach -20 to -40dB levels of backreflected intensity for lithium niobate [94] or III/V. It is not possible to get higher elimination of the return loss. Therefore it offers little interest for silica grown or ion exchanged waveguides, except in cases where no organic is accepted (see part 7).

D.Boscher and al.[90] demonstrated that a fine etching of the fibre and chip endface creates a small roughness which do not add scattering loss if some index matching material is dispensed but which eliminate a large part of the reflected light. The Return Loss is decreased of 20 to 30dB. It is lowered to less than -60dB for a 1x8 ion exchanged glass splitter. The technique is suitable for active, semi-active and hybrid flip chip passive alignment processes. It is more difficult to apply to the monolithic passive positioning although it should be possible to create the roughness during the machining step (see part 3-3).

6.4 angled facets

The most efficient way to eliminate backreflected light propagation is to tilt the optical endfaces in order that the reflected light is deflected out of the fibre. The return loss is lower than -55dB for angles of 5° or more if some index matching material is used [83]. The technique is also in use with an angle of 8° for connectors of the SC type.

It is very well suited to semi-active techniques because it is not much more difficult to polish the chip and the fibres arrays with an angle [63-66] than straight. A correct endface orientation is easy and leads to minimised insertion loss.

For the active alignment technique, where fibres are usually cleaved, the solution is to cleave fibres with an angle which is now possible on commercially available tools [83]. In order to keep the asset of mass transfer of the fibres in the case of monolithic/hybrid passive techniques, it would be necessary to use multifibre angled cleavers. They have been demonstrated but are not yet commercially available [84]. They would present a great progress because the cleaving process is faster than polishing.

One shall note that this method does not reduce reflection which may result in some high Fresnel loss. Therefore it is usually combined with an index matching technique.

6 5 conclusion

One should distinguish between the two cases of silica glass waveguides (with refractive index matched to the fibre index) and lithium niobate or III/V compounds, In the former case, gel/glue dispensing is sufficient to provide correct return loss. Some facets treatment enhance the return loss performance.

For very low (-50 -55 dB) return loss or for the later case, the most efficient solution is to tilt the optical endfaces. Until recently the only possibility was polishing. It best fits to the semi-active technique. But cleavers are nowadays an interesting alternative in terms of time consumption and flexibility in the choice of the alignment technique, turning the hybrid passive method more attractive.

7 FIXING TECHNIQUES

7 1 requirements

In former parts, we dealt with techniques to achieve high optical performances with limited connection time. We describe here the methods for securing the pigtails.

Integrated Optics components have to work in very diverse environments and each use may lead to dedicated requirements. The applications in sensor and communications systems aboard military or (space) aircrafts [18][86][88][94] demonstrate the real aptitude of IO components to stand in harsch environments , but solutions are often cost intensive.

In the field of telecomunications, the deployment of FITL trials enabled to develop and precise the truly required specifications for correct performance of the IO devices in cable systems, including long term reliability [16] [87] .

Among the tests parameters, the thermal cycling of the component provides a valuable insight on the fibre/chip hybride. The damp heat is also a critical test, particularly when organic fixing materials are used. The pulling strength test measures the force of the attachment.

At this time, several working groups are aiming to set generally accepted standards at both european (CECC WG27) and international (IEC/TC86/SC86B WG5 and ISO/TC172/SC9 WG 7) levels.

The thermal cycling test shall describe the environmental conditions of use of the IO device. Therefore one can find different tests as for example for building-sheltered (-25 +70°C) or field (-40+85°C) environments. The insertion loss variation should be below a few tenth of a dB (0,6 dB in most of published requirements) [16] [80] [91] . Some electroactive devices made of lithium niobate exhibit operating problems over 55°C [88]. They require controlled climat conditions which relaxes the requirement on connection withstand.

Typical required fibre pulling strength is of 5N [16] [80] [91]. The damp heat test usually refers to 93/95% humidity at temperatures around 40-60°C [16] [80] [91] and up to 21 days [81].

The ageing test are a real problem because the failure mechanisms have to be clearly identified while the industry does not have the long term experience. The most usual method is to extand tests over long period under conditions which are thought to be the worst [80] [91] although this is under discusion [16].

7.2 Type of fixing materials

One can distinguish between UV or thermal curable glues and metal solders.

The UV curable glue is quick cured. One can select the curing location by focusing the UV beam. This is particularly usefull for the active alignment method (se part 2 [87]). The curing process also brings limited stress because heat is minimum. Furthermore some UV glues are excellent fibre index matching materials.

However one has to check they do not deteriorate in time and that they do not soften due to their low glass transition temperature (Tg). The tests in thermal cycling and damp heat are particularly important. T.Shiota[16] proposed to improve the bonding efficiency and reliability by acid cleaning of the fibres and chip faces.

Stiffer are the thermocurable glues [88] [97], usually in addition to the UV curable material [99]. The idea is to separate the two functions of optical index matching and fixing [87]. The relatively high curing temperature of 70 - 120°C may stress the pigtails. Furthermore the fixing has to be global because the curing process is slow and can not be space limited.

An alternative to organic materials are metals which can be either soldered or welded. Ageing is better controlled and there is no outgasing risk, which is a requirement for hermetic sealing of active devices. It is commonly found for lasers pigtailing [94] [98]. All surfaces have to be metallised prior to the assembling [73]. The soldering process offers mass connection and is proposed for monolithic or hybrid passive positioning and for semi-active alignment techniques. Oxydation is a major problem. Laser welding offers relatively local attachment capabilities [73]. If organic materials are forbidden, the only index matching solution is to AR coat the optical endfaces of the chip and the fibres.

The glass welding technique does not require any additional fixing material but only applies to glass waveguides. The fibre is fused to the chip by a CO_2 laser beam [95-96] . Low return loss and low sensitivity to environment problems can be expected. But one has to overcome the different sizes of the chip and of the fibres. Therefore J.J.G.Allen [95] proposed to set the fibres in a block.

One can draw the conclusion that all techniques have their domain of applications. The cheapest and easiest method is the use of organic adhesives. It is a good candidate for low cost passive devices. The active devices do not support outgasing and they are preferably pigtailed using cost intensive soldering or welding techniques.

7 3 Fixing techniques

Let us describe how the fixing materials can be used with the alignment methods reviewed in parts 2-5.

In the case of the active alignment technique, the fibre is individually attached. Therefore UV glue or thermal well located treatment (laser welding) are the only techniques. The bond surface is small, particularly for non-buried waveguiding structure. Several authors [41] [61] propose to glue a block a top of the chip. N.Mekada and al.[61] demonstrated that the pulling strength of UV cured bonds is improved from 0,2 N up to 0,6N. They described a thermal behaviour of +/- 0,2 dB in -10+60°C cycles.

In the case of the passive positioning, the same problem may occur if the attachment of the fibre tip to the waveguide is chosen but one can also fix the fibre in the positioning structure . Thermocurable epoxies or solder apply and the fibres of one face are fixed at once. In cycling epoxy glued deviced at -40 + 70°C, S.Day and al. [56] got low (+/- 0,2 dB) insertion loss variations.

The hybrid passive alignment offers similar assets if thermal expansion coefficients of the platform and of the chip are matched. By metal soldering of lithium niobate chips to silicon platforms, M.J. Wale and al.[73] noted 1,6 dB variations in -55°C+85°C cycling. A precise design of glue bumps and glue material allowed us [75] to get variations below +/- 0,2 dB when assembling IO glass chips to silicon motherboards.

The semi-active alignment technique has similar assets and drawbacks. the fibres can be well secured in the block but one has to warrant a good matching of the thermal expansion coefficients of the chip and the block. S.Sato and al.[53] and K.Grosskopf and al.[66] demonstrated about +/- 0,1 dB variations during thermal cycles of -40 + 85 °C .

In that respect the glass fusion technique offers an interesting solution because very low variations of 0,05dB or less are measured during similar cycles. The bonds stand 10 N pulling [96] . Furthermore the technique can be applied to any alignment method.

There are very few available automats [75][98] which take in charge both alignment and fixing. They demonstrate however that the pigtailing truly turns industrial.

8 CONCLUSION

Alignment accuracy requirements for less than 0,1dB pigtailing loss are in the range of 0.1 - 0.5 microns. Among the proposed alignment method, the active technique is the only one which can fulfil them. Therefore it has some interest in the most demanding cases (either for very low excess loss or high uniformity) particularly for devices with small modes (waist of 1-3 microns). But it is intrinsically expensive and difficult to automate. The avaibility of fixing techniques is more limited.

The semi-active approach offers today a good trade-off for most of industrial fabrications. It is far less time consuming than the active alignment method and almost reaches its performances (at least for fibre mode matched waveguides) due to improved fibre and positioning structures characteristics. It offers dense connections, high return loss and good possibilities for fixing.

The fully passive technique also takes benefit from the same progress in fibre and positioning structures fabrication. The hybrid approach is particularly attractive. The connection is fast with simple equipment . It makes possible to think about very low cost pigtailing. the attachment is well secured and the whole method can benefit from more progress in both fields of the flip-chip techniques and fibre mass preparation.

Several fixing methods are proposed to meet specific requirements of either low cost passive or active devices.

Novel UV glues, possibly in combination with thermocurable epoxies demonstrate rather robust fibre attachment and long term reliability for low cost solutions.

The hermetic sealing techniques are developped to solder and weld the fibre to waveguide connection in order to warrant no outgasing.

Therefore technical paths do exist which shall allow IO components to meet the diverse requirements of their applications with either low cost or high performance solutions.

9 ACKNOWLEDGEMENTS

The author thanks Mrs. Valerie Mignot and Mr. Roland Hakoun for their contribution to the more recent results discussed in the paper, and Mr Claude Artigue for his much appreciated help and fruitful comments.

10 REFERENCES

[1] S.Miller "Integrated Optics : An Introduction" *The Bell System Technical Journal* vol 48 n°7 (1969) pp2059-2063

[2] K.W.Murphy "Components close the gap in Fibre To The Home" *Photonics Spectra* October 1991 pp95-101

[3] T.Miyashita, S.Sumida and S.Sakaguchi " Integrated Optical devices based on silica waveguides technology" *SPIE Proceedings vol993* (1988) pp288-294

[4] M.A. Powell " An ally for high speed and the long haul" *Photonics Spectra* May 1993 pp102-108

[5] F.J.Leonberger "Status and Applications of Integrated Optics" *Proceedings of ECIO'93* sect. 10 (1993) pp5-7

[[6] G.Zhang, S.Honkanen, S.I.Najafi, A.Tervonen "Integrated $1.3\mu m/1.55\mu m$ Wavelength Multiplexer and 1/8 Splitter by Ion Exchange in Glass" *Electronics Letters* vol.29 n°12 (1993) pp1064-1066

[7] J.W.Burgess, P.J.Williams, P.M.Charles, A.J.Moseley, J.P.Hall, G.Phillpott, A.Markatos "OEIC's for Local Access Applications" *Proceedings of ECIO'93* sect. 2 (1993) pp8-9

[8] T.Schwander, S.Fisher, K.Hirche, H.Hirth, M.Korn "InGaAsP/InP 3 waveguides directional couplers for WDM applications" *Proceedings of ECOC'92* MoB2-5 (1992) pp69-72

[9] C.Cremer, M.Schier, G.Ebbinghaus "Monolithically integrated grating receivers for DWDM" *Proceedings of ECIO'93* sect. 2 (1993) pp10-11

[10] P.M.Charles, G.G.Jones, P.J.Williams, R.M.Ash, P.H.Fell, A.C.Carter "Integratable high speed buried ridge DFB lasers fabricated on semi-insulating substrates" *Electronics Letters* vol.27 n°9 (1991) pp700-703

[11] P.J.Williams, R.G.Walker, P.M.Charles, R.Ogden, A.K.Wood, N.Carr, A.C.Carter "Design and fabrication of monolithically integrated DFB laser-wavelength duplexer transceivers for TPON/BPON access link" *Electronics Letters* vol.27 n°10 (1991) pp809-810

[12] C.E.Zah, F.J.Favire, B.Pathak, R.Bhat, C.Caneau, P.S.O.Lin, A.S.Gozdz, N.C.Andreadakis, M.A.koza and T.P.Lee "Monolithic integration of multiwavelength compressive-strained multiquantum-well distributed feedback laser array withstar coupler and optical amplifiers" *Electronics Letters* vol.28 n°25 (1992) pp2361-3262

[13] O.G.Ramer "Single-Mode fiber-to-Channel Waveguide Coupling" *Journal of Optical Communications* vol.2 n°4 (1981) pp122-127

[14] E.J.Murphy "Fiber attachment for Guided Wave Devices" *Journal of Lightwave Technology* JLT-6 n°6 (1988) pp862-871

[15] C.A. Armiento "Passive coupling of InGaAs/InP Laser Array and single mode fibres using silicon waferboard" *Electronics Letters* vol.27 n°12 (1993) pp1109-1111

[16] K.Grosskopf, N.Fabricius, R.Fuest "Performance of Integrated Optical Single Mode Multiport Splitters in Glass under Environmental Test Conditions" *Proceedings of ECIO'93* sect. 9 (1993) pp6-9

[17] "Telekom answers a call for 1.2 Million new lines" *OLE* May 1993 pp25-29

[18] R.Regener, H.George "Result of an 18 month qualification program and quality insurance for Integrated Optical modules" *Proceedings of ECIO'93* sect. 10 (1993) pp14-15

[19] s.Valette, P.Gidon, S.Renard and J.P.Jadot "Silicium based integrated optics technology : an attractive hybrid approach for Opoelectronics" *SPIE Proceedings vol 1128 "Glasses for Opoelectronics"* (1989) pp179-185

[20] I.R.Crostown " Silicon Optohybrids for Advanced Optoelectronics Mulitchip Modules" *Proceedings of ECIO'93* sect. 11 (1993) pp1-3

[21] M.Seki, H.Hashizume and R.Sugawara "Two-step purely thermal ion exchange technique for single-mode waveguide devices in glass" *Electronics Letters* vol.20 n°20 (1988) pp1258-1259

[22] M.McCourt and A.Cucalon "Optical and environmental performance of packaged single-mode 1xN couplers made by Ion Exchange in glass" *Proceedings of OFC'90* WE1 (1990) p 60

[23] M.Kawachi "Silica waveguides on silicon and their application to Integrated-Optics components" *Optical and Quantum Electronics* vol 22 (1990) pp391-416

[24] R.G.Walker and C.D.W.Wilkinson "Integrated Optical waveguiding structures made by silver ion exchange in glass : 2 directional couplers and bends" *Applied Optics* vol.22 n°12 (1983) pp1929-1936

[25] H.Zhenguang, R.Srivastava and R.V.Ramaswamy "Low loss small-mode passive waveguides and near-adiabatic tapers in BK7 glass" *Journal of Lightwave Technology* JLT-7 n°10 (1989)pp1590-1596

[26] I.Shanen-Duport, P.Benech, R.Rimet " Optimization of a collimation structure made by ion-exchange in glass" *Proceedings of ECIO'93* sect. 14 (1993) pp42-43

[27] D.Y.Zang and C.S.Tsai "Single-mode waveguide microlenses and microlens arrays fabrication in LiNbO3 using titanium indiffused proton exchange technique" *Applied Physics Letters* vol.46 n°8 (1985) pp703-705

[28] S.A.Reid, M.Varasi, S.Reynolds "Double Dilute Melt Proton Exchange Waveguides" Journal of Optical Communications vol.10 n°2 (1989) pp67-73

[29] C.A.Edwards "Ideal microlenses for laser to fiber coupling" *Journal of Lightwave Technology* JLT-11 n°2 (1993)pp252-257

[30] P.Doussiere "Polarisation insensitive optical amplifier with buried laterally tapered active waveguide" 3d Topical Meeting on Optical Amplifiers and their Applications Santa Fè (1992) pp140-143

[31] J.M.Verdiell, M.A.Newkirk, T.L.Koch, R.P.Gnall, U.Koren, B.I.Miller "High Performance Aspheric Waveguide Lenses in Photonic Integrated Circuits for WDM Applications" *Proceedings of ECIO'93* sect. 2 (1993) pp32-33

[32] R.Zengerle, H.Bruckner, H.Olzhausen and A.Kohl "Low -loss fibre-chip coupling by buried laterally tapered InP/InGaAsP waveguide structure" *Electronics Letters* vol.28 n°7 (1992) pp631-632

[33] G.Muller, G.Wenger, L.Stoll, H.Westermeier, D.Seeberger "Fabrications techniques for vertically tapered InP/InGaAsP Spot Size Transformers with very low loss" *Proceedings of ECIO'93* sect. 14 (1993) pp14-15

[34] T.Brenner and H.Melchior "Integrated Optical Modeshape Adapter in InGaAsP/InP for efficient fiber-to-waveguide coupling" *IEEE Photonics Technology Letters* vol.5 number 9 (1993) pp1053-1056

[35] T.L.Koch and al. "Tapered waveguide InGaAs/InGaAsP Multiple Quantum Wells Lasers" *IEEE Photonics Technology Letters* vol.2 number 2 (1990) pp88-90

[36] Y.Shani, H.Henry, R.C.Kristler, R.F.Kazarinov, K.J.Orlowsky "Integrated Optics Adiabatic Devices on Silicon" *Journal of Quantum Electronics* Vol.QE-27 number 3 (1991) p556-566

[37] J.Buus and al. "Spot Size Expansion for Laser to Fiber Coupling Using an Integrated Multimode Coupler" *Journal of Lightwave Technology* JLT-11 n°4 (1993)pp582-588

[38] *Melles Griot catalogue 1994 : datasheet on Nanotrack Products*

[39] *Newport catalogue 1994 : datasheet on Autoalign Products*

[40] R.Glass, H.Marth and P.Cornardeau "Positionnement de fibres : un nouveau concept" *Journal OPTO* number 73 (1993) pp 38-41

[41] K.H.Cameron "Simple and Practical Technique for attaching single-mode fibres to lithium niobate waveguides" *Electronics Letters* vol.20 n°23 (1984) pp974-976

[42] T.Yoshizawa, K.Takemoto, N.Takato "Novel technique for connecting waveguides and fibers" *Fiber and Integrated Optics* Vol.9 (1990) pp125-130

[43] D.J.Albares, D.B.Cavanaugh, T.W.Trask US Patent 49 30 854 (1990)

[44] J.F.Bourhis, V.Mignot, R.Hakoun " Automatic pitailing of passive components with high number of ports" *IO Symposium* Lindau SPIE vol.2213 paper number 30 (1994)

[45] M.McCourt "Status of glass and silicon based technologies" *Proceedings of ECIO'93* sect. 9 (1993) pp1-4

[46] N.Yamaguchi, Y.Kokubun and K.Sato " Low-loss Spot Size Transformer by dual tapered waveguides" *Journal of Lightwave Technology* JLT-8 number 4 (1990) pp587-594

[47] E.J.Murphy, T.C.Rice, L.McCaughan, G.T.Harvey and P.H.Read "Permanent attachment of single-mode fibre arrays to waveguides" *Journal of Lightwave Technology* JLT-3 (1985) pp795-799

[48] A.C.O'Donnell "Polarisation independant 1x16 and 1x32 lithium niobate optical switch matrices" *Electronics Letters* vol.27 n°25 (1991) pp2349-2350

[49] P.Barlier, C.Nissim and L.Dohan "Passive Integrated Optics Components for Fiber Optics Communication in Moldable Glass" *Proceedings of IOOC-ECOC'85* (1985) pp187-190

[50] H.Kragl "Race II POPCORN : A Novel Technology for Polymer Based Integrated Optics" *Proceedings of ECIO'93* sect. 9 (1993) pp24-25

[51] A.C.G.Nutt, J.P.G.Bristow, A.McDonagh and P.J.R.Laybourn "Fiber-to-waveguide coupling using ion-milled grooves in lithium niobate at 1.3µm wavelength" *Optics Letters* Vol.9 number 10 (1984) pp463-465

[52] P.S.Chung and J.Millington "Milled groove method for fibre-to-waveguide couplers" *Journal of Lightwave Technology* JLT-5 number 12 (1987) pp1721-1725

[53] S.Sato, H.Wada, H.Hashizume, S.Kobayashi, M.Seki and K.Nakama "Guided-wave Directional Couplers with PMF arrays" *Electronics Letters* vol.27 n°4 (1991) pp303-304

[54] H.W.Messenger "Excimer Laser marks LiNbO3 for fiberoptics devices" *Laser Focus World* May 1990 p227

[55] C.M.Schroeder "Accurate Silicon Spacer Chips for an Optical Fiber Cable Connector" *The Bell System Technical Journal* January 1978 pp90-97

[56] S.Day, R.Bellerby, G.J.Cannell and M.Grant " Silicon-based fibre pigtailed 1x16 power splitter"*Electronics Letters* vol.28 n°10 (1992) pp920-922

[57] G.Grand, H.Denis, S.Valette "New method for low cost and efficient optical connections between singlemode fibres and silica guides" *Electronics Letters* vol.27 n°1 (1991) pp16-17

[58] G.Grand, S.Valette, G.J.Cannell, J.Aarnio, M.del Giudice "Fibre pigtailed silicon based low-cost components" *Proceedings of ECOC'90* (1990) pp525-528

[59] J.Watanabe and T.Saito "Precision machining for micro-optical devices with powder particles collision" *Technical Digest of the optical Society of America: conference on Lasers and Electrooptics* Washington (1987) paper W04

[60] W.Ehrfeld "The LIGA Process for MicroSystems" *MICROSYSTEM Technologies 90* editor H.Reichl, publisher Springer-Verlag (1990)pp521-537

[61] N.Mekada, M.Seino, Y.Kubota and H.Nakajima "Practical Method of waveguide to fiber connection : direct preparation of waveguide endface by cutting machine and reinforcement using ruby beads" *Applied Optics* vol.29 number 34 (1990) pp 5096-5102

[62] Y.Yamada, A.Takagi, I.Ogawa, M.Kawachi and M.Kobayashi "silica-based Optical Waveguides on terraced silicon substrate as hybrid integration platform" *Electronics Letters* vol.29 n°5 (1993) pp444-446

[63] E.J.Murphy and T.O.Murphy "Characteristics of 23 Ti:LiNbO3 switch arrays" *Proceedings of ECIO'93* sect. 10 (1993) pp3-4

[64] H.M.Presby, S.Yangs, A.E.Willner and C.A.Edwards "Connectorised Integrated Star Couplers on Silicon" *Optical Engineering* Vol.31 number 6 (1992) pp1323-1327

[65] J.E.Watson, M.A.Milbrodt and T.C.Rice "A polarisation-independant 1x16 Guided-Wave Optical Switch Integrated on LiNbO3" *Journal of Lightwave Technology* JLT-4 number 11 (1986) pp1717-1721

[66] K.Grosskopf, N.Fabricius, U.Nolte and H.Oeste "Integrated Optical Multiport Splitters in Glass for Broadband Communications Networks" *Proceedings of EFOC/LAN 92* (1992) pp148-152

[67] H.P.Hsu and A.F.Milton "Flip-Chip Approach to Endfire Coupling Between Single-Mode Optical Fibres and Channel Waveguides" *Electronics Letters* vol.12 n°16 (1976) pp404-405

[68] C.H.Bulmer, S.K.Sheem, R.P.Moeller and W.K.Burns "High efficiency flip-chip coupling between single-mode fibers and LiNbO3 channel waveguides" *Applied Physics Letters* Vol.37 number 4 (1980) pp351-353

[69] E.J.Murphy and T.C.Rice "Self-alignment technique for fibre-attachment to guided wave devices" *Journal of Quantum Electronics* Vol.QE-22 number 6 (1986) pp928-932

[70] H.Kaufmann, P.Buchmann, R.Hirter, H.Melchior, G.Guekos "Self-adjusted permanent attachment of fibres to GaAs waveguide components" *Electronics Letters* vol.22 number 12 (1986) pp642-644

[71] H.Kaufmann *Proceedings of ECOC'93* Montreux September 1993

[72] G.L.Destefanis "HgCdTe infrared diode arrays" *SemiConductors Science Technology* 6 (1991) ppC88-C92

[73] M.J.Wale and C.Edge "Self-Aligned Flip-Chip Assembly of Photonic devices with Electrical and Optical connections" *IEEE Transactions on Components, Hybrids and Manufacturing Technologies* Vol.13 number 4 (1990) pp780-786

[74] J.W.Parker "Optical Interconnection for Advanced Processor Systems : A review of the ESPRIT II OLIVES program" *Journal of Lightwave Technology* JLT-9 number 12 (1991) pp1764-1773

[75] J.F.Bourhis, V.Mignot and R.Hakoun "Automat for pigtailing of passive integrated optics components with large number of ports" *Proceedings of EFOC/N* paper 85 (1994)

[76] K.P.Jackson "Optical Fibre Coupling Approach for Multi-Channel Laser and Detector Arrays" *SPIE Proceedings vol.994* pp40-47

[77] H.F.Lockwood "Hybrid Wafer-Scale Optoelectronic Integration" *SPIE Proceedings vol.1389* pp55-67

[78] R.S.Vodhanel, J.S.Ko "Reflection induced frequency shifts in single-mode laser diodes coupled to Optical Fibres" *Electronics Letters* vol.20 number 23 (1984) pp973-974

[79] N.Schunk, K.Petermann "Numerical analysis of the feedback regimes for a single mode semi-conductor laser with external feedback" *Journal of Quantum Electronics* Vol.QE-24 number 7 (1988) pp1242-1247

[80] *British Telecom Specification RC 8937*

[81] *Deutsche Bundespost Telecom , Technical Delivery Conditions* 6060-3006 May 1993

[82] M.Renaud, J.F.Vinchant, P.Barry, J.LeBris, J.G.Provost, J.A.Cavailles, M.Erman "Monolithic Integration of GaInAsP/InP Carrier Depletion Directional Couplers and GaInAs p.i.n. Detectors on Semi-Insulating InP" *IEEE Photonics Technology Letters* Vol.4 number 12 (1992) pp409-411

[83] Y.Ruello, M.Boitel, J.M.Cailleaux, D.Crespel, T.Mahé and D.Roblot "Evolution du raccordement des cables a fibres optiques : perspectives" *Proceedings of OPTO'93* Paris, (1993) pp 383-391

[84] M.Boitel, J.M.Cailleaux, D.Crespel, T.Mahé and Y.Ruello "Fracture en biais du cable ruban" *Proceedings of OPTO'93* Paris, (1993) pp 418-419

[85] Susuki and al. " HDTV photonic space-switching system using 8x8 polarization independent LiNbO3 switch matrix" *Proceedings of OSA topical meeting on Photonic Switching* Paper FE2 (1989)

[86] R.Hakoun, E.Tanguy and P.Guerin "Militarisation de multiplexeurs mulitmodes en Optique Integree sur verre par echange d'ions" *Proceedings of OPTO'90* Paris, (1990) pp 462-465

[87] "Integration replaces the twisted fibre connector" *OLE* November 1992 pp20-21

[88] M.S.Ner, C.Sharp and D.R.Gibson "Environmental performance of LiNbO3 based guided wave optical devices" *SPIE Proceedings* Vol 1180 (1989) pp183-184

[89] G.Voirin, B.Sheja and O.Parriaux "Applications of glass etching to guided wave optics" *SPIE Proceedings* Vol 1128 (1989) pp140-141

[90] D.Boscher, B.Le Marrer and G.Perrin " Traitement anti-reflexion des composants passifs a fibres et applications" *Proceedings of OPTO'90* Paris, (1990) pp 454-457

[91] Bellcore *Technical Advisory TA-NWT-001209* (12/1991)

[92] E.Friedrich, M.G.Oberg, B.Broberg, S.Nillson and S.Valette " Hybrid Integration of Semi-Conductor Lasers with Si-based Single Mode Ridge Waveguides" *Journal of Lightwave Technology* JLT-10 number 3 (1993) pp336-340

[93] J.F.Vinchant, A.Goutelle, B.Martin, F.Gaborit, P.Pognod-Rossiaux, J.L.Peyre, J.LeBris and M.Renaud " New Compact Polarisation Insensitive 4x4 Switch Matrix on InP with Digital Optical Switches and Integrated Mirrors" *Proceedings of ECOC'93* (1993) pp371-374

[94] K.R.Preston, B.M.MacDonald, R.A.Harmon, C.W.Ford, R.N.Shaw, I.Reid, J.H.Davidson, A.R.Beaumont and R.C.Booth "High performance hermetic package for LiNbO3 electro-optic waveguide devices" *IEE Colloquium on Integrated Optics,* London Digest number 1989/3 (1989) pp2/1-2/4

[95] J.J.G.Allen, S.P.Shipley and N.Nourhargh " Silica based integrated optic components for telecommunication application" *Optical Engineering* Vol.32 number 5 (1993) pp1011-1014

[96] T.Shiota, N.Taketani, K.Morosawa, T.Tokunaga, T.Fukahori "Improved optical coupling between silica-based waveguides and optical fibres" *Proceedings of OFC'94* FB-4 (1994) pp 282-283

[97] G.N.Blackie, A.D.Carr, A.Donaldson, D.D.Hall, P.K.Kimber, N.J.Parsons, I.A.Wood and R.Fedora "Recent progress at GEC Marconi in the development and manufacture of Integrated Optics travelling wave modulators"

[98] A.A.Hollander and K.J.Watkins "The Art and Scenic of Optoelectronic Joining" *Photonics Spectra* October 1993 pp110-114

[99] A.Neyer, W.Mevenkamp and B.Kretzschmann "Optical Switching System using a nonblocking Ti:LiNbO3 4x4 Switch Array" *Proceedings of the ECIO'87 conference* (1987) p32

[100] H.P.Zappe, M.Shen and H.E.G.Arnot "Hybrid Coupling and Monolithic Integration for Integrated Optical Sensors" *Proceedings of ECIO'93* section 11 (1993) pp8-9

[101] C.Duchet, C.Brot, L.Sarrabay, M.DiMaggio and D.Haux "Digital optical switching having one operational state at zero voltage" *Proceedings of OFC'93* (1993) pp478-479

Optical fiber chemical sensors

Mehdi Shadaram

The University of Texas at El Paso
Electrical Engineering Department
El Paso, Texas 79968

ABSTRACT

The theory and applications of optical fiber chemical sensors are discussed. Several transduction mechanisms for measuring the quantity of chemical species such as carbon monoxide, hydrogen sulfide, or toxic substances in the air or water are explained. A new frequency domain method which utilizes (chemo-optic) variable index material optical fiber couplers is also proposed for the application in distributed sensing.

1. INTRODUCTION

A wide variety of industrial firms are trying to develop new sensing methods for detecting physical quantities as part of their industrial and economic growth. Optical fibers have matured to the point where they can be useful in many applications ranging from high data rate communication networks to sensors of physical fields. Since the mid 1970's optical fiber sensors have been used for sensing virtually any physical parameter such as pressure, temperature, acceleration, rotation, electric field, etc[1-3]. However, the development of optical fiber sensors for detecting chemical species was initiated in the early 1980's[4-6]. The cost of chemical sensing with optical fibers can be relatively low compared to chemical monitoring with a conventional laboratory apparatus. Some advantages of optical fiber sensors are that they are rugged, portable, immune to electromagnetic fields, and can be operated in harsh environments. The development of such inexpensive and small size chemical transducers are beneficial to a variety of applications. A few potential applications for the optical fiber chemical sensors are: pollution detection, hazardous waste analysis, food processing, blood analysis, exhaust analysis of internal combustion engines, and explosive gas monitoring. Furthermore, the advances in optical fiber technology has allowed the development of multiplexed systems employing a large array of optical fiber sensors. These large arrays of fiber sensors are usually embedded in composite materials to construct smart structures of sensor networks[7-8]. The development of fiber optic smart structure technology also offers the promise of undertaking "real time" structural measurements with built-in sensor systems. The application of optical fiber smart structures can be in monitoring the distribution of chemicals in a variety of containers or boilers used in materials processing. This paper also describes the theory of operation of a new technique for discrete measurement of the distribution of the mass of chemical species.

In general, all the optical fiber chemical sensors are equipped with an optical source such as light emitting diode (LED) or semiconductor laser diode, followed by an optical fiber which is used as a transmission medium for light, a photodetector which is usually followed by a signal processing device, and a transduction mechanism that modulates the optical parameters. Like any other optical fiber sensor, optical fiber chemical sensors are classified into two categories; extrinsic and intrinsic. The extrinsic optical fiber chemical sensors are based on the fact that sensing takes place in a region outside the fiber. In essence, as light is launched into the fiber, it is guided to the point of the sensor at which the light exits from the fiber into a separate sensing mechanism and relaunched back into the fiber. On the other hand, intrinsic optical fiber chemical sensors are based on the fact that the sensing takes place within the fiber itself. Unlike the extrinsic sensors, the light being guided does not exit the fiber but is modulated within the fiber in response to an external influence. The transduction mechanism usually depends on the changing of the fundamental optical field parameters: intensity, phase, color, and polarization. In an intensity-modulation sensor, the chemical species to be detected cause a change in the intensity of the light. In a phase-modulation sensor, the chemical species to be detected affect the optical paths along the fiber which induces a phase shift in the optical signal travelling through the fiber. These sensors require the use of single-mode fibers and coherent sources. They are also considered to be the most sensitive optical fiber sensors. The operation of polarimetric sensors is based on the changes in the polarization state of the light caused by the chemical species to be detected. In color or wavelength modulated sensors, the chemical species to be detected affect the spectral distribution of the optical signal. The light sources for polarimetric sensors have a wideband spectrum and the sensing mechanism is based on either absorption of a portion of the light spectrum when it passes through the chemical species to be detected or changes in the fluorescent emission of special coatings on the optical fiber.

The development of optical fiber smart structures offers the integration of a large number of optical fiber sensors for distributed measurement systems. These systems usually consist of an array of optical fiber sensors embedded in advanced composite material structures during their fabrication. The finished product, which is known as materials with optical nerves, is viewed as a state-of-the-art sensing system for detecting and locating cracks, anomalous hot spots, radio active radiations, and excess of pressure in a variety of structures such as boilers and chemical storage containers that have potential for causing catastrophic environmental accidents. Because of difficulties in the fabrication process, conventional optical fiber chemical sensors are not assembled into an integrated sensor system. This paper explains a novel technique which makes the integration of optical fiber sensors into a system one step closer to reality.

2. REVIEW OF SENSOR DESIGNS

Several optical fiber chemical sensors are categorized and described in this section. Optical fiber chemical sensors are classified based on their structure and operation. In this paper the most common sensors; intensity, phase, and polarization modulated sensors are explained. In fact, the current commercial optical fiber chemical sensor systems are based on the intensity modulation of light[9-10]. Among these sensors are those where optical fibers are used to enable a remote sample of fluorescence or absorption. Other intensity modulated optical fiber sensors are those that utilize Raman spectroscopy and absorption spectroscopy. To this date, very little has been done to utilize phase and polarization modulating optical fiber chemical sensors. Regardless of the sensor classification, the primary configuration of any sensing system is based on a chemically sensitive material such as polymers, organic, and nonorganic semiconductor materials deposited on the end of the fiber, over the cladding, or over the core of the fiber. The operation of these chemical sensors is usually based on changes in the optical properties of the coating material when they are exposed to a specific chemical species. The variation in optical properties of the coating material is normally caused by either chemical reactions or the adsorption of chemicals on coating materials.

2.1. Intensity-modulated sensors

Figure 1 represents a typical intensity modulated optical fiber chemical sensor[11]. A variety of coating materials can be used for detecting different chemical species. For example, a thin layer of palladium (10-20 nm thick) which is acting as a micromirror deposited on the end of the fiber, can be used for hydrogen detection. Exposure to hydrogen gas converts the palladium coating to hybrid, PdH_x, which in fact reduces the reflectivity of the mirror. The absolute reflectivity of the palladium film is about 22%. The magnitude of the changes in reflectivity is directly related to the Pd film thickness, hydrogen concentration, and duration of exposure. An experiment has shown that when the sensing area is exposed to a 0 to 4% concentration of hydrogen in air, the reflectivity of the palladium film drops from 22% to 20%. This sensor may be used for the detection of a large concentration of hydrogen gas in the air. A different sensor may be obtained by using a thin polymer film such as tetrafluoroethylene (C_2F_4) or poly-tetrafluoroethylene (PTFE) on the end of the optical fiber[12]. In general, in response to a wide range of volatile organic compounds (VOCs) polymers swell. The swelling of the sensing part causes a change in the reflectivity of the fiber end. In order to provide a better reflectivity at the polymer-fiber interface, a semitransparent film of metal such as Ni is deposited on the fiber end before the polymer. The sensitivity of these types of sensors normally depends on the degree of swelling and refractive index of the polymer when it is exposed to the VOC and index of refraction the fiber core. Swelling of the polymer usually produces a decrease in reflectivity. Other film materials may be suitable for chemical sensing applications. Table 1 represents a few examples of film materials and their applications in chemical sensing.

Table 1. Examples of film materials and their applications

Film Material	**Chemical Species to be Detected**
Ni, Ti	O_2
Cr followed by Au	Hg vapor
C_2F_4 and PTFE	Trichloroethylene (TCE) and Methanol (MeOH)
Pd	H_2

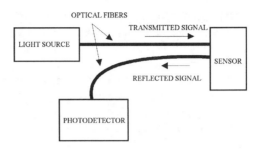

Figure 1. An example of intensity modulated optical fiber chemical sensor.

A different class of intensity modulated chemical optical fiber sensors can be realized by utilizing variable index material (VIM)[13] as cladding over the core. The variation in refractive index is caused by selective gas adsorption. The VIM of interest may include organic or inorganic semiconductor materials. The adsorbed gas on the semiconductor surface plays a role comparable to "dopant" and therefore initiates drastic changes in electrical and optical properties. The process is a consequence of the compatibility of the electronic structure of the material near its surface and the atomic structure of the gas being detected. Since the adsorption occurs at the material surface, the high surface to volume ratio of VIM films make the advantages of such structures apparent. Examples of VIMs suitable for detecting certain gases are shown in table 2.

Table 2. Examples of VIMs and their applications

VIM	Adsorbing Gas
Cobalt Acetylacetonate (TF)	NH_3
Polyvinylidene Fluoride	An
Polyfuran	H_2
Fluoenylidene Malonitrile	CO, CO_2
Polyfluoromethylene	H_2S

Figure 2 illustrates an example of an adsorption based optical fiber chemical sensing. This design employs two coupled fibers with portions of their claddings removed in order to allow a point of contact between the cores. Fiber 1 (input fiber) retains its original cladding surrounding the point of contact, while fiber 2 (sensor fiber) has a coating of VIM replacing its cladding at this point. The coupling ratio from fiber 1 to fiber 2 is directly related to the index of refraction of the VIM segment. The adsorption of some gases on these semiconductors' surfaces can produce a considerable variation in their electrical and optical properties which affects the coupling ratio. This phenomenon can eventually lead to the detection and quantification of some pollutant gasses in the atmosphere. Since the changes in the coupling ratio are based on the adsorption process, the response time of these sensors is relatively low compared to those optical fiber chemical sensors whose operation is based on chemical reaction between the coating materials and chemical species to be detected. The application of these type of sensors in distributed sensing will be discussed in the following section.

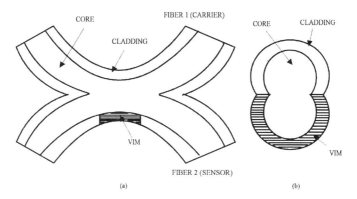

Figure 2. An example of adsorption based optical fiber chemical sensor, (a) coupled fibers, (b) cross section .

2.2. Phase-modulated sensors

A different class of optical fiber chemical sensors known as phase-modulated sensors utilizes the optical phase shift effect in a single-mode fiber as the fiber is exposed to chemical matters. In general, these sensors are the most sensitive type of optical fiber sensors. To this date, very little has been done to utilize phase-modulated chemical optical fiber sensors (interferometers). In order to achieve an optical phase modulation, one arm of an optical fiber interferometer can be modified with respect to the other. Phase-modulated sensors normally employ a single-mode fiber and a monochromic light source. The phase delay, ϕ, for a monochromatic light propagating in a fiber with a length of L and core refractive index of n is given by

$$\phi = \frac{2\pi nL}{\lambda} \quad (1)$$

where λ is the wavelength of the light in the free space. A strained or temperature varying optical fiber experiences a change in its length or refractive index which results in the phase modulation of the propagating light in the fiber. Since photodetectors are not capable of responding to optical frequencies, a mechanism for converting phase modulation to amplitude modulation is required. A typical Mach-Zehnder, Michelson, or another type of optical fiber interferometer[2] can be employed to accomplish this function by placing the modified fiber in the sensor arm of such interferometers. Figure 3 shows the structure of a Mach-Zehnder optical fiber interferometer. The laser output light is split by a 3-dB coupler/spitter with 50 percent of the light injected into the single-mode sensing fiber and the rest into the reference fiber. After passing through the interferometer the output beams from the sensing and the reference fibers are recombined by the second 3-dB coupler/spitter, and interference occurs. The interference causes the phase-modulated light to become amplitude-modulated light. The amplitude-modulated light is then photodetected.

The phase change, $\Delta\phi$, associated with the optical path modification caused by strain in the fiber core is given by

$$\Delta\phi = k\Delta(nL) = k(n\Delta L + L\Delta n) = knL(\frac{\Delta n}{n} + \frac{\Delta L}{L}) \qquad (2)$$

where $\Delta L/L$ is the axial strain S_{11} and $\Delta n/n$ is given by

$$\frac{\Delta n}{n} = -(\frac{n^2}{2})[(P_{11}+P_{12})S_{12} + P_{12}S_{11}] \qquad (3)$$

where P_{11} and P_{12} are the elements of the strain optic tensor (Pockel's coefficients), and S_{12} is the radial strain. Using the typical values of 0.27 and 0.12 for P_{11} and P_{12} respectively and n=1.48 for the fused silica fibers will lead to the following conclusion

$$\Delta\phi = knL[0.868S_{11} - 0.429S_{12}] \qquad (4)$$

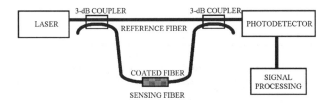

Figure 3. A typical Mach-Zhender optical fiber interferometer.

Modification of the optical path can be achieved by coating the sensing fiber with specific materials that can affect the optical properties of the fiber when the coated fiber is exposed to particular chemical matters. Tables 1 and 2 show a list of potential coating materials for sensor application. For example, palladium coated fibers can be utilized for the detection of hydrogen. It is well known that when palladium is exposed to hydrogen gas it forms a hybrid with a lattice constant which depends upon the concentration of the hydrogen gas. In other words, the applied strain on the fiber core is defined by the hydrogen concentration. Equation (4) shows that a radial strain of -10^{-6} over 30cm of the core of the fiber can generate a $\Delta\phi$ of 1.83 radian at 0.632 µm wavelength. Experiments have shown that a hydrogen concentration of a few ppb in nitrogen gas can generate a measurable $\Delta\phi$ in palladium coated fibers with a coating length of 30 cm[4]. Polymers which normally swell when exposed to various vapors can also be ideal candidates for use as coating materials. Additional interferometric optical fiber chemical sensors can be acquired by using a VIM semiconductor-cladded fiber as the sensing fiber in the interferometer.

An additional choice for phase modulation of the light can be the use of a Fabry-Perot optical fiber interferometer as shown in Figure 4. The thin layer of metallic, polymer or VIM semiconductor (with a thickness of approximately several hundred Å) deposited on the fiber end acts as a semitransparent material. The reflectivity of the semitransparent film is given by:

$$R = \left| \frac{R_1 + R_2 e^{-j2\gamma}}{1 + R_1 R_2 e^{-j2\gamma}} \right|^2 \quad (5)$$

where R_1 and R_2 represent reflectivities of fiber/film and film/air interfaces respectively and γ is the optical phase shift caused by film thickness. If we assume that the refractive index of the film, n_f, is uniformly distributed, γ can be related to the film thickness, ε, as follows:

$$\gamma = \frac{2\pi n_f \varepsilon}{\lambda} \quad (6)$$

Chemical reaction between the film and the chemical species to be detected may vary the electronic structure of the film (such as adsorption process) which modifies the refractive index of the film, or may cause swelling of the film that results in a change in the thickness of the film.

In phase-modulated sensors, the optical fibers should be firmly secured in a stationary position and kept in a temperature controlled environment to minimize noise introduced by vibration and differential thermal expansion. Although the necessity of using a monochromic light source limits the size and cost of these sensors, the fact that they measure optical phase shifts gives them the potential of being the most sensitive. Recent advances in laser diode technology and the development of integrated photonic devices may reduce the cost and size of the phase-modulated sensors to a point which will make these sensors economically competitive with the others.

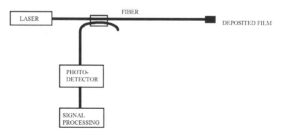

Figure 4. A Fabry-Perot optical fiber interferometer.

2.3. Polarization-modulated sensors

The transduction mechanism of a polarization-dependent sensor is based on the changes in the state of polarization of light upon the exposure of the sensor to the chemical species to be detected. Basically, any stress on the medium that light is propagating within translates into a change in its birefringence. This is mainly due to the changes in the index of refraction. In intrinsic sensors, fiber is usually coated with a specific thin polymer film which swells when exposed to the chemical species to be detected. The swelling action applies stress over the fiber core. An example of an extrinsic sensor, made of an optical fiber terminated by a small flexible balloon containing a mixture of liquid crystal coated with a thin polymer film, is illustrated in Figure 5. Liquid crystals are generally known for their electrical and thermal properties, but their optical properties are also very much dependent upon the pressure[14-15]. This influence can be optimized with proper composition and structure of the liquid crystals. Usually a mixture of two or three kinds of crystals is necessary to optimize the sensitivity. Any small pressure variations on the half spherical membrane will change the state of polarization of the light travelling through the liquid crystal. The transduction mechanism of this sensor is based on the sensitivity of the ellipticity of the light travelling in the crystal to small changes in shape and orientation of the crystal molecules. Consequently, any small variation of pressure will provide a large change in the ellipticity of the reflected light.

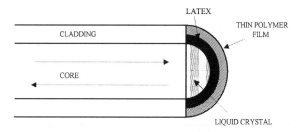

Figure 5. An example polarization optical fiber sensor using liquid crystal.

The effect of stress on the state of polarization of the light in optical fibers can be explained using electromagnetic wave theory. In general, light travels as electric and magnetic fields orthogonal with respect to each other in single-mode fibers. Because the action of light on molecules involves the redistribution of charges, and electric fields have a much greater effect of inducing these charges than do magnetic fields, only the electric fields will be considered for analysis. A polarized electric field normally travels along the fiber in the form of two orthogonal waves known as E_x and E_y. The mathematical expression for these waves are depicted as

$$E_x = E_{xo}e^{-j\phi_{xo}}$$
$$E_y = E_{yo}e^{-j\phi_{yo}}$$
(7)

where $\phi_{jo}=-jZ_o(n_i-jk_i)$, n_i=componential index of refraction, k_i=componential absorption constant, $Z_o=2\pi L/\lambda$, L=length of the fiber, λ=wavelength, and i=x or y unit vector. For ideal fibers the phase difference between the two components ($\phi_{xo}-\phi_{yo}$) is zero, thus resulting in a linear birefringence. If the difference is a multiple of $\pi/2$ then the birefringence is circular; otherwise the birefringence is elliptical. Using Jon's matrix[16] the output electric field vectors (E_{xf} and E_{yf}) can be related to the input electric field vectors (E_{xi} and E_{yi}) as follows

$$\begin{vmatrix} E_{xf} \\ E_{yf} \end{vmatrix} = \begin{vmatrix} m_{11} & m_{12} \\ m_{21} & m_{22} \end{vmatrix} \times \begin{vmatrix} E_{xi} \\ E_{yi} \end{vmatrix}$$
(8)

The characteristic matrix, m_{ij}, contains the amplitude and phase information required to represent the state of the polarization of the light traveling through the fiber. A single-mode fiber can be represented as a photoelastic (birefringent) material. When the fiber is under application of external influences the real part of its refractive index can be represented by[17].

$$\begin{vmatrix} n_x \\ n_y \\ n_z \end{vmatrix} = \begin{vmatrix} n \\ n \\ n \end{vmatrix} + \begin{vmatrix} B_1 & B_2 & B_2 \\ B_2 & B_1 & B_2 \\ B_2 & B_2 & B_1 \end{vmatrix} \times \begin{vmatrix} \epsilon_x \\ \epsilon_y \\ \epsilon_z \end{vmatrix} \quad (9)$$

where n is the index of refraction of the fiber with no external perturbance, ϵ's are the principal strains, and B_1 and B_2 are called strain optical coefficients. The stress dependence of the refraction indices are given by

$$\begin{vmatrix} n_x \\ n_y \\ n_z \end{vmatrix} = \begin{vmatrix} n \\ n \\ n \end{vmatrix} + \begin{vmatrix} C_1 & C_2 & C_2 \\ C_2 & C_1 & C_2 \\ C_2 & C_2 & C_1 \end{vmatrix} \times \begin{vmatrix} \sigma_x \\ \sigma_y \\ \sigma_z \end{vmatrix} \quad (10)$$

where σ's are the principal stresses, and

$$C_1 = \frac{1}{E}(B_1 - 2\nu B_2)$$
$$C_2 = \frac{1}{E}[B_2 - \nu(B_1 + B_2)] \quad (11)$$

The E represents the Young's modules of the fiber material, and ν is the Poisson's ratio. If we assume that only a differential stress of $\Delta\sigma = \sigma_x - \sigma_y$ is applied on the fiber with a length of L, then birefringence between the two orthogonal components in the x and y directions can be shown by

$$\phi = \frac{2\pi L}{\lambda}(n_x - n_y) \quad (12)$$

Birefringence, ϕ, can also be related to the principal stresses or strains as follows

$$\phi = \frac{2\pi L}{\lambda}(C_1 - C_2)(\sigma_x - \sigma_y) = \frac{2\pi L}{\lambda}(B_1 - B_2)(\epsilon_x - \epsilon_y) \quad (13)$$

2.4. Color-modulated sensors

Figure 6 depicts two examples of color or wavelength modulated sensors. The transduction mechanism of the sensor shown in Figure 6a is based on the absorption of a portion of the light spectrum by the chemical species to be detected. The light source has a wideband spectrum and the light is guided via the transmitting fiber to the monitoring region where the wavelength absorption (modulation) occurs. The attenuated light is then guided by the receiver fiber to the spectrometer which provides a ratio-metric measurement of the optical intensity proportional to the wavelength distribution of the modulated light. An application example of such a sensor would be the remote measurement of sulfur dioxide (SO_2) in exhaust gas from combustion engines[18]. In general, SO_2 strongly absorbs ultraviolet (UV) lights. Consequently, an air gap between the transmitter and receiver fibers can act as the modulator. The air gap behaves as an absorption filter for UV lights when SO_2 gas flows through it. The sensitivity of such sensors depends solely on the SO_2 concentration and the sensor design. The transduction mechanism of the sensor shown in Figure 6b is based on the changes in the fluorescent emission of a special coating material on the tip of the optical fiber when the tip is exposed to the chemical species to be detected. Within the medical field one of the most important applications of these sensors includes the monitoring of pH, pCO_2, and pCO in the blood. Other medical applications include measuring oximetry, dye dilution, and velocity of blood flow in vessels[19].

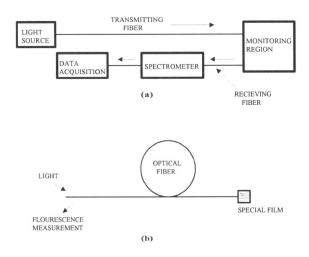

Figure 6. Examples of color modulated sensors. (a) absorption, (b) fluorescent emission.

3. A PROPOSED DISTRIBUTED SENSOR SYSTEM

Figure 7 illustrates the elements of a proposed optical fiber distributed chemical sensor system. The system consists of various electrical components, optical components and data acquisition hardware. The function of such a system is to make discrete measurements of the distribution of the quantities of chemical species. The optical transmitter periodically launches an intensity modulated sinusoidal frequency, f_o, provided by the oscillator into the optical fiber star coupler's input. The coupler then distributes the signal into N individual fibers. Each fiber is followed by a special optical fiber chemical sensor which behaves as a notch filter with a center frequency of f_o. The sensor indeed attenuates the modulated signal according to the quantity of the chemical matter to be detected. The optical signals passed through the sensors are photodetected by the pin diodes, converted to corresponding sinusoidal voltages, and then peak detected. The corresponding peak voltages are then multiplexed by an analog-to-digital converter (ADC), processed by the computer and displayed as the distribution profile of the quantities of the chemical matter affecting the sensors.

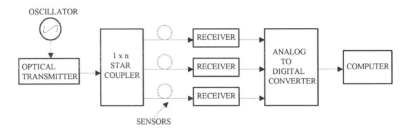

Figure 7. A proposed optical fiber distributed chemical sensor system.

Figure 8 represents an example of the sensor used in the system. The sensor consists of a fiber loop which is partially closed upon itself by means of a chemo-optic coupler (see Figure 2). Signals entering the input port of the sensor recirculate around the loop producing outputs on each transit. If we assume that the fiber loop provides a total delay time of τ, and the coupling coefficients from ports 1 and 3 of the coupler to ports 2 and 4 are k and 1-k respectively, the impulse response of the sensor can be expressed as:

$$h(t) = k \sum_{m=0}^{\infty} (1-k)^m \, \delta(t-m\tau) \qquad (14)$$

By taking the Fourier transform of both sides of Eq. 14, the transfer function of the sensor can be obtained as:

$$H(f) = k \sum_{m=0}^{\infty} (1-k)^m \, e^{-j2\pi m f \tau} \qquad (15)$$

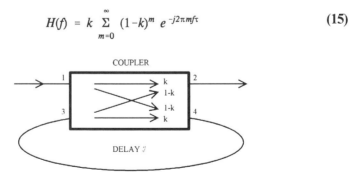

Figure 8. An example of optical fiber chemical sensor used in the distributed system.

Figure 9 shows the magnitude of H(f) versus the frequency for different values of k and τ=50 nsec which corresponds to a fiber delay length of approximately 10 meters. The sensor behaves as a notch filter with a center frequency of 10 MHz. As indicated in Figure 9-a, the attenuation at the notch frequency strongly depends on the k value.

Furthermore, Figure 9-b shows that the sensitivity of the attenuation to the k value is maximized for k values around 0.3. Figure 10 represents the magnitude of H(f) at the notch frequency as a function of k. It also shows that the maximum sensitivity can be achieved around k=0.3. Eq. 16 defines the sensitivity (S) of the sensor as the changes in the magnitude of H(f) in dB with respect to variations in k at the notch frequency.

$$S = \frac{\Delta |H|_{dB}}{\Delta k}\bigg|_{f=f_o} \quad (16)$$

Figure 11 depicts S as a function of k. The high sensitivity, utilization of intensity modulated light and multimode fibers, and simplicity of the design make these devices very attractive for use in distributed chemical sensing.

4. ADSORPTION BASED OPTICAL FIBER SENSORS

The transduction mechanism of the adsorption based optical fiber sensors (explained in section 2.1) is based on the changes in the refractive index of the fiber's cladding upon exposuer to chemical species. The variation in the refractive index of the fiber is caused by a process known as chemical adsorption or chemisorption which involves a chemical interaction between reactive sites on the surface of the cladding material (VIM semiconductor) and the adsorbed molecules (gases). The chemical interaction causes the transfer of electrons from the adsorbed gases to the VIM semiconductor such that the electrons become trapped at more or less isolated sites within

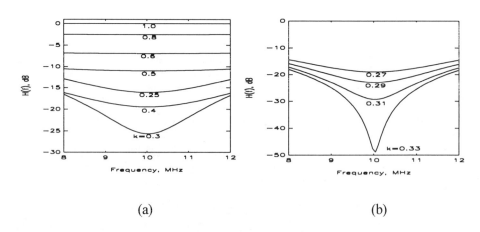

Figure 9. Magnitude of H(f) versus frequency.

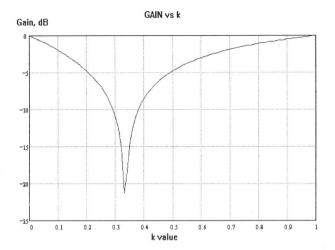

Figure 10. Gain as function of k at notch frequency.

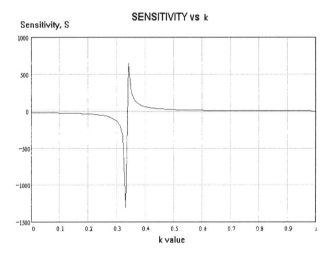

Figure 11. Sensitivity as a function of k at the notch frequency.

the VIM. The refractive index of the cladding after exposuer to gasses is given by[20]

$$\frac{n_{CL}^2 - 1}{n_{CL}^2 + 2} = \frac{4\pi}{3} N\alpha \qquad (17)$$

where N is the concentration of trapped electrons, and α is known as the electron polarizability at optical frequencies. Changes in the refractive index can be calculated by differentiating n_{CL} in Eq. 17 with respect to N:

$$\frac{dn_{CL}}{dN} = \frac{2\pi\alpha}{9n_{CL}}(n_{CL}^2 + 2)^2 \qquad (18)$$

Substituting n_{CL} from Eq. 17 into Eq. 18 leads to:

$$\frac{dn_c}{n_c} = \frac{18\pi\alpha}{(3+8\pi N\alpha)(3-4\pi N\alpha)}dN \qquad (19)$$

Assuming that the value of $N\alpha$ is much less than one, the resulting variation of refractive index can be related to a variation in N as follow:

$$\frac{1}{n_{CL}}\Delta n_{CL} \approx 2\pi\alpha\Delta N \qquad (20)$$

In order to get an idea about the typical value of Δn_{CL}, one can assume that the cladding of an optical fiber is replaced with a semiconductor VIM material such as polyfuran which has a refractive index of approximately 1.5. This cladding can adsorb hydrogen gas which leads to the electrons being trapped with a monolayer rate of $M=5\times10^{14}$ electrons/cm^2. If we assume that the fiber has cladding and core diameters of R and r respectively (R>>r), ΔN can be obtained from the following equation:

$$\Delta N = \frac{2RM}{R^2-r^2} \approx \frac{2M}{R} \qquad (21)$$

Considering a fiber with $R=6.25\times10^{-3}$ cm, ΔN will be 1.6×10^{17} electrons/cm^3. Using Eq. 20 and $\alpha=10^{-7}$ cm^3 leads to:

$$\frac{\Delta n_{CL}}{n_{CL}} \approx 10^{-3} \Rightarrow \Delta n_{CL} \approx 1.5 \times 10^{-3} \qquad (22)$$

The changes in the refractive index of the cladding can be related to the changes in the numerical aperture of the fiber which leads to variations in intensity of the light travelling through the fiber. In order to verify the amount of intensity modulation, the solid angle of acceptance of light entering the fiber is acquired using the following equation:

$$\Omega = 2\pi(1 - \cos\alpha_m) = 2\pi(1 - \sqrt{1-NA^2}) \qquad (23)$$

where α_m and NA represent the acceptance angle and numerical aperture of the fiber respectively. The relationship between NA and the cladding refractive index is:

$$NA = \sin\alpha_m = \sqrt{n_C^2 - n_{CL}^2} \qquad (24)$$

Differentiating Ω with respect to n_{CL} using Eqs 23 and 24 leads to:

$$\frac{d\Omega}{dn_{CL}} = \frac{d\Omega}{dNA} \cdot \frac{dNA}{dn_{CL}} = \frac{-2\pi n_{CL}}{\sqrt{1-NA^2}} \qquad (25)$$

Further manipulation of Eq. 25 gives the fractional change in Ω as follows:

$$\frac{\Delta\Omega}{\Omega} \approx \frac{-n_{CL}}{(\sqrt{1-NA^2})(1-\sqrt{1-NA^2})} \Delta n_{CL} \qquad (26)$$

For a fiber with NA of 0.2 and n_{CL} of 1.5, a change of 1.5×10^{-3} in n_{CL} will provide an 11.36% fractional change in Ω.

5. PROPERTIES OF POLYMER FILMS

As discussed in section 3, fibers coated with specific polymer films can provide a transduction mechanism suitable for chemical sensing. The film usually covers either a segment of the cladding or is deposited on the fiber end. The sensitivity of such sensors directly depends on the mechanical and optical responses of the film material to the chemical species to be detected. The mechanical response of the film is usually characterized by the total strain in the film due to the thermal expansion or swelling of the film[12]. The optical response of the film is assessed by the amount of fractional change in the reflectivity at the fiber/polymer interface. The sensitivity of the sensors where the film covers the cladding of the fiber depends more or less on the mechanical response, while the sensitivity of the sensors with fiber coated tips depends on the optical response of the film material.

When the film covering the fiber goes through thermal expansion or swelling, the stress-strain relationships for such a film in cylindrical coordinates are[21]:

$$\begin{aligned} \epsilon_r - \epsilon_0 &= \frac{1}{E}[\sigma_r - \nu(\sigma_\theta + \sigma_z)] \\ \epsilon_\theta - \epsilon_0 &= \frac{1}{E}[\sigma_\theta - \nu(\sigma_r + \sigma_z)] \\ \epsilon_z - \epsilon_0 &= \frac{1}{E}[\sigma_z - \nu(\sigma_r + \sigma_\theta)] \end{aligned} \qquad (27)$$

where ϵ_0 represents thermal expansion and it is normally shown by αT, where α is the linear thermal-expansion coefficient and T the temperature change. If we assume that we only have expansion in the r and z direction ($\epsilon_\theta = 0$), Eq. 27 becomes

$$\epsilon_r - (1 + v)\alpha T = \frac{1 - v^2}{E}(\sigma_r - \frac{v}{1 - v}\sigma_z)$$
$$\epsilon_z - (1 + v)\alpha T = \frac{1 - v^2}{E}(\sigma_z - \frac{v}{1 - v}\sigma_r) \quad (28)$$

The radial and axial strains modify the optical path in the fiber core as described by Eq. 2 and 3.

The optical response of the film, the reflectivity of the film, depends on the mechanical response and many other parameters such as thickness and index of refraction of the polymer. In general a variation in the reflectivity of the polymer film deposited on the fiber end can be expressed in terms of variations in the film index of refraction, Δn_f, and variations in the thickness of the film, $\Delta \epsilon$, as follows:

$$\Delta R = [\frac{\partial R}{\partial n_f}]\Delta n_f + [\frac{\partial R}{\partial \epsilon}]\Delta \epsilon + \sum_i [\frac{\partial R}{\partial z_i}]\Delta z_i \quad (29)$$

where z_i represent other parameters which effect reflectivity. Δz_i may be ignored compared to Δn_f and $\Delta \epsilon$. In Eq. 29 variation in the refractive index due to thermal expansion can be calculated from the following equation[22]:

$$\Delta n_f \approx -\frac{K}{6n_f}(n_f^4 - 1)\Delta T \quad (30)$$

where K is the volumetric thermal-expansion coefficient and is given by:

$$K = \alpha \frac{(1 + v)}{(1 - v)} \quad (31)$$

Variation in the film thickness, $\Delta \epsilon$, can be determined from the axial strain shown in Eq. 28. The derivatives of R with n_f and ϵ in Eq. 29 can be obtained from Eq. 5 by replacing R_1 and R_2 with the following equations:

$$R_1 = [\frac{n_0 - n_f}{n_0 + n_f}]^2$$
$$R_2 = [\frac{n_f - n_a}{n_f + n_a}]^2 \quad (32)$$

where n_0 and n_a represent the fiber and air index of refractions.

6. CONCLUSIONS

Different classifications of optical fiber sensors establish an appropriate technique for chemical sensing. Among all of the optical fiber chemical sensors investigated it was found that interferometric sensors exhibit the highest sensitivity and the intensity modulated sensors are the least complex to implement. The application of metal or polymer coated fibers in interferometric and intensity modulated optical fiber chemical sensors has been discussed. Several variable index materials such as polyvinylidene fluoride and cobalt acetylacetonate as potential candidates for chemical sensing have been introduced. Adsorption of gases on fibers coated with variable index materials and its effects on the fiber's optical characteristics are explained. A proposed intensity modulated optical fiber chemical sensor using a chemo-optic coupler with optical fiber delay line shows great potential for applications in distributed chemical sensor systems. It offers excellent sensitivity comparable to the phase sensors with a high degree of reliability.

6. ACKNOWLEDGMENTS

This work is supported in part by the Minority Research Center for Excellence program of the National Science Foundation, project number HRD-9353547.

7. REFERENCES

(1) T.G. Gillorenzi, J.A. Bucaro, A. Dandridge, G.H. Sigel, Jr., J.H. Cole, S.C. Rashleigh, and R.G. Priest, "Optical fiber sensing technology," IEEE J. Quantum Electron., Vol. QE-18, pp. 626-655, April 1982.

(2) C.M. Davis, "Fiber optic sensors: an overview," Optical Engineering, Vol. 24, pp. 347-351, March/April 1985.

(3) E. Udd (Editor), Fiber optic sensors, John Wiley and Sons, Inc., New York, 1991.

(4) M.A. Butler, "Optical fiber hydrogen sensor," Appl. Phys. Letters, Vol. 45, pp. 1007-1008, 1984.

(5) M.A. Butler and D.S. Ginley, "Hydrogen sensing with palladium-coated optical fibers," J. Appl. Phys., Vol. 64, pp. 3706-3712, 1988.

(6) M.A. Butler, "Chemical, biochemical, and environmental fiber sensors II," Proceedings of SPIE, Vol. 1368, pp. 46-54, San Jose, CA, 1990.

(7) R.M. Measures, "Fiber optic sensor considerations and developments for smart structures," Proceedings of SPIE, Vol. 1588, pp.282-297, 1991.

(8) G. Romero, "The development of a data acquisition and processing system for a fiber optic sensor array," Master Thesis, The University of Texas at El Paso, 1993.

(9) W.A. Chudyk, N.M. Carrabba, and J.E. Kenny, "Remote detection of ground water contaminants using far-ultraviolet laser-induced fluorescence," Anal. Chem., Vol. 57, pp. 1237-1242, 1985.

(10) T.L. Ferrell, "Fiber-optic surface enhanced raman system for field screening of hazardous compounds," First International EPA Symposium on Field Screening Methods for Hazardous Waste Site Investigations, Las Vegas, NV, 1988.

(11) M.A. Butler, "Fiber optic sensor for hydrogen concentrations near the explosive limit," J. Electrochem. Soc., Vol. 138, pp. 46-47, September 1991.

(12) M.A. Butler, R.J. Buss, and A. Galuska, "Properties of micrometer-thick plasma-polymerized tetrafluoethylene films," J. Appl. Phys., Vol. 70, pp. 2326-2332, August 1991.

(13) J. J. Robillard and H.J. Caulfield, Industrial applications of holography, Oxford University Press, New York, 1990.

(14) J.L. Fergason, "Liquid crystals in nondestructive testing," Appl. Optics, Vol. 7, pp. 1729-1737, 1968.

(15) Optics and Laser Technology, Vol. 6, p. 117, 1976.

(16) R.C. Jones, "A new calculus for the treatment of optical systems. I," J. Optics Soc. of America, Vol. 31, pp. 488-493, July 1941.

(17) P.S. Theocaris and E.E. Gdoutos, Matrix theory of photoelasticity, Sringer-Verlag, New York, 1979.

(18) L.A. Jeffers, "Fiber optic SO_2 analyzer," Proceedings of SPIE, Vol. 566, Fiber Optics Technology, San Diego, CA, 1985.

(19) W.R. Seitz, Sensors, Vol. 2, p.6, 1985.

(20) W.G. Durrer and J.J. Robillard, "VIM clad fiber optics for pollution detection," Proceedings of Materials Research Center of Excellence, The University of Texas at El Paso, Fiber Optics and Fiber Optics Materials, Vol. 3, pp. 12-29, 1992.

(21) S.P. Timoskenko and J.N. Goodier, Theory of elasticity, 3rd Edition, McGraw Hill Kogakusha, Ltd., Tokyo, 1970.

(22) C. Kitto, Introduction to solid state physics, John Wiley, New York, p. 163, 1965.